Grigori D. Shirkov
Günter Zschornack

**Electron Impact Ion Sources
for Charged Heavy Ions**

Grigori D. Shirkov
Günter Zschornack

Electron Impact Ion Sources for Charged Heavy Ions

vieweg

Produced by Hubert & Co., Göttingen

ISBN 978-3-663-09898-0   ISBN 978-3-663-09896-6 (eBook)
DOI 10.1007/978-3-663-09896-6

Preface

Presently many different types of ion sources exist worldwide for producing highly charged ions. The object of the present book is the treatment of electron impact ion sources like ECR (Electron Cyclotron Resonance) ion sources, EBIS (Electron Beam Ion Sources), EBIT Electron Beam Ion Trap) and ERIS (Electron Ring Ion Sources), which altogether are able to produce ions of high charge states. This criterion delimits the book according to classic ion sources, which as a rule can deliver high currents of low charged ions.

In the last decades there has been an intense development and building-up of sources of highly charged ions. The first impetus to the building of such sources came from heavy ion accelerator centers, since the effectiveness of a heavy ion accelerator is predominantly determined by the available ion sources. Thereby, the critical criterions for the operation of an ion source are the charge state distribution of the ions produced and the intensity of the extracted ion currents. Besides the employment of sources of highly charged ions in accelerator centers such sources increasingly are inserted separately from accelerators for basic investigations in atomic physics, surface physics and related areas.

In the present book ECR ion sources, EBIS, EBIT and ERIS are treated. Besides the fundamental theory for the description of ionization processes in electron impact ion sources the individual sources are described and discussed using the characteristics of actual sources. Thereby, laser ion sources are not discussed, because the ion production here differs greatly from those sources named above and thus their treatment would widen the frame of the existing book.

Besides the treatment of electron impact ion sources for the production of highly charged ions characteristic diagnostics methods are described for these sources, which are applied in different laboratories to diagnose the operating conditions. Thereby, besides the direct source diagnostics the diagnostic methods used often have relations to diagnostics methods in plasma physics and in particularly to the diagnostics of fusion plasmas.

Subsequently applications of sources of highly charged ions are discussed without wanting to reach here completeness. However, we examined the comprehensive work at the EBIT of the LLNL, which now makes it possible to accomplish X-ray investigations of highly charged ions into an unique way.

The authors cordially thank Dr.K.Wong from the LLNL, who kindly took over the proofreading of the english text and gave a lot of valuable references for the formation of the present book.

One of the authors (G.Z.) dedicates this book to his late wife Gisela, who contributed to the success of the work with her great understanding for the time demands of working on this book.

Dubna and Dresden, March 1996

G.D.Shirkov and G.Zschornack

Contents

Chapter 1

Elementary Processes

1.1 Collisions of Charged Particles

Positive charged ions are formed as a result of electron impact ionization of atoms or molecules in ion sources. Hereby electrons and ions are confined by external electric and magnetic fields. Multiply charged ions are created as a result of successive ionization processes.

In this chapter we will describe and classify as simple as possible the main physical processes taking place in the working area of ion sources, where ions are created and confined.

Neutral particles, ions and electrons in the plasma are in continuous motion, colliding with one another. In this context collisions mean Coulomb interactions between charged particles. Charged particles produce an electric field in their surrounding. When moving in the fields of other particles, they will change their trajectory, affecting other particles by their eigenfields. In this way energy is exchanged and redistributed between the colliding particles. These processes we shall name Coulomb collisions of charged particles .

If the internal states of electron shells change due to an electron or ion collision with the electron shell of another atom or ion, the impact is called an *inelastic collision*. Inelastic collisions can increase the total energy of the atomic electron shells. This process is named an *excitation process* and is characterized by

$$A + B \rightarrow A + B^* . \tag{1.1}$$

The impact of an electron on a neutral particle can produce an electron which energy is higher than the binding energy of the respective electron in the atom. This electron is then transferred to the continuum of unbound states and lost by the atom, e.g. it means an *ionization* of the electron. As a result of such a process the neutral atom is changed into a positively charged ion

$$e^- + A \rightarrow e^- + A^+ + e^- . \tag{1.2}$$

The ionization process of an ion of charge i caused by electron impact can be described in like manner

$$e^- + A^{i+} \rightarrow A^{(i+1)+} + 2e^- . \tag{1.3}$$

If, resulting from an interaction, there is an ionization of several electrons, it is a *multiple ionization process*. In generalized form this ionization process due to an electron impact is described as follows

$$e^- + A^{i+} \rightarrow A^{(i+k)+} + (k+1)\, e^- \ . \tag{1.4}$$

The character of interaction between electrons and between ions and neutral particles depend on their energy and the distance of interaction. In addition to processes of excitation and ionization, it is possible that the collision partners exchange one or more electrons. This process – *charge exchange* – is most likely to be found between neutral particles and ions. All interactions in which electrons are redistributed between the atomic shells of the reactances may be described in the following general form

$$A^{i+} + B^{k+} \rightarrow A^{j+} + B^{n+} + (j + n - i - k)\, e^- \ . \tag{1.5}$$

In case of ionization $(j + n - i - k)$ greater 0 is valid. If $(j + n - i - k) = 0$ holds, with $i \neq j$ and $k \neq n$ as restraints, a *charge exchange process* is involved.

Single ionization and single charge exchange are the most likely processes. As the ionization energy increases and the degree of ionization of the atom becomes larger, ionization of additional electrons from the ion will be less and less probable. So the probability of ionization decreases with an increase in ionic charge. For the same reason electron transitions between ions are not probable. On the contrary, the probability of charge exchange processes between neutral particles and ions is sufficiently high.

If the velocity of free electrons is lower than the characteristic electron velocity in the atomic orbitals, it is possible that an ion will be able to catch a free electron. This process is referred to as *recombination*.

If, as a result of an interaction, only energy or the momentum direction of the whole particle is changed without excitation of internal degrees of freedom, it will be a matter of *elastic collisions*.

On what exactly do the probabilities of collision processes depend? For an answer, a particle A will be considered. It is supposed that as a result of a collision with a particle B, particle A will be ionized. In that case the probability of ionization P of particle A by particle B at time interval Δt is dependent on

- density n_B of particle B in the volume surrounding particle A (number of particles B per unit of volume);
- relative velocity v between particles A and B;
- effective interaction cross-section σ for the process considered (here: ionization cross-section of particle A by particle B).

So, for the probability of ionization

$$P = \sigma v n_B \Delta t \tag{1.6}$$

holds.

The total effective cross-section σ of any impact process can be simply described as

$$\sigma = \pi \varrho^2 \ .$$

The quantity ϱ describes the maximum distance or the *impact parameter* at which the process considered is still possible.

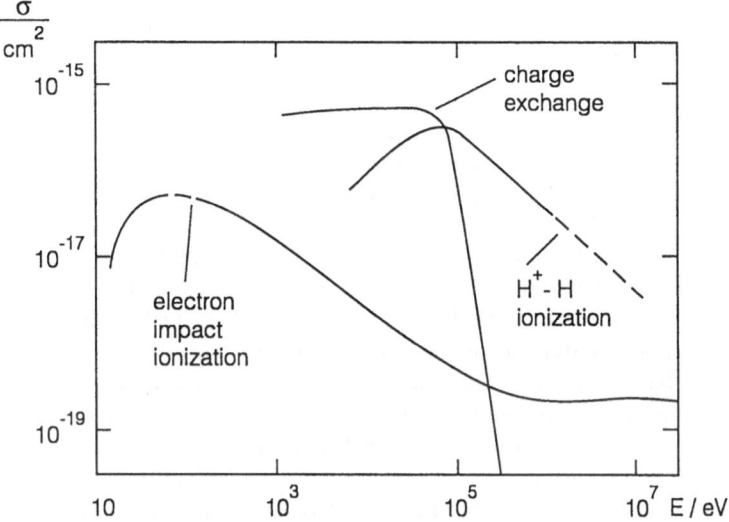

Figure 1.1   Electron impact ionization cross-sections of Hydrogen and cross-sections for processes of charge exchange and ionization between Hydrogen atoms and ions as a function of particle energy E [1].

The effective interaction cross-section σ may be understood as an area of a circle of radius ϱ. For example, the value of the impact cross-section for two hard-core spheres of radius a is $\sigma = 4\pi a^2$, since for the centres of the spheres to collide with each other they have to approach a distance which is smaller than double sphere radius.

The value of the cross-section for any inelastic processes is dependent on the energy of the colliding particles. A number of reactions show maximum reaction probabilities at different energies. An example is illustrated in Fig.**1.1** showing interaction cross-sections for ionization of Hydrogen by electron impact and for processes of charge exchange and ionization between Hydrogen atoms and ions [1].

The quantity σv is referred to as the *reaction rate*, characterizing the dependence of the considered reaction cross-sections on energy.

At particle energies which are typical for sources of highly charged ions, the most probable processes are those which show the largest reaction cross-sections and the highest rates of response. These are in particular processes of ionization of neutral particles and ions due to ionization by electron impact and processes of charge exchange of neutral particles on multiple charged ions. Another relevant matter is the redistribution of kinetic energy by elastic collisions among charged particles. These processes will be discussed in more detail in the subsequent sections.

1.2   Ionization by Electron Impact

The difficulties of producing highly charged heavy ions due to electron impact ionization have been discussed during the last few decades in particular in connection with the development of plasma physics, controlled nuclear fusion and heavy ion accelerators.

In general, processes of ionization can be divided into two fundamental groups. As a result of the interaction between an ionizing electron and an electron bound in an outer atomic subshell, a direct *continuum ionization* of the atom or ion is possible. Ionization of electrons from atomic inner-shells results in more complicated processes. Ionization or excitation of inner-shell electrons leads to inner-shell vacancies. Filling these vacancies make the whole electron shells be restructured by radiative filling through emission of characteristic X-rays and through non-radiative Auger- and Coster-Kronig transitions. If elements of high atomic number and of a complex electron shell structure are ionized, then these processes will be most essential and increase the probabilities of double and multiple ionizations considerably. Experimental investigations of the electron structure and electron transitions of multiply charged ions will be described in chapter 7.

A lot of experimental and theoretical works deal with ionization of neutral particles and ions. On the basis of different quantum mechanical approximations a great number of interaction cross-sections were obtained, they sufficiently agree with many cross-sections experimentally determined for neutral atoms within a wide range of energy. Compared to cross-sections for neutral atoms, ionization cross-sections for multiple-charged ions were relatively seldom determined by experiment. But theoretical values may fill this gap. A survey of the material available is given in [2, 3, 4, 5]. Ionization of multiple-charged ions has been discussed in a number of general publications, which dealt with producing of highly charged ions [6, 7, 8]. In the following we give a discussion of the most important characteristics to be found in ionizing neutral particles and producing highly charged ions.

The most important characteristic of the ionization cross-section is its threshold behaviour which depends on the energy of electrons E_e. It is evident that, in order to ionize an electron from an atomic subshell, the energy of the ionizing electron will have to exceed at least the *ionization energy* I_k of the bound electron. The maximum ionization cross-section is obtained at $E_e = (2 \ldots 3)\, I_k$.

The first semi-empirical formula to calculate ionization cross-sections was introduced by Tompson in 1912 [9]

$$\sigma_k^i = \frac{\pi n a_0^2}{I_k E_e} \left(1 - \frac{I_k}{E_e} \right) \tag{1.7}$$

here I_k and E_e are the energies given in atomic units (one atomic unit = 27.2 eV), n is the electron number in the outermost subshell, and $a_0 = 0.529 \cdot 10^{-10}$m is the Bohr radius.

Ionization energies have been calculated i.a. by Zschornack et al. [10] and Carlson et al. [11]. With the development of quantum mechanics it has become possible to study the behaviour of the interaction cross-sections at higher energies. The dependence of cross-sections here is of the form $\ln E_0/I_k$. For calculations of ionization cross-sections authors of present publications often use the semi-empirical formula of Lotz [12, 13], taking the correct electron binding energy and asymptotic behaviour of the cross-sections into account

$$\sigma_k^i = \frac{4.5 \cdot 10^{-14}}{E_e} \sum_{k=1}^{m} \frac{n_k}{I_k} \log \frac{E_e}{I_k} \tag{1.8}$$

with m being the number of atomic subshells occupied in the ion, n_k the number of electrons in the subshell considered, and I_k the ionizing energy for the actual subshell. Here and below all energies are given in eV.

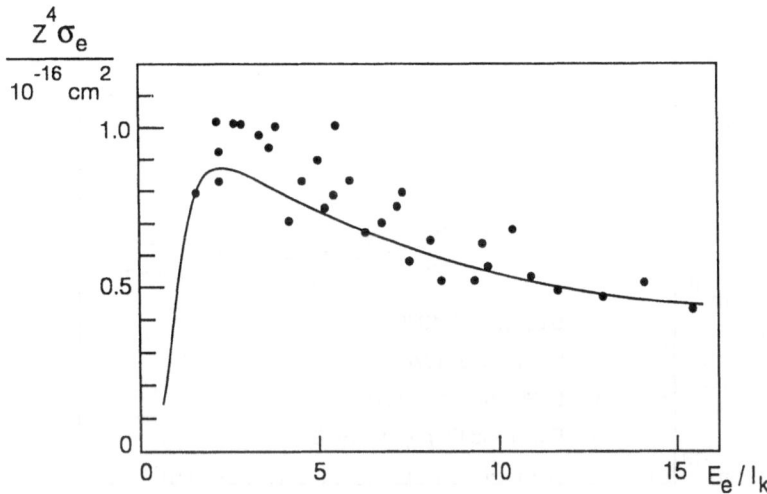

Figure 1.2 Ionization cross-sections σ_i and their universal dependence on atomic number Z and I_k/E_e. The points give experimental data from Refs.[15, 16] and the inked line calculation results from [17].

Semi-empirical equations describe cross-sections for direct ionization processes to a sufficient degree. For this reason, they are used in particular within the range of lower electronic energies where Auger and Coster-Kronig processes are not very likely to be found.

In the case of molecular ionization as a rule molecular ions are produced; they will dissociate into ionic fragments in subsequent collisions [14].

From (1.7) and (1.8) follows that the quantity $\sigma_k^i \sim I_k^{-1}$ is in essence a function of (I_k/E_e), and does not largely depend on the actual sort of ion. If one considers that I_k is proportional to k^2, for example for hydrogen-like ions $I_k = k^2/2n^2$, with n being the principal quantum number, then a determination of how ionization cross-sections depends on ionic charge will be possible

$$\sigma_{Z-1}^i \sim Z^{-2} \ . \tag{1.9}$$

Ionization cross-sections for noble gases of different charge states have been investigated by Donets [6] who examined the electron energy range up to 18 keV on the electron beam ion source KRION-1. In this connection the experimental ionization cross-sections for hydrogen-like ions up to Ar^{17+} [15, 16], which are summarized in Fig.1.2 in good accordance with theoretical values [17], are particulary interesting. This figure shows the ionization cross-section σ_i and its universal dependence on the atomic number Z, referring the electronic energy E_e to the ionization potential I. A more comprehensive summary of the dependence of ionization cross-sections on I_k/E_e has been presented by [5].

In contrast to single ionization cross-sections, there are only few known publications on double and multiple ionization cross-sections and their determination. It is evident that simultaneous ionization of two ore more electrons from the atomic shell is comparatively

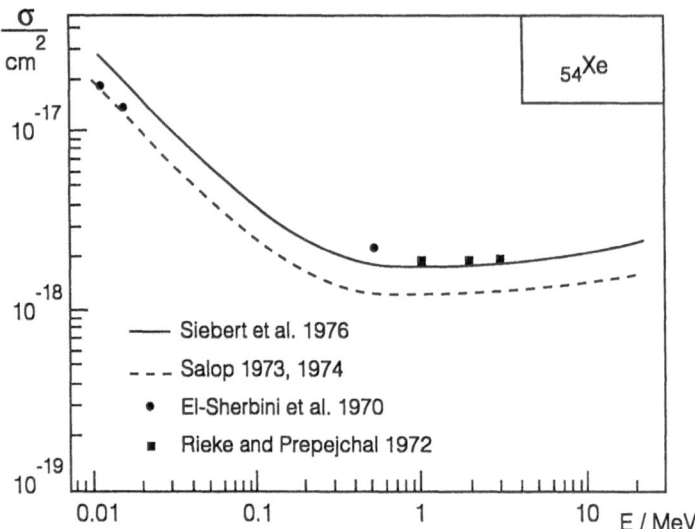

Figure 1.3   Total ionization cross-sections of neutral Xenon as a function of kinetic energy of colliding electrons. Values according to Siebert et al. [22] given as a continuous line and those according to Salop [20] as a dotted line.

not very probable. A sufficient probability will be given for *autoionization processes*, if after creation of an inner-shell vacancy, orbitals are occupied and their energetic states exceed the binding energies of the outermost electron states occupied in the energetic ground state.

Calculations of *multiple ionization cross sections* are almost not available yet. For estimates of double-ionization cross-sections the formula of Gryzinski [18] can be used. It is also possible to use cross-sections which have been obtained by an analysis of experimental data [19] and the behaviours which are in close agreement with the Lotz formula (1.8)

$$\sigma_k^{2i} = \frac{2.6 \cdot 10^{-14}}{E_e \, (I_k + I_{k+1})} \, \ln \frac{E_e}{I_k + I_{k+1}} \ .\tag{1.10}$$

The calculation of *relativistic electron-impact ionization cross-sections* for a number of gaseous elements have been done by Salop [20, 21] and Siebert et al. [22], and for Uranium by Siebert et al. [22].

Cross-sections for *direct ionization processes* can be estimated by means of the formula from Gryzinski [18]

$$\sigma_k^i = \pi n_k \, r_e^2 \, \frac{m_e c^2}{I_k} \, \ln \frac{E_e}{I_k}\tag{1.11}$$

with m_e describing the electron mass, r_e the classical electron radius, c the velocity of light and n_k is the number of electrons in the outermost occupied subshell.

In accordance with theoretical values *Auger processes* considerably contribute to the total ionization cross-section in the high energy region of electron energy. Rieke and Prejpechal [23] have taken experimental measurements of direct ionization cross-sections of neutral Krypton and Xenon atoms within the energy range between 0.5 MeV and

3 MeV. The closest correspondence between these values and calculated results has been obtained by Siebert et al. [22].   Fig.**1.3** shows the respective results, while Fig.**1.4** illustrates the total ionization by electron impact at 20 MeV achieved from adding the cross-sections for direct ionization [22] and those for Auger ionization [20, 21].

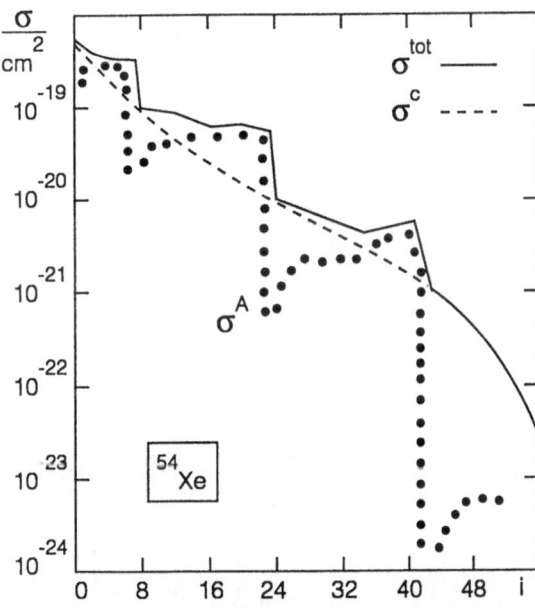

Figure 1.4 Total ionization cross-section for the electron impact ionization of Xenon atoms at an electronic energy of 20 MeV. The points describe Auger ionization cross-sections σ^A [21], the dashed lines direct ionization cross-sections σ^C [22] and the solid line total ionization cross-sections σ^{tot} [8].

As follows from calculated and experimental values (equations 1.7 to 1.11; Figs. **1.2** and **1.3**), ionization cross-sections strongly depend on electronic energy. The ionization probability described in (1.6) determines the reaction rate σv for which this dependence will appear even larger. In case of different electrons showing different kinetic energies, this means for the individual electrons to have a different degree of ionization probability, which has to be considered by all means when ion producing processes are analysed. In some cases it may be sufficient to determine mean reaction velocities σv and electron distribution functions for all ranges of different electronic energies. All this will be dealt in more detail in the following sections.

1.3   Ion-Atom Collisions

Ion-atom and ion-ion collisions are of much more complex character than collisions of electrons with ions or neutral particles. A bombarding ion produces a glaring and long-range defect in the electron shells of the target atom or ion which will loose one or more electrons. The ionized electrons get into the Coulomb field of the bombarding ion. If their relative speed is lower than the orbital velocity of the target electrons, the free electrons are very likely to be captured into highly excited states of the bombarding ion. Therefore charge exchange and electron capture processes gain in significance at relative impact velocities of $v \ll v_0 \sqrt{i}$ (here: $v_0 = 2.2 \cdot 10^8$ cm s^{-1} is the orbital velocity of the electron in the Hydrogen atom and i is the target charge found after the collision).

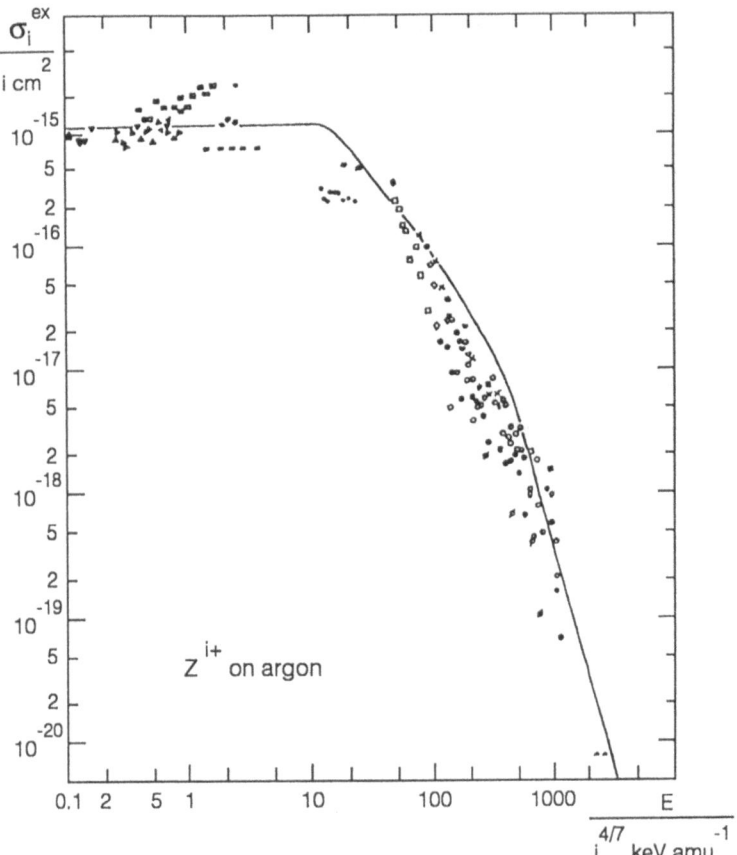

Figure 1.5 Comparison between theoretical estimations and experimental data for single-capture cross-sections of medium- and high-energy impacts of ions of charge $i \geq 4$ colliding with argon atoms. For a reference to the data see [26].

Within the range of $v/v_0 \cong i^{1/2}$, ionization and charge exchange cross-sections are of the same magnitude. For $v/v_0 \gg i^{1/2}$, ionization processes are very likely to occur, and charge exchange cross-sections will decrease very fast at an increasing impact velocity. For charge exchange processes $\sigma \cong v^{-\gamma}$ is valid here, with γ changing from 5 for charge exchange processes at very heavy atoms, where an electron capture from inner shells is predominant, to 12 for charge exchange processes at the Hydrogen atom [24, 25].

Collisions at high energies can occur at short distances of interaction; they lead to defects in the inner electron shells and consequently to the formation of vacancies there. As a result of such processes the probability of an emission of X-ray quanta and Auger electrons is increasing.

Characteristic energy dependencies of cross-sections of reactions between ions and atoms have already been shown in Fig.1.1. Experimental values for charge exchange processes are available in a great number in the literature. Figs. **1.5** and **1.6** provide examples of how charge exchange cross-sections between atoms and multiple-charged ions depend on energy.

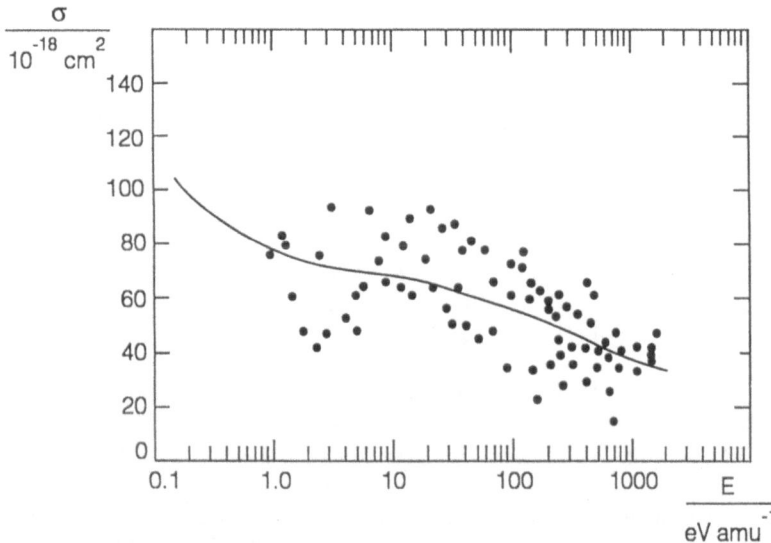

Figure 1.6   Electron capture data for O^{5+} + H and N^{5+} + H collisions in the low-energy range with experimental and theoretical data. For a reference to the data see [27].

Fig.**1.5** shows total cross-sections of charge exchange between Argon atoms and ions of the elements B, C, N, O, F, Ne, Ar, Br, I and Xe which show charge states between 4 and 17 at mean and high energies (for a reference to used data see [26]). The Figure's solid line gives results achieved from calculations according to the theory of Born and Lindhard [26]. These are given in a parameterized form, where $\sigma = \sigma/i$ and $E = E/i^{4/7}$ are valid. Fig.**1.6** shows calculations and experimental values for cross-sections of charge exchange between Hydrogen and Deuterium on the one side and N^{5+} and O^{5+} ions on the other side, at low energies (for a reference to the data used see [27]).

On the basis of the available theoretical and experimental data it is possible to conclude that within the range of about 1 eV/u and 10 keV/u ($v \cong v_0\, i^{\frac{1}{2}}$) the charge exchange cross-sections do virtually not depend on the energy of incidence and show their maximum value. This considerably simplifies the analysis of the influence of inelastic ion-ion collisions when studying electron impact processes and the storing of multiple-charged ions in ion sources. As will be shown in the following sections, the kinetic energy per ion in electron-cyclotron resonance ion sources comes to about 1 eV/u, that in electron-beam ion sources to appr. 10 eV/u, and that in the much denser relativistic electron rings of heavy ion electron ring accelerators and in electron ring ion sources to 10 keV/u. Therefore the most probable process between heavy ions in ion sources is the charge exchange of neutral atoms or molecules, which are always existent in the operational volume of the source, with ions, which have already been produced. In this connection the charge exchange cross-sections increase with increasing degree of ionization of the produced ions. All this explains why it is necessary to analyse charge exchange processes most carefully: they are among those factors which determine the ionization degree of the produced ions and increase the number of undesirable low charged ions in the operational volume of the sources considered.

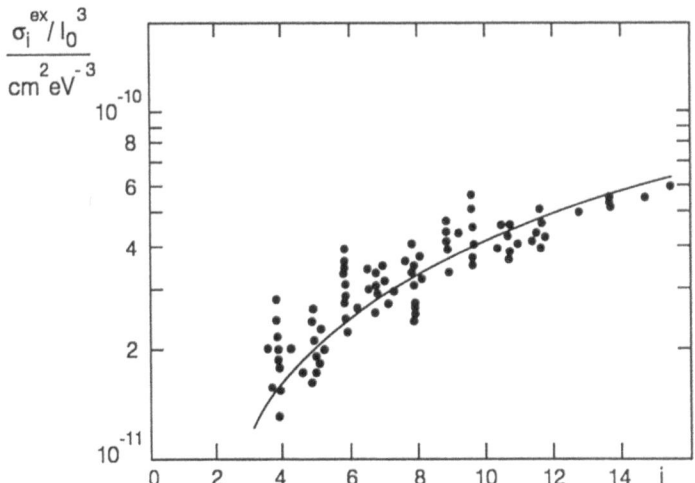

Figure 1.7   Scaling law for one electron exchange cross-section as a function of the projectile charge i [29]. Experimental data have been taken from Ref.[36]. The solid line characterizes the dependence of $\sigma \sim i$.

Here one should have a look at the main characteristics of charge exchange processes at lower energies. During the last few decades charge exchange of ions at lower energies ($v < v_0$) gained in significance in connection with the development of plasma physics and controlled nuclear fusion. Hundreds of experimental and theoretical treatises on these problems have been published since (see e.g. [1, 3, 24, 25, 28, 29, 30, 31, 32]). The known experimental values for charge exchange processes at ion energies of up to some keV give a clear idea of the charge exchange of various neutral particles.

But here is not the place to analyze and describe the available theories and calculation methods. All those who are interested in them should refer to the above mentioned papers and reviews.

If one considers that there is practically no energy dependence in the relevant energy range, the task of determining total charge exchange cross-sections, which are the sum of all quanta states of the colliding atoms and ions, can be reduced with a sufficient accuracy to determining the dependence of charge exchange cross-sections on the ion charge and the ionization potential of the atoms $\sigma_{ik} = \sigma(i, I_k)$. In many cases this dependence can be described in a simple parameterized form

$$\sigma_{ik} = A\, i^{\alpha} I_k^{-\beta} \ . \tag{1.12}$$

The majority of known models and an analysis of experimental data give for α the dependence of $\alpha = 1 \ldots 1.5$ (in some cases $\sigma \sim i \ln i$ is valid) [33].

At high impact energies, where $v > v_0$, α comes to values between 1.5 and 2 and more [34]. For β, values between 2 and 3 are usually assumed. The dependence of charge exchange cross-sections on the charge of the bombarding ion and the ionization potential of the target atom is explained in great detail in [35, 36]. Fig.**1.7** gives the charge exchange cross-sections for Carbon, Nitrogen, Oxygen, Argon and Xenon atoms

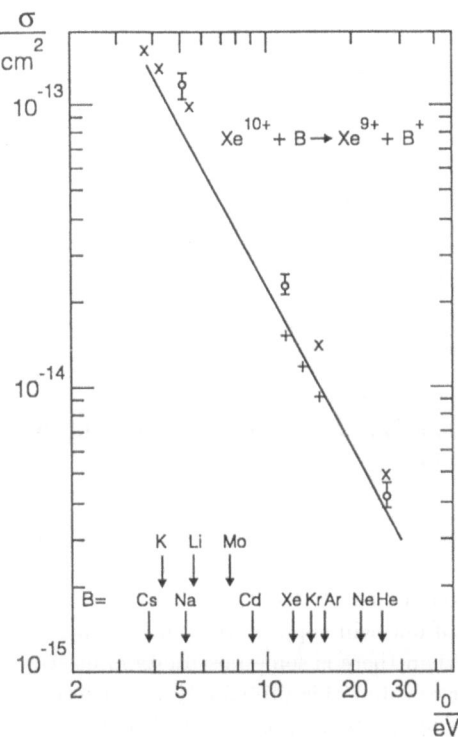

Figure 1.8 Dependence of the cross-section for the reaction Xe^{10+} + B on the ionization potential of the target atom B. The solid line is a least-square fit to experimental data by scaling $\sigma \approx I_0^{-1.94}$ [28].

in comparison with the electron capture cross-sections [36],

$$\sigma_{i,i-1} = 4 \cdot 10^{-12} \frac{i}{(I_0[\mathrm{eV}])^3} \ .$$

While Fig.**1.8** illustrates the dependence of the capture cross-section on the ionization potential of different atoms for an electron by a Xe^{10+} ion [28]. The curve in Fig.**1.8** which approximates the experimental values in the best possible way, shoes the dependence of $\sigma \cong I_0^{-1.94}$.

The cross-sections for double and multiple charge exchange processes are much smaller than those for the transition of an electron. It is possible to suppose that the general form for these cross-sections can be given by analogy with (1.12)

$$\sigma_{i,i-n} = A_n \, i^{\alpha_n} \, I_0^{-\beta_n} \ . \tag{1.13}$$

For determination of charge exchange cross-sections at noble gas atoms and ions, Table **1.1** lists the parameters A_n, α_n and β_n determined in [35] on the basis of a great number of experimental values.

In the operating range of ion sources collisions between ions occur. Therefore it is greatly important to take clear up the influence of charge exchange processes when considering ion-ion collisions. Thereby, it is difficult to obtain two intensive crossing beams of differently charged ions, there are almost no experimental data on these problems. It

Table 1.1   Parameters for an analytic determination of charge exchange cross-sections according to equation 1.13 [35].

n	A_n	α_n	β_n
1	$1.43 \cdot 10^{-12}$	1.17	2.76
2	$1.08 \cdot 10^{-12}$	0.71	2.80
3	$5.3 \ldots 5.8 \cdot 10^{-14}$	2.10	2.90
4	$3.6 \ldots 8.9 \cdot 10^{-16}$	4.20	3.00

is known however that the ionization potential of ions increases with an increase in ionic charge state and can be approximately described by

$$I_i = \frac{(i+1)^2}{2n} \tag{1.14}$$

where n is the principal quantum number of the outer shell electrons and i is the ionization degree of the ion delivering an electron. Thus if one will suppose that in this case the parametric dependence (1.12) takes place too, than there is some possibility of electron capture by multiple charged ions from another ion, but this probability is much lower than that of collisions with neutral particles. If in (1.12) the potential (1.14) is used, it is possible to estimate the probability of charge exchange of an ion at an ion.

The experimental and theoretical results achieved suggest that for very highly charged ions electron capture cross-sections at neutral atoms reach as a rule values of 10^{-13} cm^2 and more [37]. Therefore charge exchange processes can compete with processes of ionization by ion impacts even at very low concentrations of neutral particles in the operating range of ion sources, when the production of highly charged ions is concerned.

1.4   Elastic Collisions of Charged Particles

The aim of this section is to consider the general situation of electron and ion movements in sources of multiple-charged ions. The charged particles build up a long-range electric field in their surroundings; this field decreases by $1/r$. All charged particles influence one another by these fields, which effect a change in their energies and pulses. In most cases the particles meet one another only at great distances, with their trajectories being influenced in a minor way. If, in this connection, the inner degree of freedom of the particles remains unchanged, elastic Coulomb collisions are involved. In the course of time frequent elastic impacts lead to a modification of the distribution function for each particle and to redistributing the kinetic energy between the individual charged components (which – as the following will show – is of great relevance to sources of highly charged ions).

Related questions were already examined in detail some decades ago when plasma physics developed [38, 39, 40]. So they can be regarded as classical in this field of physics from our current viewpoint. On the other hand, elastic Coulomb collisions have not been

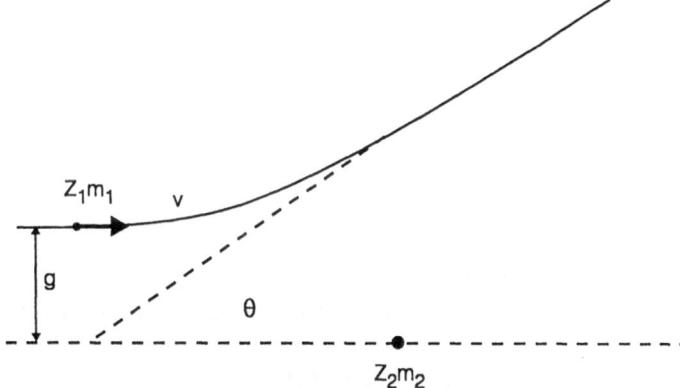

Figure 1.9   Scheme of elastic scattering of particles with the charge Z_1, mass m_1 and relative velocity v_1 to particles of charge Z_2 and mass m_2.

given the required attention when sources of highly charged ions were considered. For that reason, these questions should be examined in more detail.

For the first time in in 1984 it was investigated that it is possible to cool heavy, highly charged ions by light, low-charged ions with the aim of increasing the current of highly charged ions from an electron-beam ion source [41]. This process which is also called "ion cooling" or "mixed gas operation" has been treated in detail in [42] for different ion sources. At present ion cooling is one of the most important processes in increasing the efficiency of sources of highly charged heavy ions.

The aim of this section is to deal with the fundamental characteristics of elastic collisions of charged particles, with the papers [43, 44] being taken as a basis.

At a close look each particle moving in a cloud of charged particles simultaneously interacts with a great number of particles from its surroundings at any time, and so all these interactions have to be considered. But such a problem is difficult to solve on principle. For a solution, a number of simplifications should be introduced. The simplest method here is the approximation of two-particle collisions. In this approximation the continuous process of simultaneous interaction of each particle with many others particles leads to successive collisions between particle pairs, which results in equal energy and pulse transfers. As compared with two-particle collisions, those between three, four or more particles are very unpropable here. Therefore the two-particle collision pattern reflects the real situation sufficiently, leading to more physically correct results.

Let us consider a particle of charge Z_1 and mass m_1, which will come from infinity at velocity v and meet a resting particle of charge Z_2 and mass m_2 (see Fig.1.9). The effective scattering cross-section for a scattering at angle ϑ is described by the *Rutherford scattering formula*

$$\sigma\left(\vartheta, v\right) = \frac{Z_1 Z_2 e^2}{2\mu v^2} \frac{1}{\sin^4 \frac{\vartheta}{2}} \tag{1.15}$$

with

$$\mu = \frac{m_1 m_2}{m_1 + m_2}$$

as reduced mass and

$$\tan \frac{\vartheta}{2} = \frac{\varrho_\perp}{\varrho}.$$

Here ϱ describes the impact parameter and

$$\varrho_\perp = \frac{Z_1 Z_2 e^2}{\mu v^2}$$

is valid, where $\varrho_\perp$ is the impact parameter at 90° scattering.

If the bombarding particles are assumed to have the density n_1, the mean force $\overline{F}$, which has an effect on a resting particle of charge Z_2 from the side of the bombarding particle, yields

$$\overline{F} = 4\pi L \frac{Z_1^2 Z_2^2 e^4 n_1}{\mu v^3} \, \overline{v} \tag{1.16}$$

where L describes the *Coulomb logarithm*

$$L = \int\limits_{\varrho_{\min}}^{\varrho_{\max}} \frac{\varrho \, d\varrho}{\varrho_{\min}^2 + \varrho^2} = \ln \frac{\sqrt{\varrho_{\max}^2 + \varrho_{\min}^2}}{\varrho_{\min}} \simeq \ln \frac{\varrho_{\max}}{\varrho_{\min}} \,. \tag{1.17}$$

Determining the regions of integration on the basis of physical principles, we eliminate possible integral divergences and guarantee a finite value of the Coulomb logarithm.

Usually, ϱ is determined as the highest value of the three following quantities:

$\varrho = \dfrac{Z_1 Z_2 e^2}{\mu v^2}$ – impact parameter in case of central collisions;

a_0 – dimensions of colliding particles;

$\lambda = \dfrac{hc}{p}$ – DeBroglie-wavelength of the particle bombarding with the pulse p.

The last case is important at very high energies of the colliding particles where classical mechanics is not applicable. This will be essential if $v > Z_1 Z_2 \alpha c$ with α as fine structure constant is valid.

As the upper limit of integration $\varrho_{\max}$ either has the dimensions of the considered cloud of particles or, in the case of high particle densities, the quantity

$$r_0 = \sqrt{\frac{T}{8\pi n e^2}}$$

with T and n as temperature and particle density resp., are assumed. The quantity r_0 is referred to as the *Debye length*, characterizing the distance from a charged particle (e.g. an ion) in a surrounding medium (e.g. a plasma). Due to polarization of the medium the electric field of this particle is very small, i.e. shielded.

In the practice the relationship $\varrho_{\max}/\varrho_{\min}$ assumes very high values and the Coulomb logarithm changes very little only for a large group of tasks. Usually the Coulomb logarithm assumes values $L = 10\ldots15$. Hence the approximate determination of the regions of integration in (1.17) is sufficient and does not result in a serious error.

According to traditional terminology, the particle which is regarded separately is referred to as the *trial particle*. All other particles are considered to be a medium in which the trial particle is moving. The particles of this medium are called *field particles*.

If the mean force having effect on the trial particle is known, it will be possible to determine mean pulse and energy changes caused by the current of the field particles

$$< \frac{dp}{dt} > \ = \ -\frac{4\pi n_1}{\mu v^3} \, Z_1^2 Z_2^2 e^4 Lv \qquad (1.18)$$

$$< \frac{d\varepsilon}{dt} > \ = \ -\frac{4\pi n_1}{\mu v^3} \, Z_1^2 Z_2^2 e^4 Lv^2 \ . \qquad (1.19)$$

Let us now consider the case of a current of trial particles moving through a medium of field particles. As a result of the occurring collisions the current is broadening in the longitudinal and the transverse directions in the space of coordinates and velocities. This process is described by the equations

$$< \frac{dp_2}{dt} > = -\frac{p_2}{\tau_l^{12}} \qquad (1.20)$$

where τ_l^{12} is the mean time for longitudinal slow down of particle "1" in medium "2" and

$$< \frac{dp_2^2}{dt} > = \frac{p_2^2}{\tau_t^{12}} \qquad (1.21)$$

with τ_t^{12} is the mean time for transverse mean square deviation, and by the equation

$$< \frac{d\varepsilon_2^2}{dt} > = \frac{\varepsilon_2^2}{\tau_\varepsilon} \qquad (1.22)$$

where τ_ε is the mean time for the increase in energy spread or for the widening of the trial particle current in the longitudinal direction.

Let us assume that the field particles themselves move in a disordered way and interact with one another in an elastic way: it is most probable for real physical problems that the field particles will be distributed according to the Maxwell-Boltzmann distribution law

$$f_2 = n_2 \left(\frac{m_2}{2\pi T_2} \right)^{3/2} \exp \left[-\frac{m_2 v^2}{2T_2} \right] \qquad (1.23)$$

with T_2 being the temperature of the field particles.

By the integration of equations (1.20), (1.21) and (1.22) with the distribution function (1.23) it is possible to obtain the characteristic times of the change in parameters of a homogeneous particle beam in a medium of a Maxwell distribution.

Without explaining the derivations shown in detail e.g. in [43, 44], the results from them should be given here

$$\tau_l^{12} = \frac{\tau_1^{12}}{\left(1 + \frac{m_1}{m_2} \right) \gamma} \qquad (1.24)$$

$$\tau_t^{12} = \frac{\tau_1^{12}}{2 \left(\gamma + \gamma' - \frac{\gamma}{2x_2} \right)} \qquad (1.25)$$

$$\tau_\varepsilon^{12} = \frac{\tau_1^{12}}{\frac{4\gamma}{x_2}} \ . \qquad (1.26)$$

In this connection a symbology which is identical to that in [43] has been chosen

$$x_2 = \frac{\varepsilon_2}{T_2} = \frac{m_2}{m_1} \frac{\varepsilon_1}{T_2}$$

where x_2 defines the relation between the trial particles' energy and the field particles' temperature.

$$\tau_1^{12} = \frac{\sqrt{m_1} \varepsilon_1^{3/2}}{\sqrt{2\pi} Z_1^2 Z_2^2 e^4 n_2 L_{12}} \tag{1.27}$$

describes the *relaxation time* and

$$\gamma = \gamma(x_2) = \int_0^{x_2} e^{-t} \sqrt{t} \, dt = \gamma\left(\frac{3}{2}, x_2\right); \tag{1.28}$$

$$\gamma'(x) = \frac{d\gamma}{dx} = e^{-x} \sqrt{x} \tag{1.29}$$

are valid, where $\gamma(\alpha, x)$ is the incompletely normalized Gamma-function [45].

In addition to the above defined characteristic times, it is possible to determine the velocities of energy loss and of energy exchange between the trial particles

$$\tau_{tr}^{12} = \frac{\tau_1^{12}}{2\left(\gamma' - \frac{m_1}{m_2}\gamma\right)}. \tag{1.30}$$

Unlike the pulse (1.20) is always transmitted from the particle current to the medium with the current moderated, the change in energy of the particle current

$$\frac{d\varepsilon_1}{dt} = \frac{2\varepsilon_1}{\tau_1^{12}}\left[\gamma' - \frac{m_1}{m_2}\gamma\right] \tag{1.31}$$

can show both signs: on the one hand, the current of fast particles will heat the medium when being slowed down in it, and on the other hand, the current of moderated particles will cool the medium.

In the following, consideration will be given to concrete relaxation times in case of large energy flux values. Under these conditions $x_2 \gg 1$, $\gamma(x_2) = \sqrt{\pi}/2$ and $\gamma'(x_2) = 0$ is valid. Then, choosing

$$\tau_0 = \frac{\sqrt{m_1}}{\sqrt{2\pi}} \frac{\varepsilon_1^{3/2}}{e^4 n_2 L}$$

the values of τ_l, τ_t, τ_ε and τ_{tr} will result for the following cases:

1. for an electron beam of energy ε_e in an ion medium of temperature T_i, charge i and mass $A_i M$, where A_i defines the mass number of the ions;
2. for an electron beam of energy ε_e in an electron medium of temperature T_e;
3. for an ion beam of energy (temperature) T_i, charge i and mass $A_i M$ in an ion medium of temperature T_k, charge k and mass $A_k M$;
4. for an ion beam of energy (temperature) T_i, charge i and mass $A_i M$ in an electron medium of temperature T_e.

Table 1.2 Relaxation time values τ for different interactions between electron components and ion components of charge i and k respectively.

τ	ei	ee	ik	ie
τ_l	$\dfrac{\tau_0}{i^2}$	$\dfrac{\tau_0}{2}$	$\sqrt{\dfrac{A_i M}{m_e}}\,\dfrac{A_k}{A_i + A_k}\,\dfrac{\tau_0}{i^2 k^2}$	$\sqrt{\dfrac{m_e}{A_i M}}\,\dfrac{\tau_0}{i^2}$
τ_t	$\dfrac{\tau_0}{2i^2}$	$\dfrac{\tau_0}{2}$	$\sqrt{\dfrac{A_i M}{m_e}}\,\dfrac{\tau_0}{2i^2 k^2}$	$\sqrt{\dfrac{A_i M}{m_e}}\,\dfrac{\tau_0}{2i^2}$
τ_ε	$\dfrac{\varepsilon_e}{4\,T_i\, i^2}\,\dfrac{A_i M}{m_e}\,\tau_0$	$\dfrac{\varepsilon_e}{4 T_e}\,\tau_0$	$\dfrac{\varepsilon \varepsilon_0}{4 T_k\, i^2\, k^2}\,\dfrac{A_k}{A_i}\sqrt{\dfrac{A_i M}{m_e}}\,\tau_0$	$\dfrac{\varepsilon_i}{4 T_e\, i^2}\sqrt{\dfrac{m_e}{A_i M}}\,\tau_0$
τ_{tr}	$\dfrac{1}{2 i^2}\,\dfrac{A_i M}{m_e}\,\tau_0$	$\dfrac{\tau_0}{2}$	$\dfrac{1}{2 i^2 k^2}\,\dfrac{A_k}{A_i}\sqrt{\dfrac{A_i M}{m_e}}\,\tau_0$	$\dfrac{1}{2 i^2}\sqrt{\dfrac{m_e}{A_i M}}\,\tau_0$
range of validity	$\varepsilon_e \gg \dfrac{T_i m_e}{A_i M}$	$\varepsilon_e \gg T_e$	$\varepsilon_i \gg \dfrac{T_k A_i}{A_k}$	$\varepsilon_i \gg \dfrac{A_i M}{m_e}\,T_e$

Concrete values for the individual τ_i are summarized in Table **1.2**. The use of equations (1.24) to (1.31) and tabulated values of the function $\gamma\,(3/2, x)$ allow the calculation of relaxation times τ_l, τ_t, τ_ε and τ_{tr} at different rates of beam temperature to medium temperature.

For completion let us consider the equilibrium restoration in a two-component medium of charged particles, where each of the two components has an individual temperature. In this connection it is necessary that equation (1.31) be integrated with the *Maxwell distribution function* for the particles of first species. The result therefrom is [38, 39, 40]

$$\frac{dT_1}{dt} = \frac{T_1 - T_2}{\tau_T^{12}} \tag{1.32}$$

where

$$\tau_T^{12} = \frac{3\sqrt{\pi}}{8}\,\frac{m_2}{m_1}\,\tau_1^{12}\left(T_1 + \frac{m_1}{m_2}\,T_2\right). \tag{1.33}$$

In addition,

$$n_1 \frac{dT_1}{dt} = -n_2 \frac{dT_2}{dt} \tag{1.34}$$

is valid.

The characteristic time of establishing a temperature equilibrium between two electron components of temperature T_1 and T_2 will then come to

$$\tau_T^{ee} = \frac{3\,\sqrt{m_e}\,(T_1 + T_2)^{3/2}}{8\,\sqrt{2\pi}\,e^4 L n_2} \tag{1.35}$$

and that of establishing a temperature equilibrium between electrons and ions of temperature T_e and T_i will come to

$$\tau_T^{ei} = \frac{3 \left(A_i M T_e + m_e T_i\right)^{3/2}}{8\sqrt{2\pi}\,\sqrt{A_i m_e M}\, i^2 e^4 L n_i} \ . \tag{1.36}$$

It should be noted that the equality relation of $\tau_T^{ei}/n_e = \tau_T^{ie}/n_i$ is valid too.

For ions of different charges i and k, masses $A_i M$ and $A_k M$ and temperatures T_i and T_k holds

$$\tau_T^{ik} = \frac{3}{8\sqrt{2\pi}} \frac{\sqrt{M}}{\sqrt{A_i A_k}} \frac{\left(A_k T_i + A_i T_k\right)^{3/2}}{i^2 k^2 e^4 L n_k} \ . \tag{1.37}$$

For explanation two concrete examples should be added:

1. Let a high-energy electron beam penetrate a low-energy electron-ion medium ($\varepsilon_e \gg T_i, T_e$). According to Table **1.2** if the ions are multiply charged the scattering and moderation of the beam at ions are faster than its scattering and moderation at electrons. On the other hand, addition, energy is preferably transferred between beam and electrons, which will be heated $A_i M / i^2 m_e$ times faster than ions will be.

2. Let there be a medium which consists of electrons and ions involved in collisions. Let there also be the additional assumption that the electron energy is much higher than that of the ions. Then it follows from equations (1.35) to (1.37) that

$$\tau_T^{ee} : \tau_T^{ii} : \tau_T^{ei} \cong 1 : \sqrt{\frac{A_i M}{m_e}}\, \frac{1}{i^2} : \frac{A_i M}{m_e i^2}$$

and considering $M/m_e = 1836$ it follows that τ^{ee}, $\tau^{ii} \ll \tau^{ei}$.

The establishment of equilibrium conditions for the individual components (and hence for ions of different kinds and charge states) goes on much faster then energy transfer and heat compensation between electrons and ions, since there are great differences in mass between electrons and ions. The following will in detail consider the establishment of equilibrium conditions within one component.

Generally, the motion of particles in the co-ordinate space described by r and the velocity v taking into account the collisions among the particles is defined by distribution function $f = f(r,v,t)$ which satisfies the kinetic equation [40]

$$\frac{df}{dt} = \text{Stf} \ . \tag{1.38}$$

The symbol Stf is referred to as the *collision integral*, characterizing the change of the distribution function due to impact processes. For the present examples the collision integral is defined by the small-angle Coulomb scattering, which according to the Rutherford scattering formula (1.15) will result in the same terms as given on the right side of equations (1.18) and (1.19). Following the stationary solution of (1.38) a *Boltzmann distribution*

$$f_0 \cong \exp\{-\frac{U(r)}{T}\}\ \exp\{-\frac{M v^2}{2T}\} \tag{1.39}$$

is obtained as the equilibrium distribution function; it leads to a *Maxwell distribution* (1.23) relating the particle velocities. $U(r)$ defines the potential energy of the particles.

For an exact determination of the time required to establish an equilibrium distribution, it would be necessary to solve the related kinetic equation, which in general is not possible. A possible way is to show [43] that this time is equal to

$$\tau_i(T) = \frac{\sqrt{M}}{\pi\sqrt{2}} \frac{1}{i^4 e^4} \frac{T^{\frac{3}{2}}}{Ln} \tag{1.40}$$

in orders of magnitude. $\bar{\varepsilon}$ gives the mean energy of all particles in the considered component and $T = 2\bar{\varepsilon}/3$ is valid. The value of $\tau_i(T)$ is similar to τ_T in (1.35), (1.36) and (1.37) for one kind of particle with common average energy.

The above has dealt with elastic Coulomb collisions between charged particles when a particle beam penetrates a medium, and for media consisting of different charged particles. The comparison between characteristic relaxation times and characteristic times of ionization and charge exchange processes shows that the values of τ^{ei} are comparable to ionization and charge exchange times, but τ^{ee} and τ^{ii} are considerably smaller than ionization times $\tau^i = 1/v_e\sigma^i n_e$ and charge exchange times $\tau^{ex} = 1/v_i\sigma^{ex}n_i$. Hence it follows that elastic collisions cause a redistribution of energy between different beam components and can have a strong influence on the reaction rates of ionization and charge exchange but also – and this is of great importance – on the rate of the ion and electron losses from sources of multiple-charged ions.

1.5 Balance Equation for Ion Charge States

For developing and operating sources of highly charged ions it is important that predictions being as reliable as possible will be made in respect to the occurrence of particular ion charge states and their intensity. So a formalism which is able to describe the *ion charge state distribution* within the source at any time is required. For that it is necessary to consider all processes which in the course of time will change the number of ions for the individual charge states and all ranges of source.

The change in ion number of a defined charge state found within the source is influenced by two main processes, namely

1. ion losses and the appearance of ions from outside the source, e.g. as a result of ion injection
2. ions changing from one charge state into another by ionization and charge exchange processes.

The probabilities that the above named processes will occur at all points of the ion source operating range is dependent on the immediate coordinates and velocities of the ions. So, in general, the kinetic equation (1.38) for the distribution function $f = f(r, v, t)$ forms the basis for the equilibrium equation. The right sight of (1.38) covers all processes which are relevant to producing and destroying ions of defined charge states. On the basis of the results shown in the preceding paragraphs, the following kinetic equation

$$\frac{df_i}{dt} = n_e v_e \sum_{k=1}^{n}(\sigma_{i-k}^{ki} f_{i-k} - \sigma_i^{ki} f_i) + n_0 \sum_{k=1}^{n}(v_{i+k} f_{i+k} \sigma_{i+k}^{k,ex} - v_i f_i \sigma_i^{k,ex}) + f_i\left(\frac{1}{\tau_i^b} - \frac{1}{\tau_i^l}\right) \tag{1.41}$$

can be given for the time derivation of the distribution function of ions with the charge i. Here n_e and n_0 are the electron density and the density of neutral atoms, resp. and τ_i^b and τ_i^l are the characteristic times of producing and loosing ions of charge state i.

Equation (1.41) considers ionization processes and charge exchange processes of neutral particles to ions of an ionization degree up to n. It should be noted here too that in general, processes being of an order higher than the second one will not be considered in real calculations. In addition, the quantities τ_i^b and τ_i^l in any case should be specified. So, when calculating the charge distribution in ECR ion sources, one of the most difficult and essential problems is to determine the rate of ion losses.

In the preceding paragraph it was shown that elastic Coulomb collisions result in a Maxwell distribution of the ion velocities. After integration of the two parts of (1.41) into the velocity space, considering

$$n_i(r,t) = \int f_i(v,r,t)\, dv$$

and by analogy with the kinetic equation a relationship for ion densities is obtained

$$\frac{dn_i}{dt} = n_e v_e \left(\sigma_{i-2}^{2i} n_{i-2} + \sigma_{i-1}^{i} n_{i-1} - \sigma_i^{2i} n_i - \sigma_i^{i} n_i \right) \tag{1.42}$$

$$+ n_0 (v_{i+2} n_{i+2} \sigma_{i+2}^{2ex} + v_{i+1} n_{i+1} \sigma_{i+1}^{ex} - v_i n_i (\sigma_i^{2ex} + \sigma_i^{ex})) + n_i \left(\frac{1}{\tau_i^b} - \frac{1}{\tau_i^l} \right).$$

This only considers single and double ionization processes and single and double charge exchange processes.

The determination of the charge dispersion requires the formulation of balance equations for each ion charge component which may occur in the source. If on the assumption that only ions of one element will occur in the source (the simultaneous occurrence of various elements does not alter the fact, but considerably increases the expenditure) a set of equations for all ion charge states Z and for neutral atoms results

$$\frac{dn_0}{dt} = -n_0 \left(\sum_{i=2}^{Z} \sigma_i^{ex} n_i v_i + \sum_{i=3}^{Z} \sigma_i^{2ex} n_i v_i + (\sigma_1^i + \sigma_1^{2i}) n_e v_e \right) + n_0 \left(\frac{1}{\tau_0^b} - \frac{1}{\tau_0^l} \right)$$

$$\frac{dn_1}{dt} = n_0 \left(\sigma_1^i v_e n_e + \sigma_2^{ex} n_2 v_2 + \sigma_3^{2ex} n_3 v_3 + \sum_{i=2}^{Z} \sigma_i^{ex} n_i v_i \right)$$

$$- n_1 \left(\sigma_2^i v_e n_e + \sigma_2^{2i} v_e n_e + \frac{1}{\tau_1^l} - \frac{1}{\tau_1^b} \right)$$

$$\frac{dn_2}{dt} = n_0 \left(\sigma_1^{2i} v_e n_e + \sum_{i=3}^{Z} \sigma_i^{2ex} n_i v_i \right) + n_1 \sigma_2^i v_e n_e + \left(\sigma_3^{ex} n_3 v_3 + \sigma_4^{2ex} n_4 v_4 \right) n_0$$

$$- n_2 \left((\sigma_3^i + \sigma_3^{2i}) v_e n_e + (\sigma_2^{ex} + \sigma_2^{2ex}) v_2 n_0 + \frac{1}{\tau_2^l} - \frac{1}{\tau_2^b} \right)$$

$$\frac{dn_i}{dt} = \sigma_i^i v_e n_e n_{i-1} + \sigma_{i-1}^{2i} v_e n_e n_{i-2} + \left(\sigma_{i+1}^{ex} n_{i+1} v_{i+1} + \sigma_{i+2}^{2ex} n_{i+2} v_{i+2} \right) n_0$$

$$- n_i \left((\sigma_{i+1}^i + \sigma_{i+2}^{2i}) v_e n_e + (\sigma_i^{ex} + \sigma_i^{2ex}) v_i n_0 + \frac{1}{\tau_i^l} - \frac{1}{\tau_i^b} \right) \tag{1.43}$$

$$\vdots \quad \vdots \qquad\qquad \vdots$$

$$\text{corresponding equations for all } 3 \leq i \leq Z - 2$$

$$\vdots \quad \vdots \qquad\qquad \vdots$$

$$\frac{dn_{Z-1}}{dt} = \left(\sigma^i_{Z-1} n_{Z-2} + \sigma^{2i}_{Z-2} n_{Z-3} \right) v_e n_e + \sigma^{ex}_Z n_Z v_Z n_0$$

$$-n_{Z-1} \left(\sigma^i_Z v_e n_e + (\sigma^{ex}_{Z-1} + \sigma^{2ex}_{Z-1}) v_{Z-1} n_0 + \frac{1}{\tau^l_{Z-1}} - \frac{1}{\tau^b_{Z-1}} \right)$$

$$\frac{dn_Z}{dt} = \left(\sigma^i_Z n_{Z-1} + \sigma^{2i}_{Z-1} n_{Z-2} \right) v_e n_e - n_Z \left((\sigma^{ex}_Z + \sigma^{2ex}_Z) v_Z n_0 + \frac{1}{\tau^l_Z} - \frac{1}{\tau^b_Z} \right) .$$

The quantity Z in (1.43) is the maximum ion charge state.

As a rule, the ion densities in the operating range of ion sources are not homogenously distributed. So ions of different charge states usually show different density distributions. In this connection it is a separate and difficult task to determine the spatial distribution of ions. But in many cases, if the density distribution is known or nearly constant, one simplifies matters by integrating (1.43) above the volume which the ions take up in the source. The set of equations (1.43) then develops into a set of equations for the total number N_i of ions of each charge stage.

The set of equations (1.43) (or the analogous set of equations for N_i) is a system of non-linear differential equations of the first order, which needs initial conditions for it's solution. The values of ion densities at time $t = 0$ have to be given as initial conditions. For example, if there are not any ions at the initial moment in the source, the following values can be assumed as initial conditions

$$t = 0; \quad n_0 = n^0_0; \quad n_i = 0 \qquad (i = 1, 2, \ldots Z) . \tag{1.44}$$

The formed set of equations with the above mentioned initial conditions is called the Cauchy problem.

In order to solve the Cauchy problem for (1.43) and (1.44), which is required for determining ion rates of particular charge states and their change in time, numerical methods are used. A common technique in this connection is the Runge-Kutta-method.

For a number of cases it is necessary to know the stationary solution of the balance equations. This is of relevance e.g. for studies at ECR ion sources, since these are working in a stationary mode of operation. In this case the left sides of (1.43) and (1.44) are equal to zero and the set of equations is transformed into a non-linear set of algebraic equations. If high non-linearities occur, which is often the case in practice, the solution of sets of algebraic equations by numerical methods will prove to be more difficult than that of sets of differential equations. In connection with solving the stationary problem, it is essential to find the appropriate initial approximation for the solution.

Sets of balance equations which describe processes of successive change or transition of a particle into others have been widely used in physics. So the Cauchy problem for equations of the type (1.43) was for the first time formulated and solved by Rutherford when he described radioactive decay [46, 47]. The above set of balance equations for particular ion charge states is used in variations for calculating ion charge distributions in particular ion sources, e.g. calculations of ion charge distributions in relativistic electron rings [14, 20, 21, 48, 49, 50, 51, 52, 53], in electron beams of EBIS sources (EBIS:

Electron Beam Ion Source) [54, 55], in gas-discharge ion sources [56, 57], in ECR (Electron Cyclotron Resonance) ion sources [58, 59, 60, 61, 62, 63, 64, 65, 66] and in electron beam ion traps (EBIT: Electron Beam Ion Trap) [67, 68].

Altogether there is a great number of publications dealing with the calculation of charge distributions. They do not only include calculations for charge state distributions in ion sources but also for those in arc discharges, laser and fusion plasma, and in electron-ion devices of different nature. All these publications use a set of balance equations of different forms for the particular ion charge states. To cite all the relevant treatises published so far would mean to be beyond the scope of this book. So only the most important publications which refer to ion sources should be named here. Some of them will be considered in more detail in a subsequent place when describing the special features of the particular ion sources.

1.6   Problems in Producing Multiple-Charged Ions

The production of multiple-charged ions in ion sources plays a central role in the range of themes discussed in this book. Therefore the aim of the present section is to consider general theoretical approaches to this theme.

Let us again look at the set of equations (1.43). This set considers various processes which occur due to collisions between charged particles. The first paragraph has shown that in this context the single ionization of atoms and ions by electron impact is generally the dominant process. If only these processes are considered, the set of equations will be considerably simplified and result in a system of linear differential equations

$$\frac{dn_0}{dt} = -\lambda_1 n_0$$

$$\vdots \quad \vdots \qquad \vdots$$

$$\frac{dn_i}{dt} = \lambda_i\, n_{i-1} - \lambda_{i+1}\, n_i \qquad\qquad (1.45)$$

$$\vdots \quad \vdots \qquad \vdots$$

$$\frac{dn_Z}{dt} = \lambda_Z\, n_{Z-1}\;.$$

Equations (1.45) use $\lambda_i = \sigma_i^i v_e n_e$.

In addition, it is assumed that there are only neutral particles in the initial moment, i.e. there are no ions in the beginning

$$t = 0; \quad n_0(0) = n_0^0; \quad n_i(0) = 0\;. \qquad\qquad (1.46)$$

The solution of the Cauchy problem is comparatively easy to find in this case and has already been known (see e.g. [69])

$$n_0 = n_0^0 e^{-\lambda_1 t}$$

$$n_1 = \frac{n_0^0 \lambda_1}{\lambda_2 - \lambda_1}\left(e^{-\lambda_1 t} - e^{\lambda_2 t}\right) \qquad\qquad (1.47)$$

$$n_2 = n_0^0 \lambda_1 \lambda_2 \left(\frac{e^{-\lambda_1 t}}{(\lambda_2 - \lambda_1)(\lambda_3 - \lambda_1)} + \frac{e^{-\lambda_2 t}}{(\lambda_3 - \lambda_2)(\lambda_1 - \lambda_2)} + \frac{e^{-\lambda_3 t}}{(\lambda_1 - \lambda_3)(\lambda_2 - \lambda_3)}\right)$$

$$\vdots \quad \vdots \qquad\qquad \vdots$$

$$n_i = n_0^0 \prod_{l=1}^{i} \lambda_l \left(\sum_{j=1}^{i+1} \frac{e^{-\lambda_j t}}{\displaystyle\prod_{k=1,k\neq j}^{i+1} (a_k - a_j)} \right) .$$

The set of equations (1.45) can also be integrated under other initial conditions [48, 49, 54, 69], so for example under $n_i(0) \neq 0$.

The solution (1.47) allows the production rate of ions of any charge state to be determined. Because of the step-by-step process of ion production (single-charged ions develop from neutral atoms, double-charged ions from single-charged ions etc.), a particular charge state grows from the preceding one and so the time of producing a charge state increases then the higher its charge will be.

Developing the exponents in (1.47) according to the powers of t results in a first approximation

$$
\begin{aligned}
n_0 &= n_0^0 \left(1 - \lambda_1 t \right)\\[4pt]
n_1 &= n_0^0 \, \lambda_1 \, t\\[4pt]
n_2 &= n_0^0 \, \lambda_1 \lambda_2 \, \frac{t^2}{2}\\[4pt]
n_3 &= n_0^0 \, \lambda_1 \lambda_2 \lambda_3 \, \frac{t^3}{6}
\end{aligned}
\qquad (1.48)
$$

$$\vdots \quad \vdots \qquad\qquad \vdots$$

$$n_i = n_0^0 \frac{t^i}{i!} \prod_{j=1}^{i} \lambda_j .$$

With (1.48) it is possible to determine the characteristic time required for producing each new charge state during the time of growth of the individual ion components

$$
\begin{aligned}
\tau_1 &= \lambda_1 t_1\\[4pt]
\tau_2 &= \sqrt{2\lambda_1 \lambda_2}\,\sqrt{2 t_1 t_2}\\[4pt]
\tau_3 &= \sqrt[3]{6\lambda_1 \lambda_2 \lambda_3}\,\sqrt[3]{6 t_1 t_2 t_3}
\end{aligned}
\qquad (1.49)
$$

$$\vdots \quad \vdots \qquad\qquad \vdots$$

$$\tau_i = \left(i! \prod_{k=1}^{i} t_k \right)^{\frac{1}{i}} .$$

The equations (1.49) use $t_i = 1/\lambda_i$.

To make the task more easy all neutral particles are assumed to be changed into single-charged ions at first. The characteristic time of this process is $\tau_1 \simeq t_1$. After time t_2 the single charged ions will be changed into double-charged ones, after time t_3 triple-charged ions will develop etc. Such an assumption results in other values of ion-producing times which are usually in use in ion source physics.

The characteristic ion-producing times are

$$
\begin{aligned}
\tau_1 &= t_1 \\
\tau_2 &= t_1 + t_2 \\
&\vdots \quad \vdots \quad \vdots \\
\tau_i &= \sum_{k=1}^{i} t_i \, .
\end{aligned}
\tag{1.50}
$$

Here one can see, that these formula differ from equations (1.49).

Using the *Stirling formula*

$$
x! \approx \sqrt{2\pi x}\, x^x e^{-x}
$$

and considering the fact that the geometric average does not exceed the arithmetical average, one sees that the times from (1.49) are always shorter than those from (1.50). This result is comprehensible since the assumptions on which (1.50) has been obtained lead to excessive values due to the fact, that every new charge state $i+1$ begins to appear as soon as some of i charge states has been produced not waiting before all charge states $i-1$ are transformed to the next higher charge state. On the other hand (1.49) describe the initial stage in the production of ions of higher charge states in a correct way for $t \ll \tau_k$, but for $t \sim \tau_k$ the formulas (1.49) give the lower values of the characteristic times due to the negative powers of the exponents in (1.47). In particular when ionization cross-sections decreases fast because of an increase of the ion charge, the values obtained from (1.49) and (1.50) can differ from each other by one order of magnitude and the values of the characteristic times τ_i are between them.

Equations (1.49) give characteristic times of producing ions of a given charge state, while the values from (1.50) can serve as an estimate of the time required for achieving a mean ion charge or the ion charge at the ion charge distribution maximum. The ionization time which has been obtained using (1.50) is in better agreement with the exact solution from (1.47) and in general, it is somewhat higher than that. In addition there is a satisfactory agreement with values from experiments concerning the storage of highly charged ions [6, 15, 70] and we shall use them in future.

Equations (1.50) can also be given in another, more useable form, namely

$$
j\tau_i = \sum_{k=1}^{i} \frac{1}{\sigma_k^i}
\tag{1.51}
$$

where the product $j = n_e v_e$ describes the *electron current density*. The quantity $j\tau_i$ is referred to as *ionization factor* and depends on the ionization cross-section.

In highly charged ion source physics the ionization factor is the fundamental which determines the obtainable ion charge; in this connection it is of the same significance as is the *Lawson criterion* in nuclear fusion research which determines the necessary confinement factor $n\tau$ in the plasma for nuclear fusion reactions.

The production of ions of mean charge $\bar{i}$ depends on the ionization that an appropriate value of $j\tau_i$ is obtained. If the velocity of the electron current in ion sources changes little only, it is necessarily for the production of a given ion charge to achieve sufficiently high values of electron density and ion lifetime in order to satisfy the condition

$$j\tau_i \geq \sum_{k=1}^{i} \frac{1}{\sigma_k^i}$$

to such a degree that the electron energy will be higher than the ionization potential I_i (according to paragraph 1.2: about 2...3 times higher than the ionization potential).

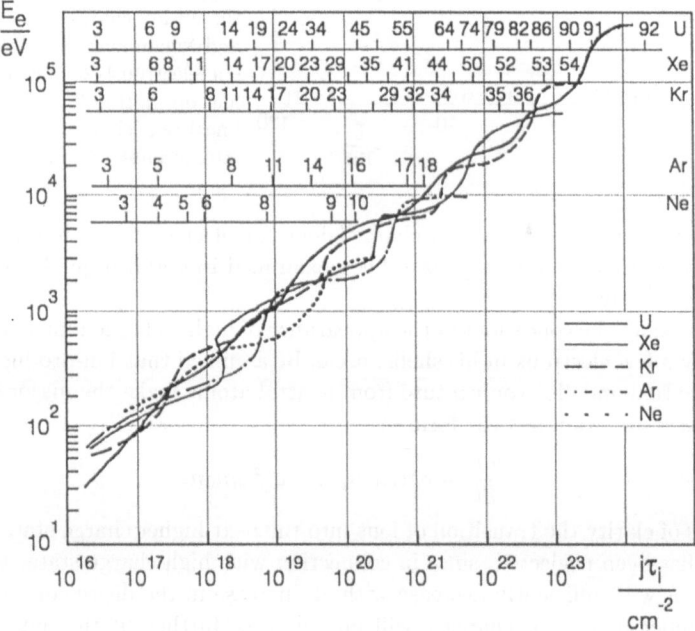

Figure 1.10 Calculated values of $j\tau_i$ required for producing Ne, Ar, Kr, Xe and U ions of a given charge state at corresponding energies of the bombarding electrons [6].

Fig.1.10 gives the values of the ionization factor to obtain respective ion charge states for Neon, Argon, Krypton, Xenon and Uranium ions at corresponding electron energies [6].

Although both the given values of the ionization factor and (1.49) and (1.50) reflect the real situation, they have been obtained for the ideal case where just single electron impact ionization by electrons is predominant. There are however a number of other processes which retard the process of producing multiple-charged ions. One exception of the rule is multiple ionization which increases the ion charge state. But the cross-section of this process is smaller than that of single ionization and for inner electron shells it can be neglected considering the whole process. All other processes, in particular ion loss and charge exchange processes, slow down the growth of ion charge or even prevent a further increase in ion charge. Ion losses depend on the nature of ion storage and ion energy; for different types of ion sources they are produced by different causes. With the presence of neutral particles within the operating range of a considered ion source the ion charge is reduced by charge exchange at the neutral particles.

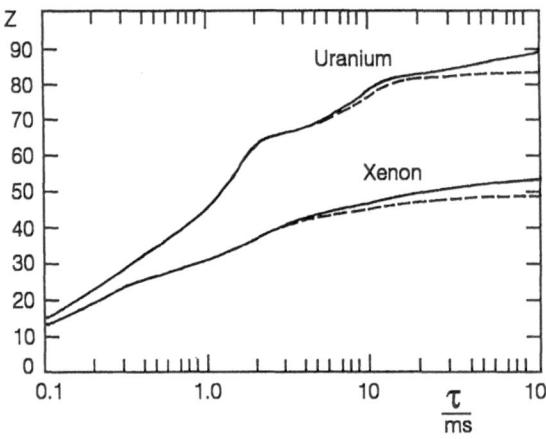

Figure 1.11 Time dependence of the mean charge state of Uranium and Xenon ions in rings with relativistic electrons[72]. The calculated values do (dotted lines) or do not (solid lines) take charge exchange processes into account.

Problems of producing multiple-charged ions for electron-beam ion sources (EBIS) and for relativistic electron rings have been examined in detail in [6, 15, 70, 71] and in [72, 73] respectively.

Let us consider the conditions of the appearance of highly charged ions. When a heavy ion has only a few electrons in his shells, it can be assumed that ion-producing processes due to ionization and electron capture from neutral atoms make the major contribution. So the balance equation has the form of

$$\frac{dn_i}{dt} = \sigma_i^i v_e n_e n_{i-1} - \sigma_i^{ex} v_i n_0 n_i \ . \tag{1.52}$$

For reasons of clarity the transition of ions into the next higher charge state $(i+1)$ due to ionization has been neglected, since in connection with high charge states the ionization cross-sections will sufficiently decrease with an increase in the degree of ionization.

The number of ions of charge i will not increase further, if the left side of (1.52) becomes zero

$$\sigma_i^i v_e n_e n_{i-1} - \sigma_i^{ex} v_i n_0 n_i = 0 \ . \tag{1.53}$$

If one abstains from a general validity in the description, $n_{i-1} = n_i$ can be assumed. This is allowed since there are not only ions of one charge state but a charge distribution for which the densities of ions of consecutive charge states generally differ from each other little only. Using the ionization and charge exchange cross-sections from (1.8) and (1.12) and the dependence given in (1.14)

$$\frac{n_0}{n_e} = 2.7 \sqrt{\frac{A}{E_i E_e}} \frac{I_0^\beta}{i^{\alpha+2}} \log \frac{E_e}{I_i}$$

results from (1.53).

If the value of electron energy is within the range of the ionization cross-section maximum, $\log\left(E_e/I_i\right) \approx 1$ will be valid. In addition, using the values of $\beta \approx 3$ and $\alpha \approx 1$ from Table **1.1** the result is a condition for the density of neutral particles, in which charge exchange processes can be neglected

$$\frac{n_0}{n_e} \ll 5n \sqrt{\frac{A}{E_i E_e}} \frac{I_0}{i^3} \tag{1.54}$$

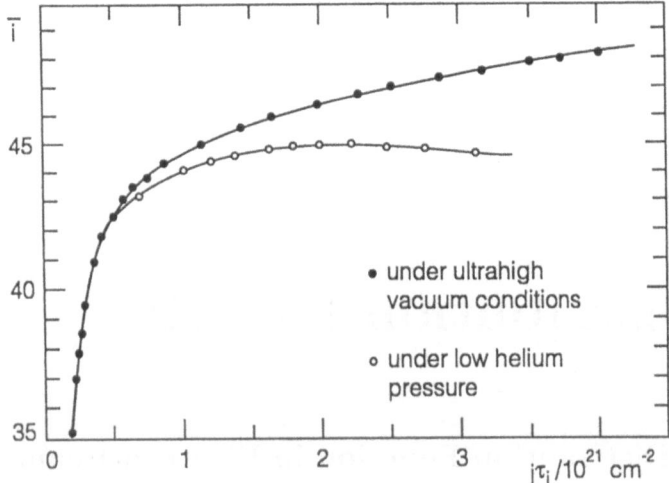

Figure 1.12   Dependence of the mean charge state of Xenon on the ionization factor $j\tau_i$ in the EBIS KRION-2 [70]

where the quantity n describes the number of electrons in the outer electron shell of the considered ion of charge state i. A is the atomic mass, E_i the ion energy and I_0 the ionization potential of the neutral atoms.

To give an example, Fig.**1.11** shows the calculated mean ion charges of Uranium and Xenon in electron rings and their parameters [73]

radius	$R = 2$ cm
radius of the ring cross-section	$a = 0.1$ cm
electron number	$N_e = 1 \cdot 10^{13}$
electron energy	$E = 20$ MeV

The solid line in Fig.**1.11** characterizes how the mean ion charge in the electron ring depends on the ion collection time in case of absent neutral particles in the ring volume. The dotted line illustrates the same dependence in regard of charge exchange of neutral particles (Nitrogen at a pressure of $2 \cdot 10^{-9}$ Torr) at Uranium and Xenon ions (Nitrogen at a pressure of $3 \cdot 10^{-9}$ Torr).

Fig.**1.12** illustrates the dependence of the mean ion charge of Xenon in an electron-beam ion source on ionization factor $j\tau_i$ at a residual gas pressure of 10^{-12} Torr (solid line) and at 10^{-9} Torr (dotted line) [70].

Summing up the conditions of producing highly charged ions for any electron-impact ion sources may be stated:

1. The ionization energy must be higher than the ionization potential of ions of a charge state to be produced.
2. An appropriate value of the ionization factor (1.51) must be guaranteed.
3. The density of neutral particles should be kept as low as possible. To estimate the admissible density of neutral particles (1.54) can be used.

For different types of electron-impact ion sources additional processes which limit the obtainable ion charge or cause ion losses may occur; but these three conditions must always be satisfied.

Chapter 2

Ion Distribution Function

2.1   Ion Distribution Function in Electron Beams

2.1.1   Kinetic Vlasov Equations

The balance equations introduced in the preceding chapter and which take electron-ion and ion-ion impact processes into consideration, are able to describe the transition of ions from one charge state into another. But as has already been shown, the probability of these processes also depends on the spatial and velocity distributions of the colliding particles. The distribution of the particles within the phase space of coordinates r and velocities v are described by the distribution function $f(r, v, t)$.

In consideration of the particles' electromagnetic eigenfields and of the impact processes which take place there, the movement of charged particles is described in a generalized form by a system of self-consistent Vlasov equations which include the kinetic equations (1.38) completed by Maxwell equations [40]. For charged particles, the kinetic equations are represented in the form as follows

$$\frac{df}{dt} = \frac{\partial f}{\partial t} + \vec{v}\,\frac{\partial f}{\partial \vec{x}} + e\left(\vec{E} + \frac{1}{c}\,[\vec{v}\vec{B}]\right)\frac{\partial f}{\partial \vec{p}} = \mathrm{Stf} \tag{2.1}$$

in considering impact processes. Here $\vec{p} = m\vec{v}$ is a particle pulse and $\vec{x}$ are the coordinates in Decarts coordinate system.

The powers of electric field $\vec{E}$ and of magnetic induction $\vec{B}$ result from the Maxwell equations

$$\mathrm{rot}\,\vec{E} = -\frac{1}{c}\frac{\partial \vec{B}}{\partial t} \qquad\qquad \mathrm{div}\,\vec{B} = 0$$

$$\mathrm{rot}\,\vec{B} = \frac{1}{c}\frac{\partial \vec{E}}{\partial t} + \frac{4\pi}{c}\,\vec{j} \qquad\qquad \mathrm{div}\,\vec{E} = 4\pi\varrho \tag{2.2}$$

with the corresponding initial and boundary conditions.

The field sources are divided into external sources and sources related to the particles of the beam, with

$$\varrho = e\int f\,d^3p$$

being valid for charge density and

$$\vec{j} = e \int \vec{v} f \, d^3 p$$

for current density.

If one regards several particles of different kinds, a system of analogous equations will result instead of (2.1). In the Vlasov equation the collision integral Stf generally describes the elastic interactions of the particles [40]. In case of an equation system for several ionic components and electrons, based on (2.1), the collision integral has to consider such terms which take inelastic processes, i.e. ionization and charge exchange processes, into consideration.

For the majority of practical cases the collision time between equal particles or with others, $\tau = 1/n\sigma v$ (σ and v are the scattering cross-section and the relative characteristic velocity of the bombarding and the scattering particles of particle density n) exceeds the lifetime of the beam. For that reason, the collision integral in (2.1) is negligible, which results in the *Liouville equation* as a continuity equation within the phase space for Hamilton systems

$$\frac{\partial}{\partial \vec{x}} \left(\frac{d\vec{x}}{dt} \right) + \frac{\partial}{\partial \vec{p}} \left(\frac{d\vec{p}}{dt} \right) = 0 \ .$$

For non-Hamilton systems the continuity equation is used in the form of [74]

$$\frac{df}{dt} + \frac{\partial}{\partial \vec{x}} \left(f \frac{d\vec{x}}{dt} \right) + \frac{\partial}{\partial \vec{p}} \left(f \frac{\partial \vec{p}}{dt} \right) = 0 \ . \tag{2.3}$$

Equation (2.3) which neglects the collision integral can also be used for describing relativistic particle movements. Relativistic Coulomb impacts can be considered in the kinetic equation for relativistic plasmas in the form given by Lifshitz and Pitaevskii [40] and Alder et al. [75].

To solve these Vlasov equations presents great difficulties. Therefore in practice some few available beam models which are exactly solvable [76, 77] are used, or various numerical methods are applied for solving the Vlasov equations [75, 78, 79]. For a number of important cases the ion distribution function and averaged characteristics of the ionic components resp. can be obtained by simplier methods.

2.1.2   Ion Distribution Function in Electron Beams

In sources of highly charged ions, such as electron beam ion sources (EBIS) or electron ring ion sources (ERIS), the ions are formed and stored in electron beams or electron rings. In the following we will in detail reflect on the ion distribution in the characteristic cases of an uniform electron density in a beam and of a beam showing Gaussian distribution of electron density in the beam cross-section. These cases have beeen studied in Refs.[8], [80] to [86], which started from the assumptions that the charge of the stored ions would be small as compared with the electron charge, and the fields produced by the ions would not influence particle movements. The results achieved that way will also be valid for sufficiently thin electron rings, if the effects related to the particles' curvature of the path are negligible.

The single charged ions in the electron beam are found in the potential gap of the Coulomb field produced by the electron charges, and oscillate there. The initial kinetic energy of the ions depends on the thermal energy of the neutral atoms and the energy transferred during collisions with electrons; it comes to a split of electron volts. In this case the depth of the Coulomb potential gap acquires a value of $U = 10 \ldots 10^5$ eV at a linear electron density of $N_e = 10^8 \ldots 10^{12}$ cm^{-1}. Thus, the formed ions remain in the beam volume, being able to experience further ionization by electron impact.

The reflections in the following will be limited to a constant electron distribution function. The velocity v_e of the longitudional electron movement is assumed to be much higher than their transversal velocity. In this case the relative impact velocity of collisions between electrons and ions, which lead to ionization processes, is equal to v_e. The transversal velocity does not influence the ion storage process and it is sufficient to know the electron density ϱ_e in the beam cross-section, which can be regarded as a permanent electron background for ions.

If the contributions of charge exchange processes, multiple ionization processes and elastic collisions is assumed to be small, the ion distribution function f_i will satisfy the following kinetic equation by analogy with the balance equations

$$\frac{df_i}{dt} = \varrho_e c \left(\sigma_{i-1} f_{i-1} - \sigma_i f_i \right) \tag{2.4}$$

where σ_i is here an electron impact cross-section for ions of the charge state i.

If the collection and storage of the ions are sufficiently long, the transition processes of ions from one charge state into another develop much more slowly than the movement of ions within the beam so that the ion distribution function can be considered to be approximately stationary. If the right sides in the system of equations (2.4) are neglected, the equations for the stationary distribution function will result

$$\frac{df_i}{dt} = 0 \ .$$

The stationary distribution function can only be a function of the integrals of motion. In this connection every ion can have two integrals of motion, namely the angular momentum relative to the beam axis $M_\varphi = m v_\varphi r$ and, if elastic collisions between ionization acts are neglected, the total energy E_i. For an azimuthally symmetric beam being circular in the beam cross-section, it is possible to assume that, if the energy of the neutral particles is relatively small $E_0 = 0$ and $v_\varphi = 0$ are valid for the angular frequency of the ions. This means that all ions which oscillate in the potential gap of the electrons cross the beam centre. So for ions of all charge states yields

$$f_i = f_i(E_i) \, \delta(M_\varphi) \tag{2.5}$$

where $\delta(x)$ is the Dirac qδ-function.

Between ionization acts the total energy E_i of every ion is maintained

$$E_i = \frac{mv^2}{2} + ieU(r) \ .$$

In addition, an ion of charge $(i-1)$ is assumed to have the distance $r = r_0$ from the beam centre and the velocity $v = v_0$ before its ionization. Then its energy will be

$$E_{i-1} = \frac{mv^2}{2} + (i-1)\,eU(r) = \frac{mv_0^2}{2} + (i-1)\,eU(r_0) \ .$$

At the moment of ionization the potential energy of the ion increases to $eU(r_0)$ and for the total energy yields

$$E_i = \frac{mv^2}{2} + (i-1)\,eU(r) + eU(r_0) = \frac{mv_0^2}{2} + ieU(r_0) = \frac{mv^2}{2} + ieU(r) . \tag{2.6}$$

Before starting the next ionization act, the ion of charge i continues to move at this energy on the energy surface $E_i(r(t), v(t)) = E_i(r_0, v_0)$.

2.1.3   Electron Beams with Constant Density over the Beam Cross-Section

An example of an electron beam showing a constant density over the beam cross-section is an electron beam ion source where the cathode of the electron gun and also the electron beam itself are located in a strong longitudonal magnetic field and the electrons are moving strictly alongside the field lines of the magnetic field.

In electron beams with a constant density over the beam cross-section the ionization probability is assumed to be the same regardless of the beam energy and the distance from the beam axis (radius). Therefore, the ion distribution function for ions of charge state i coincides with the ion distribution function for the preceding ion charge state: $f_i^0 = f_{i-1}$. After some oscillation periods in the potential gap the ions are mixed in such a way that their distribution function comes close to the stationary value f_i. But the total energy of every ion remains unchanged until the next ionization act. By averaging (2.6) over all spatial coordinates and velocities the mean energy together with the distribution function will result

$$\overline{E_i} = \frac{\overline{mv_i^2}}{2} + i\overline{eU}(r_i) = \frac{\overline{mv_{i-1}^2}}{2} + i\overline{eU}(r_{i-1}) . \tag{2.7}$$

If the potential of the electric field depends on a power of the radius in the form $U = U_0 r^\alpha$ then, according to the Virial law,

$$\frac{\overline{mv_i^2}}{2} = \alpha i \overline{\left(e\frac{U}{2}\right)}$$

will be valid; then in considering (2.6) and (2.7)

$$\begin{aligned}
\overline{r_i^\alpha} &= \overline{r_{i-1}^\alpha}\left(1 - \frac{\alpha}{i(\alpha+2)}\right) \\
\overline{v_i^2} &= \overline{v_{i-1}^2}\left(1 + \frac{2}{(i-1)(\alpha+2)}\right)
\end{aligned} \tag{2.8}$$

follows.

At constant electron density in the beam cross-section the field is linear alongside the beam cross-section and the potential represents a quadratic function of the radius. In this case $\alpha = 2$ is valid. Hence it follows that for the mean value of the square of distance from the beam axis [8, 72] yields

$$\overline{r_i^2} = \overline{r_{i-1}^2}\left(1 - \frac{1}{2i}\right) = \overline{r_1^2}\,\frac{2\,\Gamma\left(i+\frac{1}{2}\right)}{\Gamma\left(\frac{1}{2}\right)\Gamma(i+1)}$$

$$\overline{r_1^2} = \frac{\overline{r_e^2}}{2} \qquad \overline{r_e^2} = \frac{a^2}{2} \tag{2.9}$$

where a is the radius of the electron beam.

For the mean value of the square of velocity

$$\overline{v_i^2} = \overline{v_{i-1}^2}\, \frac{2i-1}{2(i-1)} = 2\overline{v_1^2}\, \frac{\Gamma\left(1+\frac{1}{2}\right)}{\Gamma\left(\frac{1}{2}\right)\Gamma(i)}$$

$$v_1^2 = a^2\, \frac{\omega_1^2}{4} \tag{2.10}$$

is valid, with ω_1 being the oscillation frequency of single-charged ions in the field of the electrons. The same results have already been achieved by other methods in Refs.[80, 81, 84].

Equations (2.9) and (2.10) allow the so-called *effective phase space volume* of the ions' transversal moment in the beam to be determined

$$F_i^2 = \overline{v_i^2}\,\overline{r_i^2} = \left(\frac{\Gamma\left(i+\frac{1}{2}\right)}{\Gamma\left(\frac{1}{2}\right)\Gamma(i)}\right)^2 \frac{a^4\omega_i^2}{4i}\ . \tag{2.11}$$

E.g. divergence and emittance of the ion beam found at the exit of the source depends on this quantity.

In the following the quantities of $\overline{r^2} = a^2$ and $\overline{v^2} = u^2$ will be introduced. Here a is the root-mean-square radius of the beam and u is the root-mean-square velocity of the ionic component.

For high ion charges ($i \gg 1$) it follows from (2.9), if ion charge increases, the root-mean-square radii of the individual ion charges with [86]

$$a_i^2 = \frac{2a_1^2}{\sqrt{\pi i}}\ . \tag{2.12}$$

The effective phase volume which slightly increases in the transition from one charge state into the next higher one comes close to the constant value

$$F_i = \frac{2F_1}{\sqrt{\pi}}\ . \tag{2.13}$$

The root-mean-square quantities of a^2 and u^2 as well as the phase space volume F sufficiently describes the properties of all ionic components of the beam; in many cases they allow the problem to be solved without exactly knowing the distribution function.

For a constant density of the electrons and neutral particles in the beam it is possible that for the stationary case the distribution function for single-charged ions will be expressed by means of the integrals of motion in the following form [83]

$$f_1 = \frac{N_i A M}{4\pi^2 a_1^2 \omega_1}\, \sigma\left(E_m - E_1\right) \delta\left(M_\varphi\right) \tag{2.14}$$

where $E_m = N_e r_e m_e c^2$ is the ion energy on the edge of the ion beam.

By integration of (2.14) in the velocity space (v, v_φ) the ion density in the beam cross-section is maintained

$$\varrho = \frac{N_1}{2\pi^2 a^2}\, \frac{\sqrt{a^2 - r^2}}{r}\ . \tag{2.15}$$

The density obtained that way shows a non-physical pecularity alongside the beam axis; its occurence is related with the fact that the velocities of the neutral particles and the energy transfer in elastic collisions between ions have not been considered, which leads to deviations of the ions from the ions' trajectory of movement running through the beam centre. In this connection one should add that this pecularity is integrable and does not result in an infinite number of ions alongside the beam axis when determining the total number of particles by integration of distribution (2.15) with respect to beam volume.

The root-mean-square radius a_1 and velocities u_1 of single-charged ions, determined for the distribution function (2.14) and density (2.15), coincide with the results achieved from (2.9) and (2.10).

2.1.4 Ion Distribution Function in Electron Beams of Gaussian Density

In the following let us regard an electron beam which is found in external focusing transversal fields changing slowly in comparison to the characteristic ion confinement time. Let the ion confinement time be much longer than the period of oscillations of the electronic and ionic components. This is true for accelerators and storage rings for electrons. For that case ion storage in an electron beam can be considered as stationary. Then the ion beam shows a Boltzmann distribution and the electron density in the beam cross-section has the value

$$\varrho_e \approx \exp\left(-eU_e(r)/2E_e\right)$$

where $eU_e(r)$ is the potential energy of the electrons in the external field and E_e is the mean energy of transversal electron movement. Considering thin beams, for a great number of cases the external fields can be assumed to be linear in a first order approximation. Thus U_e is a quadratic function of the beam radius and the electron density in the beam cross-section can be approximated as a Gaussian distribution to a sufficient degree of accuracy

$$\varrho_e = \frac{\exp(-r^2/2a^2)}{a^2} . \tag{2.16}$$

In such an electron beam which has an electron density changing alongside the beam cross-section the ionization probability depends on radius. For that reason ions of different total energies show different probabilities of further ionization. Ions of small energy, which are found in the centre of the potential gap, perform their movements in the beam centre where electron density is at its maximum, and show a high probability of being further ionized. In contrast to this, ions of high energy stay on the edge of the beam for a longer period; there the electron density is small, and so is the probability of ionization. For that reason ions of small energy are ionized more often and the mean energy of ions of every new charge stage is smaller than that obtained for beams of constant electron density. The root-mean-square dimensions and velocities are small accordingly.

For that case the stationary ion distribution function is also a function of the integrals of motion and depends only on the total energy of the ion oscillations and their angular momentum (see (2.5)).

The electron distribution (2.16) has a nonlinear electric eigenfield

$$F_e(r) = \frac{2eN_e}{r}\left[1 - \exp\left(\frac{r^2}{2a^2}\right)\right] . \tag{2.17}$$

But the majority of ions develop and move within the beam axis where the fields are approximately linear. This can be proved by means of a series expansion of the exponent in (2.17) according to powers of r. Therefore all ions are assumed to make harmonic motions at a frequency of

$$\omega_i^2 = \frac{iN_e r_e m_e c^2}{AMa^2} \ .$$

(2.18)

Then the total energy of the ions is

$$E_i = \frac{AMv^2}{2} + \frac{AM\omega_i^2 r^2}{2}$$

where the energy of single-charged ions requires the value of

$$E_1 = \frac{AM\omega_1^2 r_0^2}{2} = AM \frac{v^2 + \omega_1^2 r^2}{2} \ .$$

The initial distribution function of i-fold charged ions at the moment of their formation is proportional to the ion distribution function for the preceding charge state and the electron density

$$f_i^0 \sim f_{i-1}\varrho_e \ .$$

For an uniform distribution of the density of neutral atoms and of the electron density (2.16)

$$f_1^0 \sim \exp\left(\frac{-r_0^2}{2a^2}\right) = \exp\left(\frac{-E_1}{2AM\omega_1^2 a^2}\right)$$

will result.

Here f_1^0 is only a function of integrals of motion and does not depend on time, i.e. the function is stationary. If f_1^0 is assumed to equal f_1 and f_1 is normalized to one, it follows that

$$f_1 = \frac{AM}{\pi a_1^2 \omega_1} \exp\left(-\frac{\omega_1^2 r^2 + v^2}{2a_1^2 \omega_1^2}\right) \delta\left(M_\varphi\right)$$

(2.19)

where $a_1 = a$.

A consideration of $f_2, f_3 \ldots$ using the relation of

$$\omega_i^2 = \frac{i\,\omega_{i-1}^2}{i-1}$$

and starting from the assumption that

$$a_i^2 = \frac{(i-1)\,a_{i-1}^2}{i}$$

will result in

$$f_i^0 \sim f_{i-1}\varrho_e \sim \exp\left[-\frac{v_0^2}{2\omega_{i-1}^2 a_{i-1}^2} - \frac{r_0^2}{2a_{i-1}^2}\left(1 + \frac{1}{i-1}\right)\right] = \exp\left[-\frac{E_i}{AM\omega_i^2 a_i^2}\right]$$

and finally in

$$f_i = \frac{AM}{\pi a_i^2 \omega_i} \exp\left[-\frac{r^2 + \frac{v^2}{\omega_i^2}}{2a_i^2}\right] \delta\left(M_\varphi\right) \ .$$

(2.20)

In [87] approximated solutions considering the ion distribution of a number of low charge states are given for electron beams of Gaussian density distribution; they start from other initial conditions than those set here.

An exact solution for determining the ion distribution function can be found by integration of the kinetic equation with respect to its characteristic quantities. For ions of the first charge state such a solution has been given in [84] where a result being identical to (2.20) was obtained.

An integration of (2.20) over velocitiy space results in

$$\varrho_i = \sqrt{\frac{2}{\pi}} \; \frac{\exp\left(-\dfrac{r^2}{2a^2}\right)}{a_i r} \; . \tag{2.21}$$

The root-mean-square dimensions of the beam cross-section and the velocities of the ionic components are usually determined by the integration of distribution function (2.20) over all spatial coordinates and velocities

$$a_i^2 \;\; = \;\; \frac{a_1^2}{i} \tag{2.22}$$

$$u_i^2 \;\; = \;\; a_i^2 \omega_i^2 = a_1^2 \omega_1^2 = \text{const.} \; . \tag{2.23}$$

These dependences correspond with the assumptions made in derivation of (2.20) and concerning the dependence of the root-mean-square radius of the ion charge.

Terms (2.22) and (2.23) allow the effective phase space volume of the ions to be determined

$$F_i = a_i^2 \, \omega_i = \frac{a_1^2 \, \omega_1}{\sqrt{i}} \; .$$

The stationary ion distribution function (2.20), density (2.21) and the relations in (2.22) and (2.23) will be valid, if the distribution function changes little only when a new charge state is set up. Such kind of conditions can be satisfied e.g. for the occurence of a considerable current of ions newly formed from neutral particles. The consideration of a density which is not equally distributed in the beam cross-section leads to the fact that in a successive storage of ion charge states the root-mean-square dimensions of the ionic components will reduce more quickly than those in beams of a uniform density distribution. The effective phase space volumes will reduce simultaneously, too.

The results achieved show that multiple charged ions are mainly located around the beam axis. This effect is more felt with beams where density increases towards the centre. Under conditions where the ion charge is small as compared to electron charge, the ions will fall down to the potential gap when their charge state is increased. This effect reduces possible ion losses due to elastic ion collisions with electrons. In the case that the electron density increases towards beam axis, the reduction in the root-mean-square dimensions of the ionic components will lead to an increase of ionization velocity when ion charge increases. At the same time, a sufficiently long storage of ions in the electron beam will result in an increasing ion heating by electron impacts and an increase in transversal oscillation energy of the ions. As a result of this process the relations (2.12) and (2.22) can be strained and the oscillation amplitudes of the ions can increase in such a way that ions from the beam can be lost. Neutralization of the electronic space charge by stored ions can have analogous effects.

Except for the fact that the ion distribution functions (2.14) and (2.20) and the relations (2.9) to (2.13), (2.22) and (2.23) have been maintained within a greatly simplified model, these results are not only of a methodical interest, but they are also of a great importance for understanding physical processes in the formation and storage of ions in electron beams and sources of highly charged ions.

2.2   Momentum Method and Distribution Function

The method of complete moments of the distribution function is one of the most successful current approaches to solving the system of Vlasov equations; it has been elaborated by E.A.Perelstein and his staff. The fundamentals of this method are explained in [88]. In [8] the momentum method is considered for problems which are related to multicomponent electron-ion beams.

In analyzing beams of charged particles, in particular for sources of multiple-charged ions, especially the averaged beam characteristics, such as mean velocity, root-mean-square dimensions and temperature are of interest. The advantage of the momentum method is to give a description of beam dynamics, being short as compared to the Vlasov equations, and so to allow the principal laws to be followed. The relations introduced in the following will be guided mainly by the Refs.[8, 88].

2.2.1   Selfconsistent Description of the Movement of Beams of Charged Particles

There are several models for describing time behavior with respect to envelopes and the root-mean-square dimensions of beams of charged particles in electromagnetic fields. To determine envelopes, $a_{1,2} = \max(x_1, x_2)$ is valid; the maximum within the whole phase space which the beam achieves is unknown. For monoenergetic beams which show an equally distributed density in an elliptical beam cross-section and non-correlated transversal particle motions, the model of Vladimirskii-Kapchinskii [76, 77, 78] is often used.

The solution of (2.1) is an arbitrary function of the integrals of motion not considering collision processes. A general integral is known for the movement of particles in fields deflecting in a linear transversal direction with respect to x_1, x_2 and showing an arbitrary time-dependence (longitudional coordinate z). In that case the equation of motion assumes the following form

$$\frac{dx_i}{d\tau} = v_i; \qquad \frac{dv_i}{d\tau} = -\omega_i^2(\tau)\, x_i \qquad i = 1, 2, \dots \tag{2.24}$$

Depending on the formulation of a problem the variable τ is identical to time t or is connected with the coordinate of beam cross-section z.

By introducing the matrices

$$Y_i = \begin{pmatrix} x_i \\ v_i \end{pmatrix} \qquad \text{and} \qquad P_i = \begin{pmatrix} 0 & 1 \\ -\omega_i^2(\tau) & 0 \end{pmatrix} \tag{2.25}$$

the equations of motion take the form of

$$\frac{dY_i}{d\tau} = P_i Y_i \tag{2.26}$$

with

$$Y_i|_{\tau=0} = \begin{pmatrix} x_{i0} \\ v_{i0} \end{pmatrix} = Y_{i0} \tag{2.27}$$

being valid for the initial conditions.

For a ring-like beam e.g. in a weakly focusing magnetic field

$$\omega_r^2 = \omega_0^2 (1 - n) ; \qquad \omega_z^2 = \omega_0^2 n \tag{2.28}$$

is valid, where ω_0 is the angular frequency of the particles and n is the magnetic field index.

The solution of the Cauchy problem in (2.26) and (2.27) is a linear function with the initial conditions as follows

$$Y_i(\tau) = R(\tau) Y_{i0} \tag{2.29}$$

where matrix $R(\tau)$ satisfies (2.26) and its initial conditions $R(0) = I$ with I as the unity matrix.

Let an arbitrary symmetrical square matrix A_0 ($A_0 = \tilde{A}_0$ with A_0 as the transpose matrix) of dimension 2x2 be used for developing the quadratic form

$$F(Y_0) = Y_0 A_0 Y_0 . \tag{2.30}$$

By construction of the matrix for the initial conditions as obtained by putting the results from (2.29) in (2.30) the integral of motion for the set of equations (2.24) is the result; its form is as follows

$$F(Y) = \tilde{Y} A(\tau) Y = F(Y_0) = \text{const.} \tag{2.31}$$

with

$$A(\tau) = \tilde{R}^{-1} A_0 R^{-1} \tag{2.32}$$

where R^{-1} is the inverse matrix of R.

If the elements of matrix A are expressed by values of envelope a and their derivation, then the known equation [76, 77, 78] results from the integral of motion for the envelope

$$\frac{d^2 a_i}{d\tau^2} + \omega_i^2 a_i - \frac{\varepsilon_i}{a_i^3} = 0 \tag{2.33}$$

with $\varepsilon = F(Y_0)$ is the beam emittance divided by π. The envelope (2.33) is valid for the movement of charged particles in electromagnetic external and eigenfields being linear with respect to x_1 and x_2.

The model of Vladimirskii-Kapchinskii has been generalized by Yarkovoi who was successful in solving the problem with regard to the transversal motion and the inertial forces (mass change of the particles due to acceleration) [91].

In order to partially remove from the model character of the problem, an approach was developed in the early 70's; instead of the envelope it uses the root-mean-square beam dimensions for a large class of phase distributions and obtains self-consistency by approximation [92, 93, 94].

In this connection the result achieved by Sacherer [93] is of particular interest. This is based on the fact that for the two-dimensional case with an elliptical geometry (or the three-dimensional case with an ellipsoidal geometry) of charge distribution in the beam cross-section, the equations for the root-mean-square dimensions do not depend on the form of the distribution.

Another solution of the self consistent problem (equations (2.1) and (2.2)), which has been given by Mirer [95], uses a model distribution depending in particular on the quadratic form (2.31). The equations for the elements of the matrix A are obtained by minimization of the functional

$$\Phi(f) = \int \left(\frac{df}{d\tau}\right)^2 dY \tag{2.34}$$

where integration is done over the whole phase space. The exact solution corresponds with $\Phi(f) = 0$.

2.2.2   Moments of the Distribution Function

Into the class of distributions functions integrated by an arbitrary power of the weighting function, in [89] has introduced moments of the distribution function for the entirety of all phase space coordinates. In this connection it should be noted that in the hydrodynamics and gasdynamics only spatially infinite distributions are treated and for that reason only velocity moments are discussed.

Let us consider the first moments of the distribution function. The moment of zeroth order

$$N = \int f(Y, \tau) \, dY \tag{2.35}$$

is an integral of motion according to (2.1) and equals the total number of particles in the beam, with Y describing the vector for all phase space coordinates. If Y characterizes the vector of transversal phase coordinates and refers to a monoenergetic beam N will indicate the linear particle density in the beam.

The moments of the first order are described by the vector

$$\overline{Y} = \frac{1}{N} \int Y f(Y, \tau) \, dY \tag{2.36}$$

where the spatial vector components of $\overline{Y}$ indicate localization of the mass centre and the velocity components determine mean velocities. For fields being linear with respect to the phase space coordinates where the equations of motion for the particles assume a form according to (2.26), the moments of the first order will satisfy equation

$$\frac{d\overline{Y}}{d\tau} = P\overline{Y} \tag{2.37}$$

the solution of which is found according to (2.29).

The central moments of the second order can be written in form of a symmetrical square matrix

$$M = \frac{1}{N} \int \left(Y - \overline{Y}\right) \left(\tilde{Y} - \tilde{\overline{Y}}\right) f(Y, \tau) \, dY \ . \tag{2.38}$$

The diagonal elements of Matrix M provide the root-mean-square dimensions and the root-mean-square deviations of the particle velocities in the beam. Then a differentiation of (2.38) with respect to τ, using the continuity equation in the phase space for the case of fields being linear regarding the coordinates, will result in

$$\frac{d\mathrm{M}}{d\tau} = \mathrm{PM} + \mathrm{M}\tilde{\mathrm{P}} \ . \tag{2.39}$$

2.2.3   One-Dimensional Non-Relativistic Case

For reasons of illustration let us consider a band-shaped non-relativistic beam of particles [89]. The equations for transversal motion of the particles in the beam are of the form

$$\frac{dx_1}{dt} = x_2 = F_1 \qquad \text{und} \qquad \frac{dx_2}{dt} = F_2(x_1, t) \ . \tag{2.40}$$

The distribution function satisfies the Liouville equation

$$\frac{\partial f}{\partial t} + v_0 \frac{\partial f}{\partial z} + \sum_i F_i \frac{\partial f}{\partial x_i} = 0 \ . \tag{2.41}$$

Furthermore, let an infinite system of moments which is connected with the distribution function through the relations

$$\mathrm{M}^{p,q}(t, z) = N \, \overline{x_1^p x_2^q} = \int_\Omega x_1^p(t) \, x_2^q(t) \, f(t, z, x_1, x_2) \, dv(\Omega) \tag{2.42}$$

be constructed. Here N is the linear beam density which equals $\mathrm{M}^{0,0}$, and $dv(\Omega)$ is the volume element Ω of the phase coordinates of the beam.

If the force $F_2(x_1, t)$ is represented as powers of x_1

$$F_2(x_1, t) = \sum_k a_k(t) \, x_1^k \tag{2.43}$$

will follow.

If equation (2.42) is differentiated with respect to time and using equations (2.40), (2.41) and (2.43)

$$\frac{d\mathrm{M}^{p,q}}{dt} = p \, \mathrm{M}^{p-1,q-1} + q \sum_k a_k(t) \, \mathrm{M}^{p+k,q-k} \tag{2.44}$$

will be obtained. This differentiation with respect to time takes the form of

$$\frac{d}{dt} = \frac{\partial}{\partial t} + v_0 \frac{\partial}{\partial z} \ .$$

For linear fields ($F_2 = -\omega^2(t) \, x_1$)

$$\frac{d\mathrm{M}^{p,q}}{dt} = p \, \mathrm{M}^{p-1,q+1} - q\omega^2(t) \, \mathrm{M}^{p+1,q-1} \tag{2.45}$$

is valid for the set of equations (2.44).

So the coupled set of equations (2.44) is divided into subsystems of equations for moments of order $n = p + q$.

For $n = p + q = 0$ the linear beam density is obtained from (2.45)

$$\frac{d\mathrm{M}^{0,0}}{dt} = 0 .$$

From the set of equations for moments of the first order an equation for free oscillations of the mass centre of the beam is obtained

$$\frac{d^2\mathrm{M}^{1,0}}{dt^2} + \lambda^2(t)\,\mathrm{M}^{1,0} = 0 . \tag{2.46}$$

From the set of equations for second order moments the equation for the root-mean-square beam dimension $\mathrm{M}^{2,0} = a^2$ can be derivated

$$\frac{d^3\mathrm{M}^{2,0}}{dt^3} + 4\omega^2\frac{d\mathrm{M}^{2,0}}{dt} + 2\mathrm{M}^{2,0}\frac{d\omega^2}{dt} = 0 . \tag{2.47}$$

For the beam envelope this equation is equivalent to (2.33) included in the model of Vladimirskii-Kapchinskii, and leads to [89, 96]

$$\frac{d^2a}{dt^2} + \omega^2 a - \frac{E^2}{a^3} = 0 \tag{2.48}$$

with $E^2 = \det \mathrm{M}^{p,q}$ and $p + q = 2$.

In the following let us consider the spectrum of fundamental oscillations (2.45) in case of constant fields ($\omega^2 = $ const.). Let the actual values of coordinates x_1 and x_2 be expressed by the initial conditions (2.29) and the result be put in (2.42). The result is that the fundamental frequencies λ in case of odd $n = 2k + 1$ will equal

$$\lambda = \pm i\omega\,(2j + 1)$$

and in case of even $n = 2k$ equal

$$\lambda = \pm 2i\omega j$$

with $j = 1, 2, \ldots k$.

2.2.4   The Two-Dimensional Non-Relativistic Problem

Let us start from the assumption that the particles in the beam cross-section will move in linear external fields. Then the equations for transversal motion will have the form of [96]

$$\frac{d\mathrm{X}}{d\tau} = \mathrm{V}; \qquad \frac{d\mathrm{V}}{d\tau} = b^{\mathrm{ext}}\,\mathrm{X} + a\mathrm{V} + \mathrm{F}^S(\mathrm{X}, \tau) \tag{2.49}$$

where X and V are two-dimensional vectors of the coordinates and velocities and $\mathrm{F}^S(\mathrm{X}, \tau)$ is the vector depending on the electromagnetic eigenfield of the beam. The quantity b^{ext} relates to the influence of external fields. Then the block form of the matrix for moments of the second order can be represented as

$$\mathrm{M} = \begin{pmatrix} \mathrm{M}_{xx} & \mathrm{M}_{xv} \\ \tilde{\mathrm{M}}_{xv} & \mathrm{M}_{vv} \end{pmatrix} \tag{2.50}$$

where matrix M_{xx} of dimension 2x2 corresponds with the moments of the spatial coordinates, e.g. $M_{xx}^{ij} = \overline{x_i x_j}$.

The moments of the second order satisfy the system of differential equations (2.39) which can be represented in the form of

$$
\begin{aligned}
\frac{dM_{xx}}{d\tau} &= M_{xv} + \tilde{M}_{xv} \\[2mm]
\frac{dM_{xv}}{d\tau} &= M_{vv} + M_{xx}\tilde{b}^{\text{ext}} + M_{xv}\tilde{a} + \tilde{F}_{sx} \\[2mm]
\frac{dM_{vv}}{d\tau} &= b^{\text{ext}}M_{xv} + \tilde{M}_{xv}\tilde{b}^{\text{ext}} + aM_{vv} + \tilde{M}_{vv}\tilde{a} + F_{sv} + \tilde{F}_{sv} \ .
\end{aligned}
\tag{2.51}
$$

The square matrices of the second order F_{sx} and F_{sv} are determined by vector $F^s(X,\tau)$

$$
F_{sx}^{ij} = \overline{F_i^s x_j}; \qquad F_{sv}^{ij} = \overline{F_i^s v_j} \ .
\tag{2.52}
$$

The system of equations (2.51) can be considered as closed if matrices F_{sx} and F_{sv} are expressed by moments of the second order. For inherent linear forces the closure is performed automatically as in the one-dimensional case. For the three-dimensional case the problem is discussed in [98]. Non-linear inherent forces of the system lead to an infinite coupled system of equations for moments. A linearization of the electromagnetic eigenfields of the beam showing an elliptical symmetry of charge densities on the condition of a minimum root-mean-square linear deviation from real forces is given in [92, 93, 96]. As a result of this, the matrices take the form

$$
F_{sx} = b^s M_{xx}; \qquad F_{sv} = b^s M_{xv}
\tag{2.53}
$$

where matrix b^s is represented by matrix M_{xx}

$$
b^s = \frac{Nmc^2 r_e Z^2 M_{xx}^{-1/2}}{\beta^2 \gamma^3 \text{Sp}\left(M_{xx}^{1/2}\right)} \cdot \begin{cases} 1 & \text{for a linear geometry} \\ \omega_0^2 R_0^2 & \text{for a ring geometry} \end{cases}
\tag{2.54}
$$

In (2.54) N is the linear beam density, r_e is the classical electron radius, $\text{Sp}(M_{xx}^{1/2})$ is the trace of matrix $M_{xx}^{1/2}$ and R is the ring radius of the annular beams.

Equation (2.53) describes the transformation of electromagnetic eigenfields into effective linear fields

$$
F^s(x,\tau) = b^s X \ .
\tag{2.55}
$$

For Hamilton systems covering linear forces the root-mean-square phase space volume *emittance* is known to be an integral of motion. If the forces are linearized then the increase in the root-mean-square phase space volume, caused by non-linearities, will no longer be taken into consideration. How the nonlinear electric eigenfield is considered by moments of the first orders of the charge density is described in [97] for a band-shaped beam and for a charged cylinder of a circular cross-section.

After putting (2.53) and (2.54) in (2.51) a system of nonlinear equations results for the root-mean-square dimensions of the beam. For a beam of a circular cross-section an equation for the root-mean-square radius R of the beam is resultant

$$
\frac{d^2 R}{dt^2} + \left(\Omega_1^2 + \omega_\lambda^2\right) R - \frac{2 N_e r_e c^2}{\gamma^3 R} - \frac{F_1^2}{R^3} = 0
\tag{2.56}
$$

where Ω_1^2 is the square of the frequency which is connected with an external focusing (e.g. of electrons by ions), and $\omega_\lambda = \omega_c/2$ describes the Lamor frequency. The constant F_1 is determined from relation [96]

$$
\begin{aligned}
F_1^2 &= 4\varepsilon^2 - J^2 + M_3^2 \\
&= 4\left\{ \overline{(x_1 - \overline{x_1})^2}\ \overline{(v_1 - \overline{v_1})^2} - \left[\overline{(x_1 - \overline{x_1})(v_1 - \overline{v_1})}\right]^2 \right. \\
&\quad \left. - \left[\overline{(x_1 - \overline{x_1})(v_2 - \overline{v_2})}\right]^2 \right\}\Bigg|_{t=0} + M_3^2 \ .
\end{aligned}
\tag{2.57}
$$

The quantity J characterizes the mean value of the azimuthal mechanical moment of pulse and $M_3 = J + \omega_c R^2 = \text{const}$ the generalized azimuthal moment of the constants of motion. Equation (2.56) was derived by Lee and Cooper in [99] .

Taking into consideration the space forces of ions localized in the beam, for the motion of an electron beam in a transversal magnetic field holds

$$
\frac{d^2 R}{dt^2} + \omega_\lambda^2 R - \frac{2 N_e r_e c^2}{\gamma R}\left(\frac{1}{\gamma^2} - f\right) - \frac{F_1^2}{R^3} = 0
\tag{2.58}
$$

with f as *neutralization factor* of the beam, corresponding to the relation between the linear charge densities of electrons and ions. If the Coulomb repulsion in the beam is compensated in such a way, that $f > 1/\gamma^2$ holds, so we get in accordance with (2.58) a decreasing of the beam cross-section without magnetic field. This effect in the literature is called *ion focussing* or *beam self-focussing*.

2.2.5   Moments of Second Order for Annular Beams

Usually two models are used for investigating the equilibrium characteristics of annular particle beams and their behaviour at adiabatic changes of the external electromagnetic field. The Yarkovoi model [91] does not take into account the energy spread and the radial dimensions of the beam, which are determined by the amplitudes of the betatron oscillations. In the model by Rubin and Yarkovoi [100] the radial phase space volume is assumed to be zero and the radial deviations are determined by the energy spread in the beam. In the following, we will use the method of complete moments of the distribution function to examine stationary states, free oscillations, and adiabatic changes of the mean square deviations of annular beams of charged particles with non-zero energy spread and a final radial phase space volume [101, 102].

Let us examine an azimuthally symmetrical ring of charged particles of defined energy spread which moves within a magnetic field $\vec{B} = B(B_r, 0, B_z)$.

At sufficiently long confinement times of relativistic electron rings there are important effects which are connected with energy losses by the emission of synchrotron radiation or with the scattering of electrons on atoms of the residual gas or on ions stored in the ring. The influence of the emitted synchrotron radiation on the formation of electron rings was investigated in [103].

The linearized equations of motion in a cylindrical coordinate system (R, θ, Z) can be represented in the following form

$$
\ddot{x} + \left(\frac{\dot{\gamma}}{\gamma} + \frac{P}{\gamma m c^2}\right)\dot{x} + \omega_r^2 x + F_x + \delta F_x = \frac{\omega_0 W}{\gamma m R_0}
$$

$$\ddot{z} + \left(\frac{\dot{\gamma}}{\gamma} + \frac{P}{\gamma mc^2}\right)\dot{z} + \omega_z^2 z + F_z + \delta F_z \ = \ 0 \tag{2.59}$$

$$\dot{W} + P\left((1-2n)\frac{x}{c} + \frac{2W}{\gamma me^2}\right) \ = \ 0$$

where R_0 is the radius of the annular beam, n designates the magnetic field index, $x = R - R_0$, $W = M_\theta - M_\theta^0$, $\omega_{r,z}$ is the frequency of the betatron oscillations and $P = 2e^4 B_z^2 \gamma^2 / 3m^2 c^3$ is the power output by radiation losses of equilibrium particles.

The force $\vec{F}$ is caused by the effect of the electromagnetic eigenfield of the beam. The stochastically ocurring forces $\delta \vec{F}$ are connected with the scattering of electrons on ions or neutral atoms [99].

To achieve completeness for the system of equations (2.59) an equation describing the change of the radius of the annulus and the coupling of the electron energy to the magnetic induction is added

$$\frac{dR_0}{dt} + \frac{R_0}{(1-n)B_z}\left(\frac{dB_z}{dt} - \frac{1}{2}\frac{d\overline{B}_z}{dt}\right) + \frac{R_0 P}{(1-n)\gamma mc^2} \ = \ 0 \tag{2.60}$$

$$\gamma mc^2 \ = \ -eB_z R_0$$

with $\overline{B}_z$ as the mean value of the induction field B_z on the circle with the radius R_0.

If the coordinates and velocities of the particles are combined to form a column vector Y, the equation of motion (2.59) can be represented in matrix form

$$\frac{dY}{dt} = AY + F + \delta F \ . \tag{2.61}$$

The matrix A may be determined as

$$A = \begin{pmatrix} 0 & \Sigma \\ b & a \end{pmatrix} \ .$$

The matrices Σ, a and b can be determined from the system of equations (2.59)

$$\Sigma = \begin{pmatrix} 1 & 0 & 0 \\ 0 & 1 & 0 \end{pmatrix} \qquad b = \begin{pmatrix} -\omega_r^2 & 0 \\ 0 & -\omega_z^2 \\ -\dfrac{(1-2n)P}{c} & 0 \end{pmatrix}$$

$$a = \begin{pmatrix} -\dfrac{P}{\gamma mc^2} - \dfrac{\dot{\gamma}}{\gamma} & 0 & \dfrac{\omega_0}{\gamma m R_0} \\ 0 & -\dfrac{P}{\gamma mc^2} - \dfrac{\dot{\gamma}}{\gamma} & 0 \\ 0 & 0 & -\dfrac{2P}{\gamma mc^2} \end{pmatrix} \ . \tag{2.62}$$

The column vector F is constructed from components of the Lorentz forces of the electromagnetic eigenfield of the beam.

The matrices M_{xx}, M_{xv} and M_{vv} are determined as in the preceding relations, e.g. M_{xx} as matrix of the mean square deviations of the beam

$$M_{xx}^{ij} = \overline{x_i x_j} \qquad i,j = 1,2 \ . \tag{2.63}$$

Unlike equation (2.38), these matrices show different orders: M_{xx} and M_{vv} are quadratic symmetrical matrices of 2nd and 3rd order and M_{xv} is a matrix of the dimension 2x3.

The moments of second order satisfy a system of differential equations

$$\frac{d\mathrm{M}}{dt} = \mathrm{AM} + \mathrm{M}\tilde{A} + \overline{\mathrm{F}\tilde{Y}} + \overline{\mathrm{Y}\tilde{F}} + \overline{\delta\mathrm{F}\tilde{Y}} + \overline{\mathrm{Y}\delta\tilde{F}} \ . \tag{2.64}$$

If the radial and axial motions are separated, separate solutions of the system of equations (2.64) can be found. Then for the moments of second order follows

$$
\begin{aligned}
\dot{\overline{x^2}} &= 2\overline{xv_r} \\
\dot{\overline{v_r^2}} &= 2\omega_r^2 \overline{xv_r} - 2\left(\frac{P}{\gamma mc^2} + \frac{\dot{\gamma}}{\gamma}\right)\overline{v_r^2} + \frac{2c}{\gamma mR_0^2}\,\overline{Wv_r} + 2S_r \\
\dot{\overline{xv_r}} &= \overline{v_r^2} - \overline{x^2}\omega_r^2 - \left(\frac{P}{\gamma mc^2} + \frac{\dot{\gamma}}{\gamma}\right)\overline{xv_r} + \frac{c}{\gamma mc^2}\,\overline{Wx} \\
\dot{\overline{xW}} &= \overline{v_r w} - \frac{1-2n}{c}\,P\overline{x^2} - \frac{2P}{\gamma mc^2}\,\overline{xW} \\
\dot{\overline{v_r W}} &= -\omega_r^2 \overline{xW} - \left(\frac{P}{\gamma mc^2} + \frac{\dot{\gamma}}{\gamma}\right)\overline{v_r W} + \frac{c\overline{W}^2}{\gamma mR_0^2} - \frac{1-2n}{c}P\overline{xv_r} - \frac{2P}{\gamma mc^2}\,\overline{v_r W} \\
\dot{\overline{W^2}} &= -\frac{2(1-2n)}{c}P\overline{xW} - \frac{4P}{\gamma mc^2}\,\overline{W}^2 \\
\dot{\overline{z^2}} &= 2\overline{v_z z} \\
\dot{\overline{zv_z}} &= \overline{v_z^2} - \omega_z^2\overline{z^2} \\
\dot{\overline{v_z^2}} &= -2\omega_z^2\overline{zv_z} + 2S_z
\end{aligned}
\tag{2.65}
$$

with

$$S_{r,z} = \frac{1}{\gamma m}\iint \gamma\delta F_{r,z}v_{r,z}f_i f_e\, dY_i\, dY_e \ .$$

The dot above the individual quantities in (2.65) means the full derivation with respect to time and the bar designates averaging with the distribution function. Here, f_i, f_e, dY_i and dY_e designate the distribution functions and phase space elements for electrons and ions.

By the use of the results from [99, 102] and from section 1.4 there follows

$$S_r = \frac{2I^2 r_e^2 c^3}{\gamma^2 N}\iint \frac{(x_i - x_e)^2\, \varrho_e\varrho_i}{[(x_i - x_e)^2 + (z_i - z_e)^2]^2}\, dx_e\, dx_i\, dz_e\, dz_i \tag{2.66}$$

for arbitrary electron and ion densities in the cross-section of the annuli $\varrho_e(x_e, z_e)$ and $\varrho_i(x_i, z_i)$ for changes of the mean transverse electron velocities which were caused by scattering processes on ions of the charge i.

For the quantity S_z an expression analogous to (2.66) can be given in which the scattering in z-direction is described.

The scattering of ions of the charge i is calculated by integration of (2.66) for two space regions:

1. for impact parameters smaller than the ion radius, corresponding to scattering at a nucleus of the charge Z and at $(Z - 1)$ electrons of the ion examined;

2. for impact parameters greater than the ion radius but smaller than the beam diameter, corresponding to scattering at the ion as a whole.

For the equilibrium density of the electrons and ions then follows

$$S_r = S_z = \frac{r_e^2 N_i c^3}{2\pi a b \gamma^2} \left[(Z^2 + Z - i) \ln \left(\frac{a_0 \gamma m c}{\hbar (Z - i)^{1/3}} \right) + i^2 \ln \left(\sqrt{\frac{2}{a^2 + b^2}} \frac{ab}{a_0} \right) \right] \tag{2.67}$$

with a_0 as the Bohr radius, $\hbar$ as the Planck constant, a and b as mean square radii of the beam ($a = \sqrt{\overline{x^2}}$ and $b = \sqrt{\overline{z^2}}$) as well as $v_r = \sqrt{\overline{v_r^2}}$ and $v_z = \sqrt{\overline{v_z^2}}$.

A consideration of the eigenfields leads to a Coulomb shift of the frequency

$$\nu_r^2 = 1 - n - \frac{r_e N_e R_0^2}{\gamma^3 \beta^2 a (a + b)} \tag{2.68}$$

with $\beta = v_e / c$.

The equation system (2.65) completely describes the change of the annulus cross-section geometry and their spread in the moments of the constants of motion of an axisymmetric beam, taking into account the emission of radiation and the increase in phase space volume caused by scattering processes of electrons on ions stored in the beam.

If the parameters of the equation system (2.65) change with time only slightly, the relations

$$\frac{\dot{\gamma}}{\omega_0 \gamma} \sim \frac{\dot{R}_0}{\omega_0 R_0} \sim \varepsilon \ll 1 \tag{2.69}$$

are satisfied, the first derivatives with respect to time in (2.65) being quantities of first order in ε. If contributions of higher order are neglected, one will get

$$\frac{d}{dt} \left(\frac{\gamma v_r^2}{\omega_r} \right) = \frac{\gamma}{\omega_r} \left[\frac{P v_r^2}{\gamma m c^2} \left(\frac{(1 - 2n) c^2}{R_0^2 \omega_r^2} - 1 \right) + S_r \right] \tag{2.70}$$

$$\frac{d}{dt} \left(\frac{\gamma v_z^2}{\omega_z} \right) = \frac{\gamma}{\omega_z} \left(S_z - \frac{P v_z^2}{\gamma m c^2} \right) \tag{2.71}$$

$$\frac{d \overline{W^2}}{dt} = -\frac{2P \overline{W^2}}{\gamma m c^2} \left(\frac{(1 - 2n) c^2}{R_0^2 \omega_r^2} + 2 \right) \tag{2.72}$$

with the mean square velocities

$$\begin{aligned} v_r^2 &= a^2 \omega_r^2 - \frac{c^2 \overline{W^2}}{\gamma^2 m^2 R_0^4 \omega_r^2} \\ v_z^2 &= b^2 \omega_z^2 \,. \end{aligned} \tag{2.73}$$

For the case that the emission of energy by synchrotron radiation and the electron scattering yield only insignificant contributions, from equations (2.70) to (2.72) the adiabatic invariants [102]

$$\gamma \omega_r \left(a^2 - \frac{c^2 \overline{W^2}}{\gamma^2 m^2 R_0^4 \omega_r^4} \right) = E_r = \text{const}$$

$$\gamma \omega_z^2 b^2 = E_z = \text{const} \qquad (2.74)$$

$$\overline{W^2} = \text{const}$$

result with E_r and E_z as effective phase space volumes of the particle motion in the beam cross-section.

As is usual in the accelerator physics, the first of the equations (2.74) can be rewritten by use of the betatron and synchrotron deviations

$$a^2 = a_s^2 + a_b^2 .$$

Here holds

$$a_b^2 = \frac{E_r}{\gamma \omega_r} \quad \text{and} \quad a_s = \frac{c^2 \overline{W^2}}{\gamma^2 m^2 R_0^4 \omega_r^4} .$$

For $a_s \gg a_b$ the results of the model follow from (2.71) [91]. In the inverse case, equations coinciding with the results from [100] are obtained.

If the contribution by the eigenfields are neglegted, the relation

$$\frac{d\overline{W^2}}{dt} = 2 \left(\frac{3 - 4n}{1 - n} \right) \overline{W^2} \qquad (2.75)$$

results from (2.72) [103].

Estimations show that for electron energies of $E = 20$ MeV ($\gamma = 40$) and a ring radius $R = 4$ cm the characteristic processes in connection with the emission of synchrotron radiation [73] and with the scattering of electrons on stored ions [104] are in the range of $10 \ldots 100$ ms. In the case of a sufficiently long life of the electron-ion rings, as required for generating ions of high charge states, the synchrotron radiation leads to a reduction of the ring cross-section and at $0 \leq n \leq 3/4$ to a spread for the generalized constants of motion. The scattering of electrons on ions increases the dimensions of the ring cross-section.

It should be stated here that also the passage of a linear electron beam through a gas or through a foil leads to an increase of the beam cross-section. That may be taken into account in (2.56).

During the passage of an electron beam through a thin foil, if changes of the electron energy and of the beam cross-section can be neglected, the resultant mean square emittance of the beam over

$$\varepsilon = \sqrt{\varepsilon_0^2 + \Delta \varepsilon^2}$$

with ε_0 as the initial value of the emittance may be determined. Here the increment amounts to

$$\Delta \varepsilon^2 = \frac{4\pi a^2 r_e^2 n Z^2 d}{\beta^4 \gamma^2} \ln \frac{137\gamma}{Z^{1/3}} \qquad (2.76)$$

with a as the root-mean square deviation of the beam cross-section, n as the number of atoms having the charge Z in the unit volume of the foil and d as the foil thickness [105].

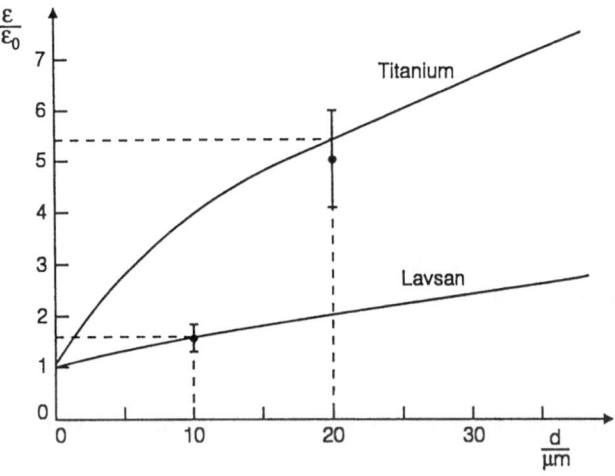

Figure 2.1 The relative change of the emittance of an electron beam in dependence on the thickness of a Titanium foil (curve 1) and a Lavsan foil (curve 2) during the passage of the beam through the foil [105]. • – experimental values.

For example, Fig.**2.1** shows calculated values of the relative increase of emittance $\varepsilon/\varepsilon_0$ for a beam with the parameters $a=1$ cm and $\gamma=4.3$ during its passage through a Titanium or Lavsan foil in dependence on the foil thickness and also results of experimental measurements of emittance, which well coincide with the calculated values, as it is obvious from Fig.**2.1** [105].

2.2.6   Multicomponent Beams of Charged Particles

The momentum method allows the modelling of the collection and successive ionization of ions in electron beams as well as the description of the change of the deviations of each charge component, taking into account the mutual influence of the individual charge components.

For the existence of several beam components of charged particles the motion of the electrons and of ions having the charge i can be described in the following form

$$
\begin{aligned}
\ddot{x}_e + \frac{\dot{\gamma}}{\gamma}\dot{x}_e + \omega_x^2 x_e - \frac{1}{m\gamma}\left(F_x^{ee} + \sum_{i=1}^{z} F_x^{ei}\right) &= 0 \\
\ddot{z}_e + \frac{\dot{\gamma}}{\gamma}\dot{z}_e \,\omega_z^2 z_e - \frac{1}{m\gamma}\left(F_z^{ee} + \sum_{i=1}^{z} F_z^{ei}\right) &= 0 \qquad\qquad (2.77) \\
\ddot{x}_i - \frac{1}{AM}\left(F_x^{ie} + \sum_{j=1}^{z} F_x^{ij}\right) &= 0 \\
\ddot{z}_i - \frac{1}{AM}\left(F_z^{ie} + \sum_{j=1}^{z} F_z^{ij}\right) &= 0; \qquad i,j = 1,2,\ldots,z \, .
\end{aligned}
$$

The index e indicates that the given quantity refers to electrons, the indices i and j, however, refer to one of z ion charge states. The quantity x gives the deviation of the particle position from the equilibrium radius. The frequencies of the eigenoscillations of electrons in the annular beam are

$$\omega_x^2 = \frac{(1-n)\,c^2}{R_0^2} \quad \text{and} \quad \omega_z^2 = \frac{n\,c^2}{R_0^2}\;.$$

The forces F^{ei}, F^{ie}, F^{ee} and F^{ij} are due to the effect of the eigencharges of the annulus components. For instance, F^{ei} designates the force acting on an electron from an ion having an i-fold charge.

The ion distribution function satisfies the kinetic equations (1.41). If only single-ionization processes are taken into account, we get

$$\frac{df_i}{dt} = n_e v_e \left(\sigma_{i-1}^i f_{i-1} - \sigma_i^i f_i \right)\;. \tag{2.78}$$

With results from Refs.[84, 106] the equations for the mean root-square deviations yields

$$\ddot{a}_e + \frac{\dot{\gamma}}{\gamma}\dot{a}_e + \omega_{xe}^2\,a_e - \frac{E_{xe}^2}{\gamma^2\,a_e^2} = 0$$

$$\ddot{b}_e + \frac{\dot{\gamma}}{\gamma}\dot{b}_e + \omega_{ze}^2\,b_e - \frac{E_{ze}^2}{\gamma^2\,b_e^2} = 0 \tag{2.79}$$

$$\ddot{a}_i + \omega_{xi}^2\,a_i - \frac{E_{xi}^2}{a_i^2} = 0$$

$$\ddot{b}_i + \omega_{zi}^2\,b_i - \frac{E_{zi}^2}{b_i^2} = 0 \qquad i = 1,2,3,\dots,z$$

with ω as the averaged eigenfrequency of the particles examined and E as effective phase space volumes of the electron and ion components.

With an ion collection that is slow as compared with the period of the eigenoscillations of the particles the effective phase space volumes remain constant. For annular particle beams with a mean square spread of electrons with respect to the generalized moments $\overline{W^2}$ the equation (2.79) change into a system of algebraic equations analogous to (2.74)

$$\gamma\,\omega_{xe}\left(a_e^2 - \frac{\overline{W^2}\,c^2}{\gamma^2\,m^2\omega_{xe}^4\,R^4}\right) = E_{ex}$$

$$\gamma\,\omega_{ze}b_e^2 = E_{ze} \tag{2.80}$$

$$\omega_{xi}\,a_i^2 = E_{xi}$$

$$\omega_{zi}\,b_i^2 = E_{zi} \qquad 1 = 1,2,3,\dots,Z\;.$$

Here, we have a closed system of equations if the averaged frequencies of the particle eigenoscillations ω are expressed by moments of second order. This requires averaging over the forces acting between the individual particle components in the beam cross-section. The result of the averaging depends on the spatial distributions of each particle type.

Let us assume that the electrons have a Gaussian density distribution (2.16) in the beam cross-section. It was shown already in section 2.1. that during the formation of ions in such an electron beam the ion density shows an integrable peculiarity in the beam centre (2.21). Elastic ion collisions lead to a gradually developing equaldistribution of the ion density. The characteristic times for the occurence of a Boltzmann distribution of energy and a Gaussian density distribution corresponding to this process correspond to a time scale of microseconds, as already shown in section 1.4. As a rule, these times are considerably lower than the life-time of the beam and the rate of ion accumulation. If it is assumed that the electron and ion densities in the beam cross-section should have a Gaussian distribution with elliptical symmetry, then follows

$$\varrho_{e,i} = \frac{N_{e,i}}{2\pi a_{e,i}\, b_{e,i}} \exp\left(-\frac{x^2}{2a_{e,i}^2} - \frac{z^2}{2b_{e,i}^2}\right) \tag{2.81}$$

with $N_{e,i}$ as linear particle densities. The individual oscillation frequencies then have the form

$$\omega_{xe}^2 = \frac{c^2}{R_0^2}(1-n) - \frac{e^2 N_e}{\gamma^3 m a_e(a_e+b_e)} + \frac{2e^2}{\gamma m}\sum_{i=1}^{z}\frac{iN_i}{a_e^2+a_i^2+\sqrt{(a_e^2+a_i^2)(b_e^2+b_i^2)}}$$

$$\omega_{ze}^2 = \frac{nc^2}{R_0^2} - \frac{e^2 N_e}{\gamma^3 m b_e(a_e+b_e)} + \frac{2e^2}{\gamma m}\sum_{i=1}^{z}\frac{iN_i}{b_e^2+b_i^2+\sqrt{(a_e^2+a_i^2)(b_e^2+b_i^2)}} \tag{2.82}$$

$$\omega_{xi}^2 = \frac{2ie^2}{AM}\left(\frac{N}{a_e^2+a_i^2+\sqrt{(a_e^2+a_i^2)(b_e^2+b_i^2)}} - \sum_{j=1}^{z}\frac{jN_j}{a_i^2+a_j^2+\sqrt{(a_i^2+a_j^2)(b_i^2+b_j^2)}}\right)$$

$$\omega_{zi}^2 = \frac{2ie}{AM}\left(\frac{N}{b_e^2+b_i^2+\sqrt{(a_e^2+a_i^2)(b_e^2+b_i^2)}} - \sum_{j=1}^{z}\frac{jN_j}{b_i^2+b_j^2+\sqrt{(a_i^2+a_j^2)(b_i^2+b_j^2)}}\right).$$

Other distributions of particle densities show a different dependence of the eigenfields on the distance to the beam centre, therefore leading to different results for the eigenfrequencies. Without a considerable expenditure, in this way the oscillation frequencies for electron beams with constant density can be found which are characteristic, e.g., of electron beam ion sources (EBIS). Corresponding frequencies for the case of constant density for electrons and ions in the beam cross-section are given in [106].

An integration of (2.78) with respect to the space coordinates and the velocities leads to the system of balance equations

$$\frac{dN_i}{dt} = \varrho_e e\,(\sigma_{i-1}N_{i-1} - \sigma_i N_i) \tag{2.83}$$

if the ion density (2.81) is used for the linear ion density N_i in the beam.

2.2.7   Numerical Modelling with the Momentum Method

The equations obtained allow the description of ion accumulation processes, taking into account different distributions for electrons and ions of different charge as well as effective active forces acting on the charged particles from other components of the electron beam. Equations for describing multicomponent clouds of charged particles with non-identical local mass centres were derived on the basis of the momentum method. Those equations constitute the basis for the modelling of the acceleration of electron-ion rings on the basis of the collective method of particle acceleration [84, 106] and are used for the calculation and optimization of accelerating and transfer channels for multiply charged ions, taking into account the eigencharge of the beam [94, 108] and for other tasks connected with the dynamics of multicomponent electron-ion beams or ensembles.

For example, let us examine a calculation variant for the adiabatic compression of an electron-ion ring in a weakly focussing azimuthally symmetrical field growing with time, as was realized at the heavy ion collective accelerator of the Dubna research centre [84].

The calculations proceeded from the following parameters of the electron rings [83]
initial radius of the electron ring $R_0 = 35$ cm
initial radius of the ring cross-section $a_e = b_e = 2$ mm
number of electrons $N_e = 1 \cdot 10^{13}$
The gas pressure (Nitrogen) in the compressor chamber was assumed to be $4 \cdot 10^{-6}$ Pa. During the ring compression ($\approx 2 \ldots 3$ ms) the mean ion charge grew up to $\bar{i} = 5$ and the neutralization of the electron charges amounted to $40 \ldots 50\%$ (f=0.4...0.5). The dispersion of the ion charge distribution was about 0.6.

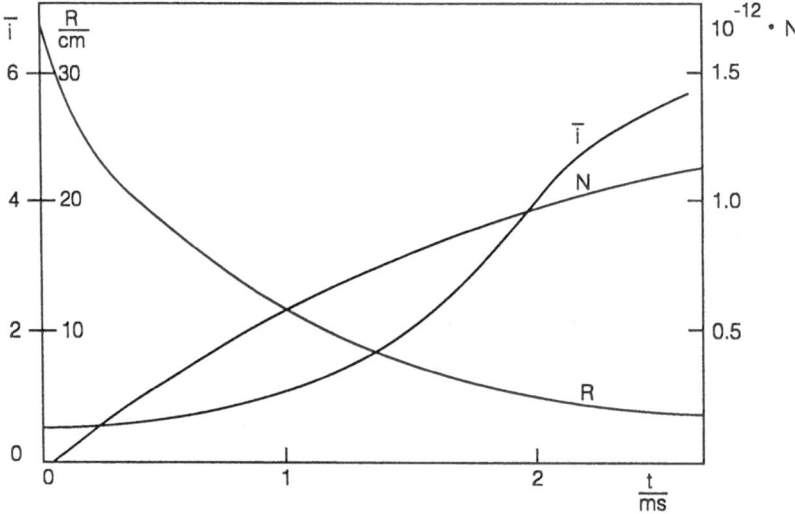

Figure 2.2   Change of the electron ring radius R, of the relativistic factor of the electrons γ and of the total number of ions N in the ring as a function of time [84].

In Fig.2.2, time controlled values of the electron ring radius R, of the relativistic factor of the electrons γ and of the total number N of ions in the ring for the example considered are shown.

In Fig.2.3, time controlled values of the root-mean cross-sectional dimensions of the electron component (a_e) and of the ion component (a_i) of the ring are demonstrated in the final phase of ring compression. Axial ring dimensions $b_{e,i}$ approximately coincide with the radial dimensions. As shown by the figures, in the final compression phase the root-mean dimensions for ions of charge 5 (component a_5) coincide with those valid for electrons. This is explained by the fact that increasing neutralization of electron charges results in a decrease of ion oscillation frequencies.

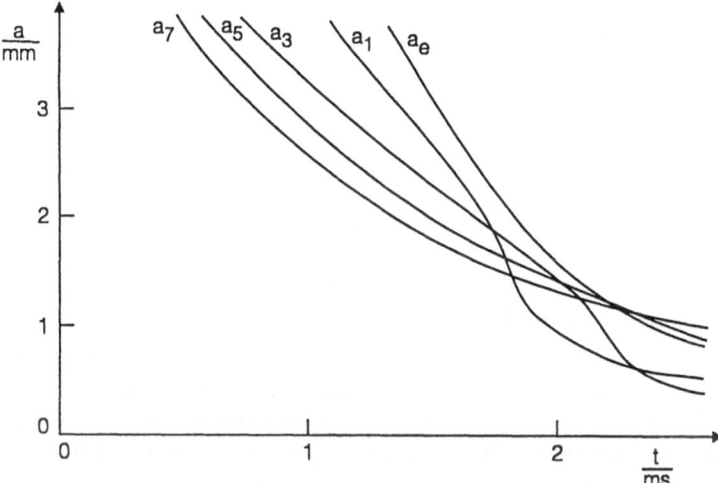

Figure 2.3   Change of the root-mean-square dimensions for the electron (a_e) and ion components $(a_1 \ldots a_5)$ in the final state of compression of the electron ring [84].

Summarizing, it is pointed out, that until today wide-spread use was made of the momentum method for the solution of various physical tasks and calculations in the theory of accelerators, in magnetooptics and in dynamics of intense particle beams. In this connection, second order moments applicability of the root-mean-square beam characteristics was worked out most extensively. Equations for second order moments were obtained by linearization of electromagnetic fields (paraxial approximation) and obviously coincide with equations known from strongly self-consistent beams models. Equations obtained are applicable for a great class of particles distributions in the phase space and are of a high degree of generality.

It is an established advantage of the method of second order moments to take into consideration such essential physical aspects such as divergence of a beam passing through a gas or a solid-state film, ionization of atom or molecular beams and accumulation of ions in the beam, the interaction of multicomponent beams and non-identical cross-sectional dimensions and various charge states of ions in the beam, synchrotron radiation effects for ring-shaped electron beams and many other phenomena.

Often the momentum method is the only applicable means to be used on the computer for a numerical treatment of optimization problems of accelerator structures, transport channels, the adaption of various systems taking into account the beam space charge and many other problems. Using the momentum method, in general much less time is required for computing than by using other alternatively applicable methods.

At the same time, however, the method of second order moments is restricted by the fact that important effects are not taken into consideration which are linked with the infringement of paraxiality of electromagnetic forces influencing the particles and which, for example, cause an increase of the root-mean-square beam emittance. This aspect may be avoided by the further development of the theory so that the infinite chain of coupled equations for the individual moments are adequately breaking-off and, self-consistent equations for moments of higher order will be obtained.

2.3   Method of Finite Particles

The momentum method of distribution functions represents an adequate means to discuss the dynamics of beams or of charged particles ensembles, among them multicomponent ensembles, as well. The finite particles method is successfully used in the investigation of detailed beam characteristics, the distribution function of particles, their change in the ion transitions from one charge state into another and when considering the nonlinearity of the eigenfields of electrons and ions. This method will be used for tasks linked with the description of the movement of a dense medium, in electrodynamics of dense media and when modelling beams of charged particles [75, 79].

To describe a dence medium, the Euler or Lagrange access is used. In the Euler access for a dense medium the systems state at definite points of the physical space, e.g. at fixed mesh points is considered. Here, the coordinates of the mesh points are independent Euler variables. In time course, the mesh points pass various parts of the medium. The direct solution of the kinetic equation may be used here as an example for the Euler access. In the Lagrange access, the movement of various selected particles is observed. The entirety of phase trajectories of these particles informs us about the process taken into account. The number of particles or their initial position in the phase space is used as an independent variable (Lagrange variable) for this process. It is assumed, that the initial or actual phase space positions are uniquely linked with the trajectories equations. The method of macroparticles or finite particles may be used as an example for the application of the Lagrange access to analyze beams of charged particles. In many cases, both accesses yield the same results are achieved. For example, in the finite particles method, the movement of particles in Lagrange variables is described. Charge density produced by the particles, and the current and the field where the particles move are determined in a stationary network, i.e. in Euler variables.

In order to get adequate information on the process considered, it is necessary to examine a sufficiently great number of particles and to statistically average the findings of the examinations. The initial positions of the particles, i.e. the starting points of the phase space trajectories are selected by statistical methods in line with the statistics accepted.

In the initial moment, the phase space volume of each component of the ensemble subdivides into a certain number of non superimposing elementary cells and, the movement of each of these volumes by the movement of any of its particles may be identified by the respective integral charge and mass. In the following, to simplify matters, model particles or macroparticles are called particles.

The application of the finite particles method is investigated for the modelling of processes of ion accumulation in azimutally symmetric electron beams. For instance, for this purpose in Ref.[109] a program was developed which for every charge component takes into account up to 2000 finite particles. For each time, in a given net of mesh points, the particle density ϱ is determined. The potentials U of the eigenfields result from the solution of the Poisson equation in cylindrical coordinates

$$\Delta U = -4\pi\varrho \qquad\qquad (2.84)$$

with boundary conditions as

$$U|_{r=R} = 0 \qquad \text{and} \qquad \left.\frac{\partial U}{\partial r}\right|_{r=0} = 0$$

and R as boundary of the range considered when dealing with this task.

The Poisson equation can be solved numerically by fast Fourier transformation [110]. The potentials of the eigenfields U are found by linear interpolation between the knots of the net used. In the ring cross-section, the particles movement is of the type

$$\frac{dr}{dt} = v_r; \qquad \frac{dv_r}{dt} = \frac{M_\Theta^2}{m^2 r^3} + \frac{Ze}{m}\left(F + F_{\text{ext}}\right) \tag{2.85}$$

where by $M_\Theta = m v_\Theta r$ the angular momentum is described as integral of movement, by v_r and v_Θ the radial and angular velocities of the particles movement, by Ze and m the charge and mass of the particles and by F and F_{ext} the forces effecting from the eigenfields U and from external fields U_{ext}.

To determine the particles trajectory, the integral of the system (2.85) is used, i.e. the conversation of energy E is

$$\frac{dr}{dt} = \pm \sqrt{\frac{2}{m}\left(E - \frac{M_\Theta^2}{2mr^2} - Ze(U + U_{\text{ext}})\right)}. \tag{2.86}$$

This equation, in the environment of a vanishing radius is integrable with numerical methods than the second equation in the system of equations (2.85). In addition, using the energy conversation law instead of the movement equations to determine the particles coordinates, the number of the equations to be solved is reduced by the factor two. The equation (2.86) may be integrated with the Runge-Kutta method of fourth order.

For the initial arrangement of the finite particles in the phase space of the coordinates and velocities, according to the given distribution function f, an algorithm will be used which is described in the following.

For the factorization of the initial distribution function

$$f(x_1, x_2, \ldots, x_i, \ldots, x_k) = f_1(x_1) f_2(x_2) \ldots f_i(x_i) \ldots f_k(x_k)$$

independently the distribution for each coordinate is considered. Then, for example, the initial values of the coordinates x_{in} of the finite particles of the number n $(n = 1, 2, \ldots, N)$ are of the values

$$x_{in} = F_i^{-1}(R(n)), \qquad x_i \in [a_1, b_1]$$

with $R(n)$ as random number generator of the interval $[0, 1]$. Here, by the function F_i^{-1} the inverse value with respect to

$$F_i(x_{in}) = \frac{1}{N}\int_{a_i}^{b_i} f_i(y) g_i(y)\, dy$$

is shown. The g_i are weight functions, determined by the metric of the space. For example, for the distribution according to the radii in polar coordinates (φ, r), the relation $g(r) = r$ follows.

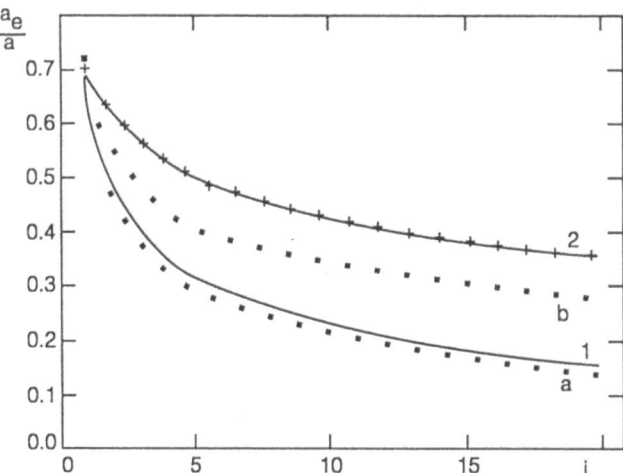

Figure 2.4 Change of the root-mean-square dimensions of ion components in dependence on the ion charge [86]. Curve 1 – beam of Gaussian density, determined according to (2.22); Curve 2 – beam with uniformly distributed electron density, determined according to (2.12). The points a and b describe calculations using the finite particles method for two regimes of transition of ions from one charge state into another in an electron beam of Gaussian like density. Crosses correspond to calculations by the finite particles method for an electron beam of uniformly distributed electron density.

Now, basic results of calculations are dealt with which are described in detail in [85, 86, 109, 111]. Dynamics of second order moments of ions as a function of ion charge will be investigated by the method of finite particles. To describe each ion component, 1000 particles were used, respectively. Electrons were assumed to be distributed in the ring cross-section according to a Gaussian distribution (see (2.16)) and, in (2.17), the nonlinear electrical eigenfield was described. Fields produced by ions were not taken into account.

Calculations were made for two cases. In the first case (case a in Figure **2.4**) valid for the conditions for which (2.20) was derived, it was assumed that, when new ions were formed, for the previous charge state the distribution function remains unchanged. In case b, the change of the distribution function was taken into account.

In Fig.**2.4** the change of the relations between root-mean-square dimensions of ions of charge state i and root-mean-square dimensions of the electron beam a_i/a_e is indicated. Points for a Gaussian distribution of density of the electron beam are illustrated as they were calculated following (2.22). In addition, for the cases a and b, calculation results are indicated taking into account nonlinear fields (2.17) for the cases a and b.

Accuracy of numerical calculations can be scrutinized by means of calculations of ion collection in an electron beam of constant density. Results of these calculations (characterized by crosses in Fig. **2.4**) differ from the accurate values obtained by (2.12) by $1\ldots 2\%$. They are described by the black lines 2 and the difference for single charged ions is less than 0.5%. Calculations were made for identical root-mean-square dimensions and linear beam densities with constant electron density or Gaussian-like densities. Based

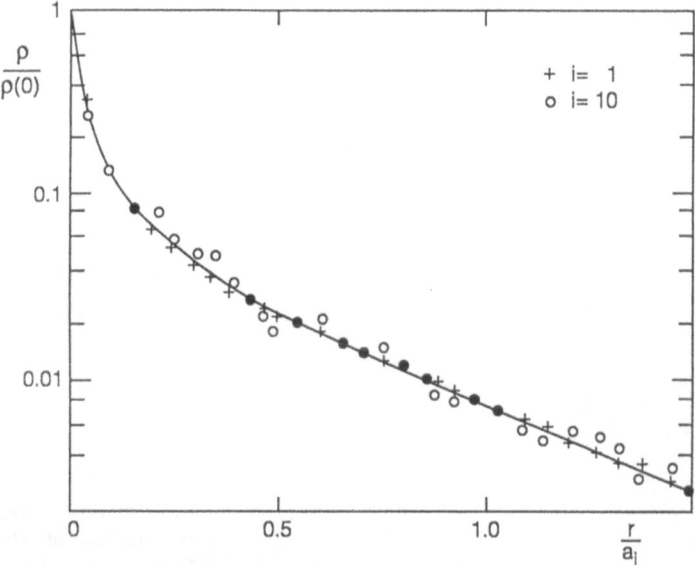

Figure 2.5   Some dependencies of the ion densities on the radius of the first (+) and tenth (o) charge state in an electron beam of Gaussian electron density distribution [111].

on these results, it is concluded that nonlinearity of the eigenfields of Gaussian electron density distribution considerably influences the ion distribution function for the first two charge states.

In Fig.**2.5**, calculated density distributions averaged according to the thickness $h_r = a_i/20$ of ring discs, alongside the ring radius are indicated, which refer to the averaged density in the central range ϱ_0 with the radius $h_r/2$ for the first and 10th charge state in case a. In the same figure, the black line indicates ion density averaged in the same way (2.21) and electron density (2.16) for $a_e = \sqrt{2}a_i$ and $N_e = 10N_i$.

The described numerical results coincide well with the model distribution function (2.20), hence, their relevance is substantiated. Density ϱ_0 in the beam centre is determined by the value h_r and can exceed electron densities by one order of magnitude. As already stated by (2.21), and analogously by (2.15), for electron beams of constant density in the beam centre, infinitely high density is obtained which can not be realized by physical means.

Taking into account angular moments of ions [81] different from zero determined by initial energies of ions ε_0 and connected with the thermal motion of primary particles and the momentums transmitted by ionization processes, entails that ions do not arrive the beam centre and, the above mentioned singularity do not occur. Minimal distance, in which the ion approaches the beam centre, is $r_{min}/a_i = \sqrt{\varepsilon_0/\varepsilon_i}$. This value is for electron-ion rings $10^{-3}\ldots10^{-4}$ and for $\varrho_e/\varrho_0 \sim 10^{-2}\ldots10^{-3}$. Up to now, the problem of possible ion-electron instabilities in the centre of these beams which may bring about a modification of the distribution function (2.20) and an increase in the transversal phase volume of ions has not yet been investigated.

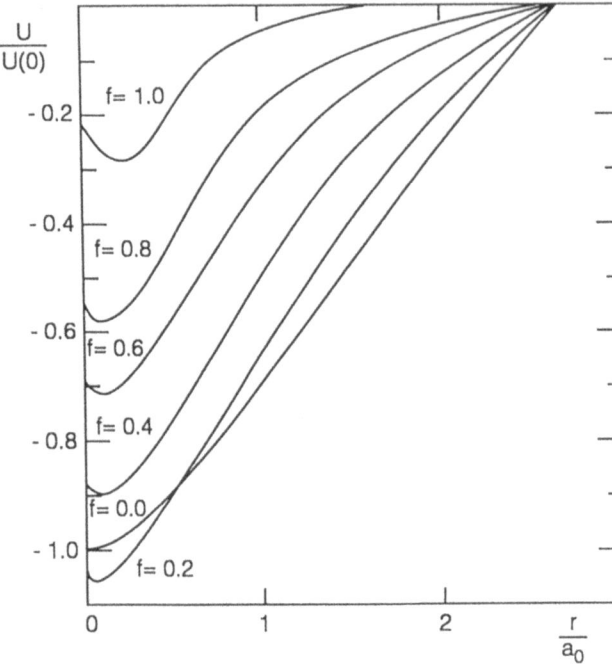

Figure 2.6 Dependence of the electrical potential U on the radius of the electron beam r for various values of the neutralization factor f of the electronic charge by ions [111].

It should be noted that due to the restricted number of ions in the beam centre, considerable changes of eigenfields and frequencies do not occur. Coinciding with the estimations made in [111], the ion distribution (2.20) in the beam centre changes when

$$f \gg \sqrt{\frac{\pi}{2}} \frac{1}{4\bar{i}}$$

where by $\bar{i}$ the averaged ion charge is described. At a sufficient confinement time of ions in the beam, collisions of electrons and ions in the beam centre result in the transformation of the distribution function into a Boltzmann distribution which shows no singularities.

In Fig.**2.6** the dependence of the electrical potential U on the radius r for various neutralization factors f of the electron charge is represented [109, 111, 112]. The occurrence of a local maximum in the beam centre is explained by the ion density in the beam centre according to the distribution function (2.20). At increasing ionization factor up to complete charge compensation ($f = 1$), ions are kept only weakly by the electrons in the beam. Their root-mean-square dimensions are greater than those of the electrons and the potential gap does not completely disappear.

In the Figures **2.7** and **2.8**, phase space densities of electrons and ions are presented in relative coordinates r and velocities v_r [111]. Thereby, it is shown how the dynamics on ion distribution functions is formed. The successive collection of four different ion charges in the electron beam was considered. The charges of individual ion components are 1:2:5:10. When the neutralization factor is appoximately 1, ion losses began and the growth of the neutralization factor was stopped.

In Figure **2.9**, the change of the root-mean-square dimensions of a_i as a function of the neutralization factor f is illustrated. Hence it appears that with increasing neutra-

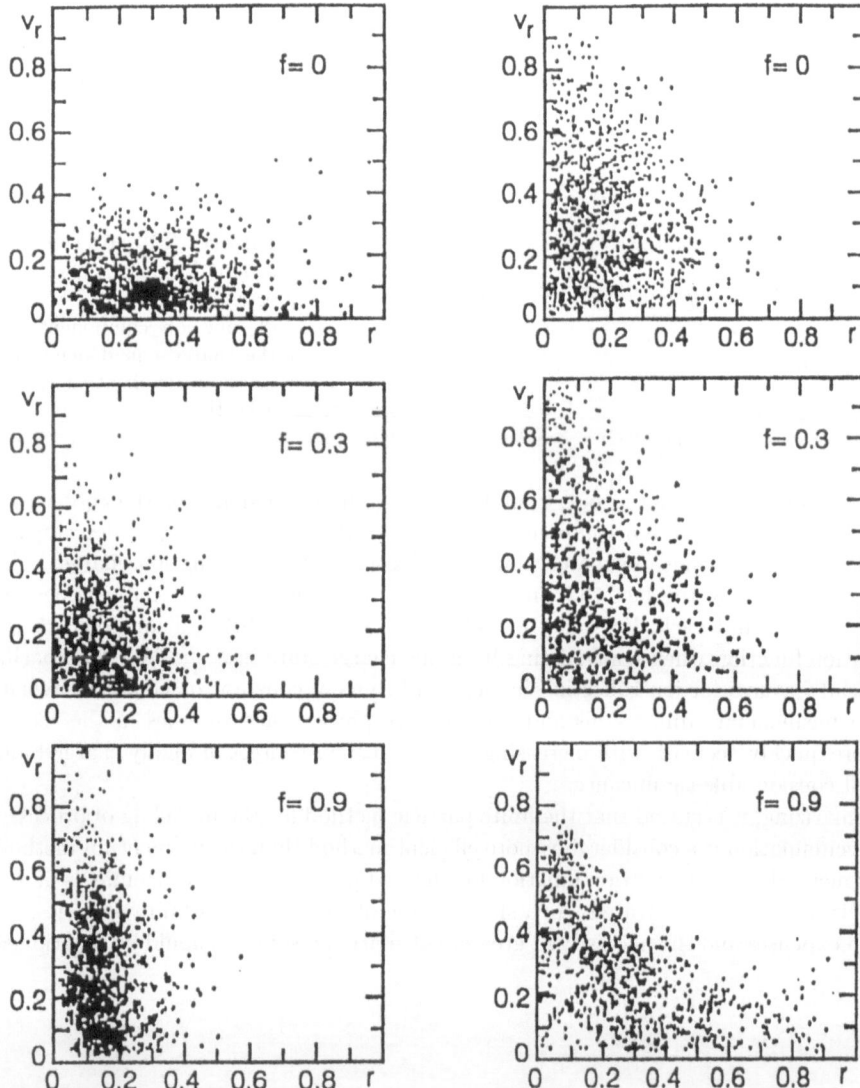

Figure 2.7 Dynamics of the electron distribution function in the phase space of the coordinates and velocities at increasing neutralization factor f [111].

Figure 2.8 Dynamics of the ion distribution function in the phase space of the coordinates and velocities at increasing neutralization factor f [111].

lization factor, the root-mean-square dimensions of the electrons rapidly decrease. This phenomenon is known as *ion focussing*. At the same time, ion dimensions increase and become greater than electronic ones. Analogous results were obtained on the basis of the momentum method for the distribution function (Figure **2.3**) [84]. At high values, growth of the neutralization factor is reduced when ion losses begin and, the root-mean-square dimensions for the first charge state begin to decrease.

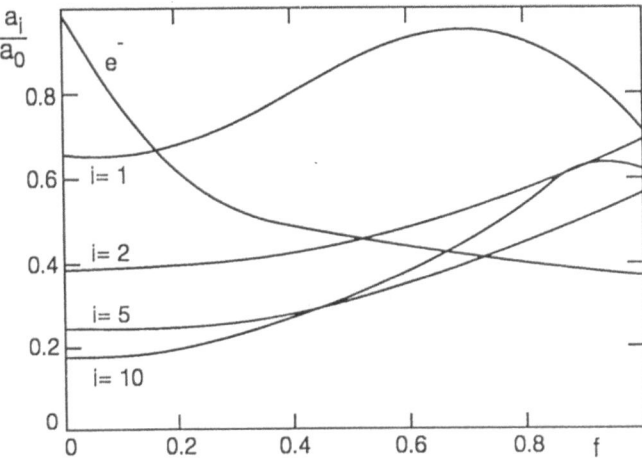

Figure 2.9 Root-mean-square dimensions of the electron component (e^-) and of four ion components of the charge i in dependence on the neutralization factor f [111].

Here, it should be pointed out, that for $f \ll 1$ the dependence of the root-mean-square dimensions of the ion components differ from the law $a_i = a_e/\sqrt{2i}$ described in [85], whereas the root-mean-square velocities increase with increasing ion charge. This may be explained by the fact that in the quoted calculations the ion charge does not increase by one unit each, but more quickly $(1 \Rightarrow 2 \Rightarrow 5 \Rightarrow 10)$ and ions disturb the distribution function when transmitting from one charge state into another. Comparing the special features for ion collection in electron beams of constant or Gaussian density the root-mean-square dimensions and the effective phase space volumes in the second case more quickly decrease with increasing ion charge. For sources of highly charged ions this is of considerable significance.

Summarizing, it is stated that the finite particle method for the modeling of processes of ion accumulation is a considerably more efficient method than the momentum method. By this method, the distribution function for charged particles and the eigenfields of ions and electrons may be illustrated. As a shortcoming of the method of finite particles, the required expenses and the necessity of greater calculation resources should be mentioned.

Chapter 3

Production of Multiply Charged Ions in Relativistic Electron Rings

3.1 Collective Ion Acceleration in Electron Rings

3.1.1 The Collective Method of Ion Acceleration in Electron Rings

Collective ion acceleration is marked as a whole group of particle acceleration methods. The collective method of particle acceleration became known as a result of investigations which were accompanied with the realization of the coherent method of acceleration proposed by V.I.Veksler [113].

For traditional accelerators the use of outer electromagnetic fields for particle acceleration is characteristic. Thereby the acceleration of charged particles depends on the intensity of the accelerating fields and on the charge to mass ratio of the particle to be accelerated. Since these values are limited by technical and physical reasons, the achievable final energy of the accelerated particles through increasing the accelerator size is limited. Thus, the further progress was done in acceleration techniques with new acceleration principles, which make it possible, to increase the effective field intensity of the accelerating electric and magnetic fields, which influences on the particles.

At coherent acceleration methods the accelerating field arises as a result of the interaction of groups of particles or also of the particles mutually, by the interaction with plasma currents or by the interaction with electromagnetic radiation. These different interactions were considered in the works of V.I.Veksler [113].

The collective ion acceleration through electric eigenfields of an electron cloud is an example for the realization of the coherent acceleration method proposed by Veksler. The electric field, which interacts with the space charge within the cloud, can achieve extreme values at comparatively small electron charge density.

As an example the estimation for a field of an endless cylindric electron beam with linear particle density N should be accomplished here. If the charge is distributed homogeneously over the ring cross-section – a circle with the radius a – the intensity of the electric field at the ring border yields

$$F_e = \frac{2eN}{a} \, . \tag{3.1}$$

At $a = 0.1$ cm and $N \approx 10^{11} \ldots 10^{12}$ cm^{-1} the field intensity according to (3.1) is estimated to $E \approx 10^5 \ldots 10^6$ V/cm.

Thus it becomes possible to increase with the use of the eigenfields of electron clouds for ion acceleration the rate of ion energy per acceleration length.

However, if the production is striven for high eigenfields in electron clouds first the existence and maintenance of such electron clouds must be guaranteed. This is complicated by high electric eigenfields that lead to a strong electric repulsion of the electrons. To compensate this repulsion, outer fields are necessary, which are comparable in their intensity with that of the eigenfields which still exceed these. Here the well known attenuation effect of Coulomb repulsion in relativistic beams can be taken advantage of at the formation of the corresponding electron ensembles. Due to the magnetic attraction of similar charges, which move in the same direction, the repulsion of these charges is mutually weakened in a direction orthogonal to the moving direction in comparison with the resting charges around the factor γ^2, whereby γ is the relativistic factor of particle movement. Therefore, on an electron moving in a linear relativistic beam, which is located at the boundaries of the beam, acts the force eF_e/γ^2, whereby F_e determines itself from (3.1). The repulsive force can be compensated at sufficiently relativistic electrons through comparatively small outer fields. Additionally it is possible, to compensate the repulsion forces acting between the electrons through a slight amount of ions in the electron beam. The ions themselves, which dispose of no directed movement, thereby become held by the large fields, produced by the electrons electric fields. In this way there arises selffocussing electron-ion-beams, their existence was already pointed out by Bennet [114].

A simple realization of an electron beam with spatial limitation of relativistic motion is a ring. Thereby, the electrons circulate with relativistic velocities to such an extent, that their repulsion forces are weakened in the ring cross-section around the factor γ^2. The ions are immovable in the azimuthal direction and are held in the electron ring by high electric eigenfields of the electrons. For the characteristic case of the collective acceleration method, the radius of the small ring cross-section is much smaller than the mean ring radius, e.g., the repulsive forces can be estimated for a linear beam by (3.1).

It is assumed, that such an electron-ion ring is accelerated by outer electromagnetic fields into the perpendicular direction to the ring plane. Such a ring acceleration without separation of ionic and electronic components should be marked as acceleration of a compact ring. The maintenance of the compactness of the ring should be guaranteed by the eigenfields of the electron-ion ring.

The principle of collective ion acceleration by electron rings requires, that the interaction of the ring with outer accelerating fields is determined essentially by the electrons. This means that for the acceleration in the electric field the summary ion charge must be considerably smaller than the entire electron charge.

A qualitative picture of the collective ion acceleration can be given in the following way: An electron-ion-ring is polarized in the presence of outer accelerating fields so that the electron component is localized spatially before the ion component in the acceleration direction. Therefore the electron eigenfields hold the ions during the acceleration in the ring, i.e. the effectiveness of the ion acceleration is determined by the force of the electron ring eigenfields. Since these fields are very strong, an ion acceleration results, which occurs in the effective accelerating fields which exceed the usual acceleration in otherwise modern accelerators considerably.

Further more, we will give an explanation of the operation principle for the collective ion acceleration. However, as already mentioned, the interaction of the electron-ion ring with outer accelerating fields is determined by the summary electron charge eN_e with N_e as electron number in the ring. The total mass of the ring is then immediately

$$m\gamma N_e + AMN_i$$

with m and M as electron and ion masses, A as the mass number of the accelerated ions.

The relationship of the entire electronic charge of the ring to the ring mass has then the value

$$\frac{eN_e}{m\gamma N_e + AMN_i} .$$

This relationship for the collective acceleration method is more important than the relationship of the ion charge to the ion mass. Thus, at the acceleration of a compact electron-ion ring the effective charge to mass relationship increases. Therein lies the most important advantage of the collective ion acceleration method in contrast to the classic ion acceleration in outer fields.

If the ion number is so small that their contribution to the total mass is negligible, then the ring is accelerated like a pure electron ring. The ions are thereby held by the Coulomb potential of the electrons in the ring, so they have equal velocity like the electrons (in acceleration direction) and they receive a about the factor $AM/m\gamma$ higher energy than the electrons.

For the collective ion acceleration in electron rings such conditions are characteristic, at which the acceleration of the electron-ion ring is determined by an outer field through the electron component. The ions remain in the electron ring by virtue of this high eigenfield.

The ion energy can be estimated for the collective acceleration method using equation (3.1). For an ion of the charge Z and mass number A the energy increase per length unit of the acceleration path yields

$$E \approx k\,\frac{10^{-11}\,N_e}{2Ra}\,\frac{Z}{A}\,\frac{\text{MeV}}{\text{nucleon m}} \qquad (3.2)$$

with R and a as mean radii of the electron ring and of the ring cross-section and k as coefficients, which are determined from the condition for the acceleration of the electron-ion-ring as compact ensemble. The numeric value of k depends on the acceleration conditions and lies between $0.25 \ldots 0.75$.

Ions are formed or further ionized in the electron rings by electron impact ionization. If ions are held in the electron rings for some milliseconds it is possible to produce multiply charged ions of heavy elements.

To obtain a numerical estimation of the energy of a heavy ion, in (3.2) the following values are inserted: $N_e = 1 \cdot 10^{13}$, $R = 4$ cm, $a = 0.2$ cm, $Z/A = 0.1$ and $k = 0.75$. With these values result an energy increase of about 5 MeV/(nucleon m). Thus, it becomes

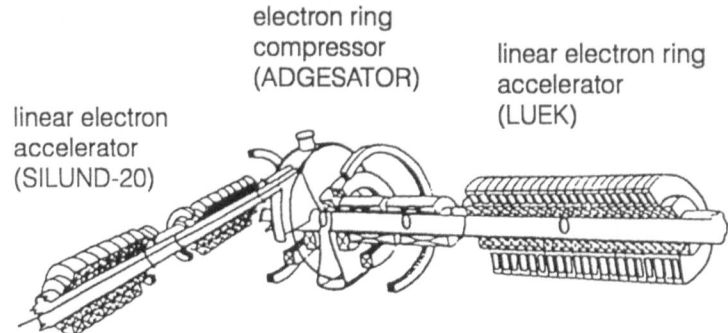

Figure 3.1   Function diagram of the heavy ion collective electron ring accelerator KUTI-20 [116].

in principle possible to accelerate on an acceleration length of 1 m ions up to energies, which are necessary for nuclear physics experiments.

Works on the collective ion acceleration with electron rings began in the sixties in Dubna under the direction of V.I.Veksler and continued in the seventies and eighties through the group around V.P.Sarantsev. The theoretical principles of the collective ion acceleration with electron rings were worked out by the group around E.A.Perelstein. Fundamental results of these investigations are summed up in [83].

Following the works in Dubna, corresponding investigations began in Germany (Garching and Karlsruhe), the USA (Berkeley and Maryland) and in Russia (Moscow and Tomsk).

The books of V.P.Sarantsev and E.A.Perelstein [83] and by U.Schuhmacher [115] are dedicated to the collective acceleration method with electron rings, in which the most essential theoretical principles of this method and experimental results, in the above mentioned scientific centers, are presented. Likewise abundant quotations are indicated to original works.

However, experimental works to the collective acceleration method had to be given up soon, since a series of technical problems appeared, which were connected with the formation of electron rings with such parameters, as they are necessary to the collective ion acceleration. Only in Dubna experimental works were continued to end of the Eighties. Here the most essential results would receive also for the collective acceleration method, on which the later project of a *electron ring ionizer* or ERIS (electron ring ion source) is based.

3.1.2   The Heavy Ion Collective Electron Ring Accelerator

Here we will describe some particularities of the accelerator construction at the layout of the Dubna accelerator. For the realization of the collective ion acceleration it is necessary to produce electron rings with high charge density and with electrons, which have sufficiently high relativistic velocities in reference to their rotation movement in the ring. The electron current in the ring should reach values of 10^3 A or more highly at

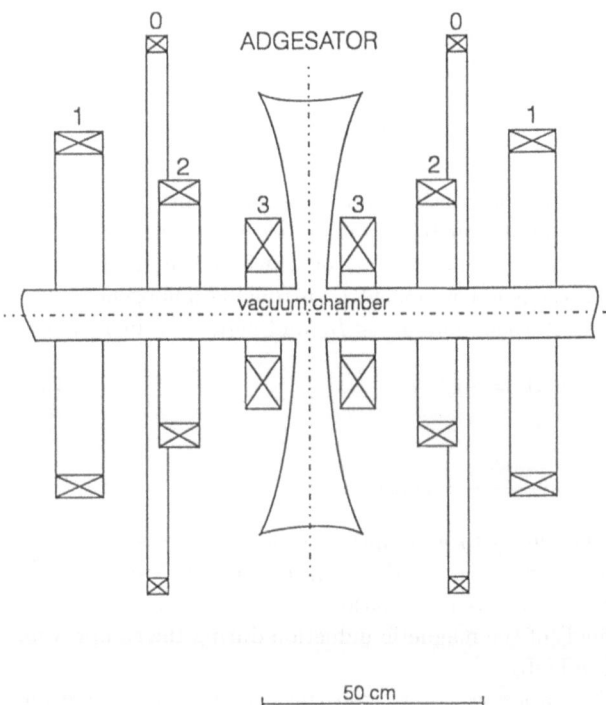

Figure 3.2 Diagram of the magnetic compression system of the ADGESATOR. The numbers 0, 1, 2 and 3 mark the coils for the individual compression stages [83].

values of the ring cross-section radius of about 0.1 cm. The factor for the relativistic movement of electrons in the ring must be chosen thereby in the region $\gamma \approx 20 \ldots 50$.

The formation of an electron ring with the needed parameters is connected with a series of difficulties through the injection of a linear electron beam in a suitable magnetic field configuration. Thus, different compression schemes are applied to the production of dense electron rings, which make it possible, to increase the initially small electron charge density at growing relativistic factor of the electron rings. The ring compression results in a timewise or spatially growing magnetic field in devices, which are called as compressor or ADGESATOR (Russian: adiabatic generator of charged toroids).

The function diagram of the accelerator KUTI-20 is presented in Fig.**3.1**. In the linear induction accelerator SILUND-20 an intense electron beam is accelerated [116], which is characterized by an energy up to 2 MeV, a current of approximately 500 A, an impulse length of 30 ns and a repitition frequency of 20 Hz.

Using the electron accelerator SILUND-20 and the compressor chamber ADGESATOR an electron ring is formed in the compressor chamber [83]. The compressor, as is shown in Fig.**3.2**, consists of a thin-walled Titan chamber and a system of current coils (compression stages). The injection of the electron beam in the compressor chamber results at presence of a weakly focussing magnetic field, which forms from the linear electron beam a ring with a radius $R \approx 35$ cm. Through the successive switching of further compression stages the magnetic field grows timewise and compresses thereby creating the electron ring.

In the course of the compression the azimuthal moment of movement of the electrons is conserved, which leads to the integral of movement

$$M_\Theta = mcR\beta\gamma + \frac{e\overline{B}_z}{2c}\,R^2\;. \tag{3.3}$$

Thereby, in (3.3) $\gamma = 1/\sqrt{1-\beta^2}$ is the relativistic factor of the electrons, $\beta = v_e/c$ and B_z is the mean value of the magnetic induction B_z within the electron ring.

We will use a cylindrical coordinate system (r, φ, z), in which the direction of the z-axis coincide with the ring axis. If the field component B_z is roughly homogeneous, i.e. it changes only weakly with the radius, then holds $B_z \approx \overline{B}_z$ and under consideration of

$$mc\beta\gamma = -eB_z R \tag{3.4}$$

we get

$$M_\Theta \approx \frac{mcR\beta\gamma}{2} = \text{const.} \; .$$

A growing magnetic field B_z thus leads to a compression of the electron ring with simultaneous growth of the electron energy. During the compression β will receive 1 and the product $R\gamma$ will be approximately preserved. Deviations from a constant value in $R\gamma$ appear in radial inhomogeneous fields of the magnetic induction during the compression of the electron ring in the ADGESATOR.

Also known are the compression processes of electron rings in stationary, spatially inhomogenous magnetic fields [117, 118].

The metallic compressor chamber of the KUTI-20 is a weld construction from Titan sheet of thickness 0.8 mm. The surfaces of the side walls of the ADGESATOR chamber show a spherical form. The chamber is supplied with additional flanges for electron injection from the linear accelerator, for introduction of different gauges, for devices for capturing of the electron beam in an equilibrium path, for mounting of a corrector for the electron trajectories in the chamber and for the vacuum system. In the chamber center two pipes are mounted for injection of neutral gases in the compressor chamber and for the extraction of the electron-ion-ring from the chamber.

The magnetic system of the compressor chamber (Fig.**3.2**) is composed of:

- coils for a constant (slowly changeable) magnetic field;
- coils for four compression stages, which produce pulsed magnetic fields;
- a solenoid, in that field the acceleration of the ring results.

The geometry of the compression stages and of the solenoids was chosen under consideration of the screening of the magnetic fields by the metallic chamber walls. The vacuum system of the ADGESATOR was dimensioned for receipt of a vacuum of approx. 10^{-7} Pa. The magnetic field of the compressor chamber produces electron rings with up to 10^{13} electrons and with diameters of the ring cross-section of $2\ldots3$ mm at a mean ring radius of $R_0 \approx 4$ cm in the compressed state. The formation time of an electron ring totals thereby of about $2\ldots3$ ms.

Ions are produced in the electron rings either from atoms of the residual gas or from a specially injected beam of neutral atoms, which cross the ring cross-section. For this purpose different sources of neutral atoms were developed. At the accelerator KUTI-20

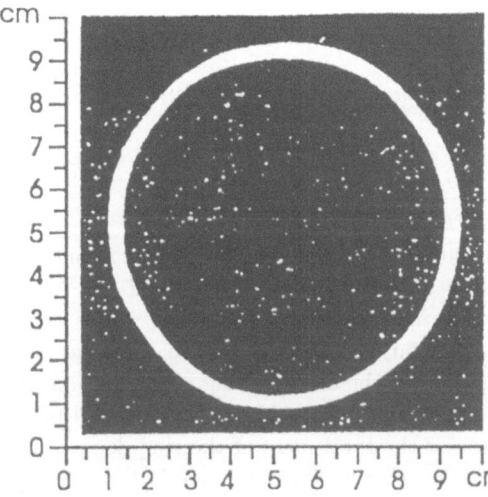

Figure 3.3 Photograph of the compressed electron ring in the accelerator KUTI-20 [122].

a pulsed gas source for pulses of neutral, gaseous media [119] as well as laser sources for condensed materials [120] were both installed.

For extraction of the electron-ion rings from the compressor chamber the currents are varied in the coils of the last compression stage in such a way, that the magnetic potential gap, in which the electron ring is located, moves it out from the mean compressor chamber plane. As a result the electron ring loaded with ions is extracted from the ADGESATOR in a solenoid with a magnetic field gradient, where its preliminary acceleration results in the magnetic field (magnetic acceleration).

The magnetic field gradient producing solenoid is composed of a system of current coils, which produce a field sloping slowly in reference to the axis of the magnetic field with a gradient

$$\frac{1}{B_z}\frac{\partial B_z}{\partial z} \approx -0.15\,\text{m}^{-1}\ .$$

The further acceleration of the electron-ion-ring results in a linear induction accelerator for electron rings (LUEK) [121].

In Fig.**3.3** a photograph of an electron ring in the last phase of the ring compression is shown [122]. The ion acceleration in electron rings is a threshold process. For the acceleration of ions by electron rings the ring radius parameter, radius of the ring cross-section, electron number, number and charge of the ions, spatial distribution and dynamics of the magnetic and electric fields, pressure in the compressor chamber and a series of other parameters must move in well defined boundaries. If the demanded parameters are not observed, the electron ring is destroyed and ions are lost from the ring.

The maximum permissible acceleration, at which the ions are held through the eigenfields of the electron ring, is proportional to the relationship of ion charge i and mass A. Now we assume, in the ring are found only ions of an element with the mass number A and on the electron ring with the radius R and the mean square radius of the ring cross-section a_e acts a constant accelerating force F. Through such a ring ions are accelerated, which have a charge [107]

$$\frac{i}{AM_i} > \frac{2\pi R a_e}{k m_e^2 c^2 N_e r_e \gamma_e (\xi + 1)} \tag{3.5}$$

with

$$\xi = \frac{AM_i N_i}{m_e \gamma_e N_e}$$

whereby m_e and r_e describe the electron mass and the classic electron radius, k a constant, which indicates a certain reserve for the factor i/AM_i, c the velocity of light and N_i the ion number.

For presently realized methods of ion accumulation a sufficiently broad charge spectrum emerges [53, 104]. Besides the desired ions in the electron ring ions also from the residual gas are collected [104]. Ions with small values of i/AM_i can not be held in the ring and get lost so. As a result of this process the total mass of the ring decreases at growing acceleration and ions of the next higher charge stages become lost etc. Under certain circumstances these processes can lead to avalanches in ion loss processes. After leaving the electron ring the ions move along the axis of acceleration. Simultaneously, the electron eigenfields decrease under loss of the ability of the ring, to accelerate ions, i.e., the process of ion acceleration comes to the succumbing. Along the acceleration direction the electrons then hurries ahead of the ions, which only due to their inertia move away.

This means, for the modelling of the electron-ion ring acceleration the existence of ions of different charge states must be taken into account. Especially important is the maximum permissible ring acceleration for concrete ion charge distributions. The values of ion charge states, which are not lost due to ion loss cascades, have been determined for some cases in [50].

Particularities of the ion storage in relativistic electron rings, which have a direct cover to the theme of the existing book, are treated in the next paragraph.

In the development of a heavy ion-electron-ring accelerator beside technical problems, such as the development of magnetic field coils, which operate under extreme electrical, mechanical und thermical conditions, the feed of these coils with pulsed currents of defined values and duration, the development of a thin Titan chamber of an extended volume ($\approx 10^5$ cm^3), the production of a high vacuum better than 10^{-8} Torr in the chamber and a series of other problems, also some physical problems occur.

The physical difficulties result from the limitation of the electron number in the ring, which appear due to the interaction of the electron charges with the thin metallic chamber; the conditions of penetration of pulsed magnetic fields in the metallic vacuum chamber of the compressor, which acts by limiting the electron and ion number in the ring and the possible appearance of transversal and longitudinal instabilities of the electron component and of the electron-ion-ensemble. These and other difficulties were the reasons, why the work was not continued at the collective ion acceleration. So the accelerator described here KUTI-20 was not inserted as an injector for a larger ion accelerator or as a separate accelerator. However, investigations from Dubna become known by techniques to accelerate ions in electron rings in a decreasing magnetic field [123, 124] and in a pulsed electric field [121].

It is appropriate to point out the analogies between the problems, which appear in the development of a heavy ion collective accelerator on the basis of electron rings and those due to thermonuclear synthesis. In both cases electron-ion ensembles are found in pulsed electromagnetic fields. There appear characteristic instabilities and simultaneously all necessary system parameters must be realized, whereby the realization of each of these parameters lies at the borders of present possibilities. In this sense both the collective ion acceleration as well as the thermonuclear synthesis are *treshold processes*. Their realization depends on the simultaneous achieving of a set of parameters. The deviation from one of the necessary system parameters from the necessary boundaries leads then to the extinguishing of the process.

3.2 The Storage of Ions in Electron Rings of the Heavy Ion Collective Accelerator

In the development and operation of a heavy ion collective accelerator there are fundamental problems of production and storage of ions in electron rings to be solved. Thus with the beginning of the development of such an accelerator a series of theoretical works, which deal with the modelling of the ion storage in electron rings, (see [14, 20, 21, 22], [48] to [50] and [125] to [128]) became known.

The theoretical works describing the ion assembly in electron rings proved to be stimuli for the determination of ionization cross-sections of atoms and ions at relativistic electron collisions [20, 21, 51, 129], since they are characteristic for electrons in collective accelerators. Particularly it was shown in this connection, that at the ionization of inner atomic shells by energetic electron collisions cause Auger transitions [20, 21] and multi-ionization processes [20, 21, 51]. These are essential contributions to the total cross-section and are to be considered for calculations to the ion storage.

In later works of ion storage the creation of ions from molecules of an originally neutral gas was included [14] with the charge exchange of a neutral particle with the ions [52, 53]. With the development of the momentum method for the ion distribution function differences in the spatial distribution in the ring cross-section both for the electron as well as for the ion component were considered [84, 101, 102, 106].

In addition, later works of the ion storage in relativistic electron rings are treated, which were published in Dubna. Here we will consider this works in more detail.

3.2.1 Ion Storage from the Residual Gas

The storage of Nitrogen ions from the residual gas under consideration of single- and double charge exchange processes of neutral particles at the formed ions during the compression of the electron ring in the compressor chamber of the KUTI [52] were considered. Thereby the charge exchange cross-sections are calculated according to formulae indicated in [33].

During the ring compression the ring dimensions (ring radius and radius of the ring cross-section) change in the course of time (t = 1.8 ms) about the factor ten. The dependence on the mean ring radius R as a function of the time is shown in Fig.**3.4**. The ring volume as function of the compression time can be expressed in the form

$$V(t) = A + B\,e^{-ct}$$

with a precision of about 10 % .

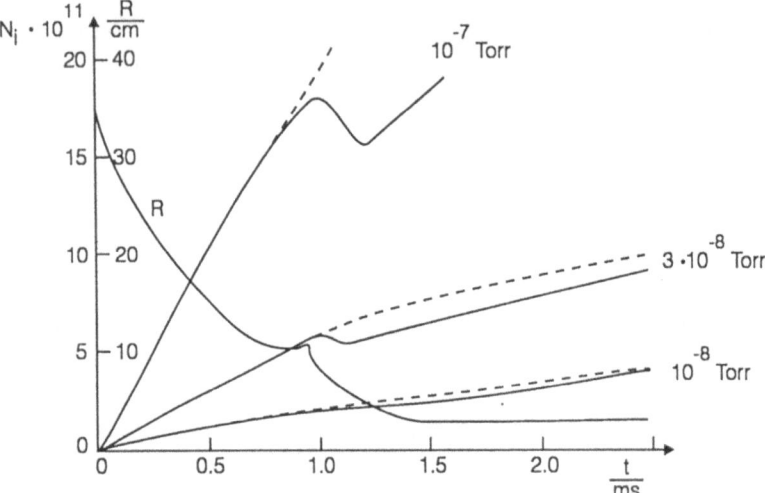

Figure 3.4   Radius of the electron ring R and ion number N_i in dependence of the residual gas pressure in the compressor chamber of the KUTI at different electron numbers in the ring [52]. The dotted lines correspond to calculations without consideration of the charge exchange of neutral atoms at already formed and stored ions.

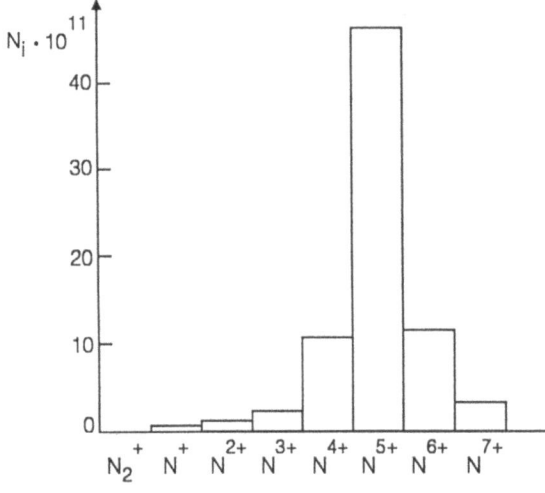

Figure 3.5   Nitrogen ion distribution in the last compression phase of the electron ring [52].

Results of the modelling of the ion storage are shown in Figs.**3.4** and **3.5**. The indices a and b indicate ring compression regimes of final radii $R = 3.2$ cm and $R = 3.7$ cm. The cross curves represent an electron number $N_e = 4 \cdot 10^{12}$ and smooth lines $N_e = 1 \cdot 10^{13}$. Dotted curves correspond to analogous calculations without consideration of ion-ion interactions. In Fig.**3.4** the change of the ion number in the ring during the compression time to 1.8 ms and the subsequent ion collection in the compressed electron ring to 2.5 ms is presented. At a pressure of $P = 10^{-7}$ Torr the net ion charge at the end of the compression time exceeds the net electron charge and the ions are no longer held in the electron potential gap.

In this case the consideration of the ion-ion interaction influences the calculation results considerably. Charge exchange processes between lowly charged ions can neutralize these ions and they can leave the ring volume. This leads to a slowdown of the growth of the ion number in the ring and at $\bar{i} = 1\text{-}2$ ($t = 1.2 \ldots 1.3$ ms) and even to a reduction of the ion number in the ring. Double charge exchange processes increase the mean ion charge. Fig.**3.5** shows the ion charge distribution at the end of the storage process.

3.2.2   Approximation Methods to the Calculation of the Ion Number and Ion Charge State Distribution in Electron Rings

In the first paragraph it was already pointed out at low ion charge stages the ion-ion interaction plays no essential role in the collection of ions with intermediate and high atomic numbers in electron rings. This simplifies considerably the determination of the mean charge $\bar{i}$ and of the ion number N. For some cases the problem can be solved even analytically. An exact solution is a differential-equation system, which describes the collection of ions in an electron ring with constant geometric dimensions [49]. Of likewise considerable meaning for the development and for the operation of collective accelerators is the determination of parameters of ions stored in the ring during the formation time of the electron rings. This problem was treated in [53].

As a rule in the compression of electron rings the change of the ring radius R can be approximated by expressions of the structure [83, 115]

$$R \sim e^{-\alpha t} \quad \text{and} \quad R \sim (1 + \alpha t)^{-1} \ .$$

If selffocussing effects are neglected, yields here $V \sim R^3$.

The total ion number N in a relativistic electron beam of changing volume and neglecting of ion-ion interactions results in the following system of equations

$$\frac{dn_0}{dt} \;=\; \frac{S\bar{u}}{V}\,(n_0 - n) - \frac{\sigma_1\,N_e\,c\,n}{V} \tag{3.6}$$

$$\frac{dN}{dt} \;=\; \sigma_1\,N_e\,c\,n \tag{3.7}$$

with n and n_0 as densities of the neutral particles in the beam and outside the beam, S as beam surface, $u = u_t/\sqrt{6\pi}$ as mean value of the velocity component of the neutral particles in perpendicular direction to the beam surface, u_t as thermal velocity of the neutral particles, σ_1 as ionization cross-section for electron impact ionization of neutral atoms by relativistic electron collission and c as electron velocity, equal to the velocity of light in the relativistic case.

Yields

$$V = V_0\,(1 + \alpha t)^{-3} \quad \text{and} \quad S = S_0\,(1 + \alpha t)^{-2}$$

and the initial conditions

$$V = V_0\,; \quad S = S_0\,; \quad n = n_0\,; \quad N = N_0\,; \quad t = 0$$

then the integral (3.6) has the form

$$n_1 = n_0\,\sqrt{\frac{a_1}{b}}\;e^{-x^2}\int\limits_0^x e^{x^2}\,dx + e^{x_0^2 - x^2}\left(n_1^0 - n_0\,\sqrt{\frac{a_1}{b}}\;e^{-x_0^2}\int\limits_0^{x_0} e^{x^2}\,dx\right) \tag{3.8}$$

with

$$a_1 = \frac{S_0 \bar{u}}{\sigma_1 N_e c} \, ; \quad b = \frac{\alpha V_0}{S_0 \bar{u}}$$

$$x^2 = \frac{[(1 + \alpha t)^2 + a_1]^2}{4 a_1 b} \qquad x_0^2 = \frac{(1 + a_1)^2}{4 a_1 b} \, .$$

For the solution of the integrals (3.8) tables can be used , so e.g. [45]. For many concrete applications yields $x^2 - x_0^2 \gg 1$ and thus

$$n = \frac{n_0 \, a_1}{[(1 + \alpha t)^2 + a_1]} \, .$$

For this case (3.7) can be written as

$$N = \frac{V_0 \, n_0}{\sqrt{a_1} \, b} \left[\arctan \frac{1 + \alpha t}{\sqrt{a_1}} - \arctan \frac{1}{\sqrt{a_1}} \right] \, . \tag{3.9}$$

With this equation it becomes possible, to determine the total number of ions, which have been formed in the ring in the course of his compression. For an estimation of the mean ion charge the knowledge of the total charge of all ions in the beam is necessary.

The total charge of the ions in the beam can be determined by

$$\frac{di}{dt} = \frac{d(N\bar{i})}{dt} = \left[\sigma_1^s n + \sum_{i=1}^{i} \frac{\sigma_{i+1}^s \, N_i}{V} \right] N_e \, c \tag{3.10}$$

with N_i – number of the ions of charge i and $\sigma_i^s = \sigma_i^i + 2\,\sigma_i^{2i} + 3\,\sigma_i^{3i} + \ldots$ – sum of all multiple ionization cross-sections for an ion of the charge i.

Numerical calculations have shown, that for the ion collection in the electron ring only ions of fewer charge states are present. Further it is known, that the ionization cross-sections for medium and heavy elements change little within an atomic subshell [20, 21]. Thus the calculation of the charge spectrum in the boundaries of an electronic subshell gives

$$\frac{di}{dt} = \left[\sigma_1^s n + \frac{\sigma N}{V} \right] \, . \tag{3.11}$$

The quantity σ gives here the mean value of σ_i^s for a considered subshell.

Under application of (3.7) and (3.8) and the initial conditions at $t = 0$ the dependence of the mean charge by time can be found

$$i = i_0 = N^0 \, \bar{i}_0$$

$$\bar{i} = \frac{N^0 \, \bar{i}_0}{N} + \left(1 - \frac{N^0}{N} \right) \left(\frac{\sigma_1^s}{\sigma_1} - \frac{\sigma a_1}{4\sigma_1 b} \right) + \frac{1}{4ab} \left((1 + \alpha t)^4 - \frac{N^0}{N} \right)$$

$$+ \frac{n_0 V_0}{n N a_1 b^2} \left(3 a_1 \alpha t - (1 + \alpha t)^3 + 1 \right) \tag{3.12}$$

$$a = \frac{S_0 \bar{u}}{\sigma N_e c} \, .$$

An estimation for the time Δt for the complete ionization of an electron shell yields according to (3.12)

$$\Delta t = \frac{(4abk)^{1/4} - 1}{\alpha} \tag{3.13}$$

if k describes the electron number in the considered subshell.

For some simple cases the ion number and their mean charge can be determined still more simply. Dense electron beams can not compensate for the entering current of neutral particles which decrease due to ionization and charge exchange processes. As a result the density of neutral particles in the beam is considerably lower than in the surrounding area. If we proceed on the assumption, that practically all neutral particles, which arrive from the residual gas in the beam, are ionized and remain so in the beam, the linear density of the ions can be estimated, which become formed in the courses of the time t

$$N = \sqrt{\frac{2\pi}{3}}\, u_t\, d\, n_0\, t \qquad (3.14)$$

where d is the beam diameter or the diameter of the ring cross-section.

In beams of low density ($n_e \ll u_t/\sigma_1 d v_e$) the density of neutral particles is practically the same as n_0 and it yields

$$N = \sigma_1\, v_e\, n_0\, N_e\, t \qquad (3.15)$$

whereby N and N_e describe either the linear ion and electron densities or the total number of ions and electrons. The mean ion charge can estimated from (1.51) by

$$c\, n_e\, t = \sum_{k=1}^{i+1} \frac{1}{\sigma_k} . \qquad (3.16)$$

Thereby, equation (3.16) for the unknown size i is solved, whereby consideration finds, that the electron density n_e can change in the course of time , so e.g. at the compression of the electron ring.

As an example we consider the accumulation of Argon, Krypton and Xenon ions from the residual gas during the ring compression at the KUTI-20. For comparison purposes the calculations were accomplished with and without consideration of charge exchange processes according to (3.9) and (3.12). Charge exchange cross-sections were considered in the form (1.12). Calculations were accomplished with different parameters α and β in (1.12):

1. $\alpha = 1$; $\beta = 2$;
2. $\alpha = \beta = 1.5$.

The coefficients A_1 and A_2 were determined in accordance with measured cross-sections from reference [130]. For ionization cross-sections values were used from a combination of Auger ionization cross-sections [20, 21] and by cross-sections for direct ionization processes [22]. For double ionization cross-sections values are assumed likewise from [22]. The total ionization cross-sections for Xenon ions are presented in Fig.1.4. The ring volume V and the ring surface S became approximated through equal dependencies, which became used at the derivations of equations (3.8) and (3.11) [53]. Left out were the values of the

initial radius of the electron ring R = 35 cm;
final radius of the electron ring R = 3.5 cm;
compression time $t \sim 2.3$ ms;
electron number N_e = $1 \cdot 10^{13}$.

For Argon a residual gas pressure of $4 \cdot 10^{-8}$ Torr, for Krypton $2 \cdot 10^{-8}$ Torr and for Xenon $1 \cdot 10^{-8}$ Torr were assumed.

Table 3.1 Ion number N, mean charge $\bar{i}$ and charge dispersion D at the storage of Argon, Krypton and Xenon ions from the residual gas [53]. For all ion numbers N counts the factor 10^{11}.

	N_{1i}	N_{2i}	N	N_A	$\bar{i}_{1i}$	$\bar{i}_{2i}$	$\bar{i}$	$\bar{i}_A$	D_{1i}	D_{2i}	D
Ar	8.87	8.94	8.82	8.77	6.71	7.13	6.49	6.50	1.21	1.04	1.77
Kr	5.48	5.50	5.12	5.06	11.08	11.29	11.18	10.56	3.30	2.50	3.69
Xe	3.77	3.77	3.68	3.59	16.06	16.30	16.07	15.45	8.58	6.26	8.85

Numerical results for rare gases are summed up in Table **3.1**. The sizes N_{1i}, $\bar{i}_{1i}$, D_{1i} and N_{2i}, $\bar{i}_{2i}$, D_{2i} correspond to calculations under consideration of ion-ion collisions with cross-sections, which were calculated according to the above indicated parameter sets 1 and 2. The size D describes the *charge dispersion*, i.e. the mean square deviation of the ion charge from the mean value of the charge. For the calculation a system of balance equations (1.43) was used. The values for N, $\bar{i}$ and D were calculated without consideration of ion-ion-interactions. The sizes N_A and $\bar{i}$ are calculated according to equations (3.9) and (3.12).

In Table **3.1** numerical values are presented. The contribution of ion-ion interactions is small for the accumulation process of comparatively low charged ions themselves at heavy atoms. Charge exchange processes considerably reduce the charge dispersion, that particularly becomes clear at Argon.

the in Table **3.1** summed up results lead to the conclusion, that with (3.9) and (3.12) calculated results are sufficiently exact. Thus it becomes possible, to accomplish estimations without complicated numerical calculations for any interesting cases.

3.2.3   Ion Accumulation from a Beam of Neutral Particels

The storage of ions from gaseous elements is in principle possible from an atmosphere of the corresponding gas in the compressor chamber. The production of ions of such elements like Copper, Lead, Uranium etc. requires against a source for beams of neutral atoms of the considered elements.

For an efficiently working source of neutral particles and for a minimal deterioration of the vacuum in the compressor chamber it is necessary, to inject a pulsed particle beam directly in the compressed electron ring.

In the storage of ions from pulses of neutral particles, the pulse duration time must be considerably smaller than the ion storage time, because of high vacua in the compressor chamber results in no permanent new formation of lowly charged ions from neutral particles. Thus in the electron ring there exist no ions, of charge of which is considerably lower than the mean ion charge in the electron ring for this moment in time. This leads to a reduction of the ion charge dispersion. Thus the ion storage from pulses of neutral particles is considerably better suitable for the production of highly charged ions than the production of ions from a residual gas atmosphere.

During the loading of the electron ring with pulsed neutral particle beams a part of the atoms injected in the ring is not ionized or do not arrived the ring volume and so

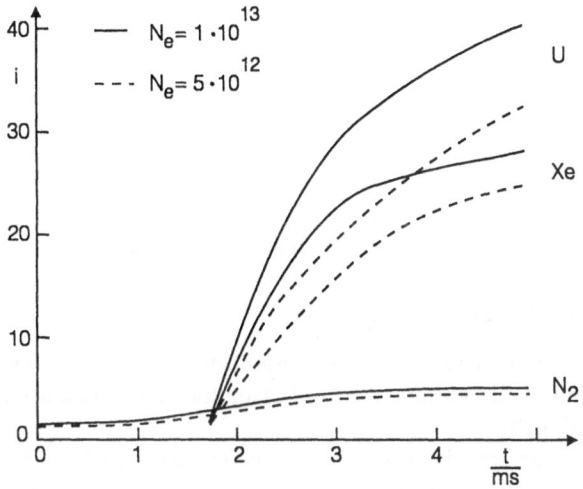

Figure 3.6 Change of the mean charge $\bar{i}$ of Nitrogen, Xenon and Uranium ions as a function of the storage time t in the electron ring [104].

worsens the vacuum in the compressor chamber. For the description of this process a mean free path l of the neutral particles in the ring is introduced. For a ring of the volume V and a velocity of the gas jet u yields

$$l = \frac{Vu}{\sigma N_e c} \ .$$

If the particle source produces an axial symmetric, azimuthal homogeneous beam of a mean density n_0 with a radial width $d \geq 4a$ and an impulse length τ, so the mean particle number, which is not included by the electron ring, can be estimated to [104]

$$N \approx 2\pi R u n_0 \tau \left(d - 4a\,e^{-l/\pi a} \right) \ .$$

Here a is the half root-mean-square dimension of the ring cross-section. For more complicated source geometries (for instance an ensemble of several individual gas jets) the number N can be estimated likewise.

For an efficient source operation as many as possible neutral particles should be ionized in the ring, to not worsen the vacuum in the compressor chamber. On the other hand the demand stands, that the ion production should result as evenly as possible over the ring cross-section, to reduce the radius of the ion component in the ring cross-section [84]. If ions are produced only in the outer zones of the electron ring, increase the mean square dimensions of the ion component in the ring cross-section, which can lead to an increase of ion losses from the ring at the ion storage and acceleration. Thus the weakening of the neutral particle beam in the ring cross-section must be held small. This demand is fulfilled at $l \sim 4a$.

If the efficiency of ion production and the necessary ion number for an optimal load of the electron rings is known, the number of neutral particles for an impulse can be estimated. Calculations are known for the storage of Xenon and Uranium ions from gas jets of the length 0.25 ms in a residual gas atmosphere (Nitrogen) in the compressor chamber of the KUTI-20 [104]. These modellings did not consider ion-ion interactions, since these do not influence the result at the storage of ions to high charge states. The compression time of the electron ring totaled thereby 2.5 ms, after that the ring dimensions were assumed as constantly. Further it was proceeded on the assumption, that the source of neutral particles begins to work 1.75 ms after beginning of the ring compression at a ring radius of $R = 5$ cm.

The most important results of the described calculations are presented in Fig.**3.6**. Presented becomes the change of the mean ion charge $\bar{i}$ from Nitrogen, Xenon and Uranium ions in the electron ring. From the comparison of both curves follows, that the mean ion charge $\bar{i}$ considerably depends on the electron number N_e in the ring.

3.2.4　Diagnostic Methods for Electron-Ion Rings

During the developing of electron ring accelerators also a series of diagnostics methods was developed for the determination of the parameters of electron-ion-rings. Essential methods should be described here shortly.

The *ion number* in the electron ring can determined over the evaluation of by the ring emitted *electron bremsstrahlung* [8]. Bremsstrahlung emerges as a result of Coulomb scattering of electrons at the atomic nuclei of ions or neutral particles located itself in the ring. In this case the measuring size describes the product $N_e(N_i + N_0)$.

The *electron number* in the ring can be received by the spectroscopy of the electron bremsstrahlung at the beginning of the electron ring formation, where still no ions were formed and only atoms of the residual gas, whose pressure is known, are available. Over the knowledge of N_e the ion number available after the ring compression in the ring can be determined from the spectrum of the electron bremsstrahlung.

The *synchrotron radiation* of the relativistic electrons is used likewise to diagnostic purposes. The ring of relativistic electrons themselves presents an intense source of synchrotron radiation. The emitted synchrotron radiation power of the electron ring with N_e electrons, an electron energy E_e and a ring radius R yields

$$P = -\frac{2}{3}\frac{N_e\,E_e^4\,r_e}{m_e^3\,c^5\,R^2} \tag{3.17}$$

with m_e as electron mass. Thereby P reaches 1 kW in the visible and infrared region of the electromagnetic radiation.

Of particular interest is the theoretically predicted and experimentally shown widening of the emission cone of the synchrotron radiation from the electron ring at the storage of ions in the ring [131, 132].

Synchrotron radiation of a relativistic electron becomes emitted in a small angle area $\Phi \sim 1/\gamma$ (γ – relativistic factor) in the movement direction of the electrons. Since the electrons in the ring execute Betatron oscillations around their equilibrium paths, it comes to oscillations of the direction vector of the electron velocities around the general movement direction of the electron beam and with it to a widening of the angle distribution of the synchrotron radiation. The full width at half maximum (FWHM) of the angle distribution in axially direction can be received from the expression

$$\Theta_\lambda = \sqrt{\Phi^2 + \Theta_z^2}$$

with Θ_z as the caused through Betatron oscillations mean square angle deviation of the particles from the mean ring plane [134]. The frequency of the Betatron oscillations is

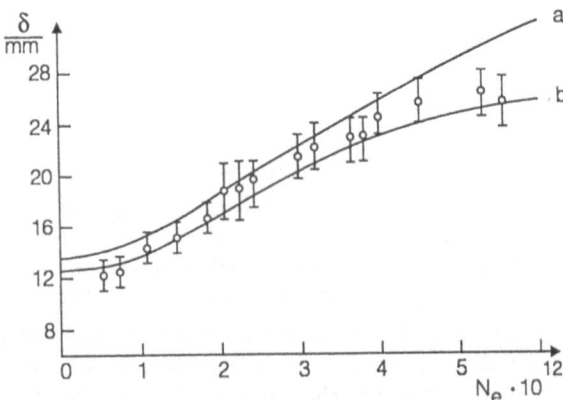

Figure 3.7 Change of the full width at half maximum δ of the intensity distribution of the electron synchrotron radiation in dependence of the electron number N_e in the ring [131]. 1 – calculation without consideration of the ion focussing for an initial radius of the ring cross-section $a = 1.7$ cm; 2 – calculation with consideration of the ion focussing and $a = 1.5$ cm. The indicated points describe experimental values.

determined thereby by the acting outer focussing and by the ion self-charge in the ring. It can be shown, that the opening angle of the angle distribution of the synchrotron radiation increases with increasing ion number in the ring. Simultaneous the ion number is proportionally to the electron number available in the ring. A widening of the synchrotron radiation distribution at growing electron number became experimentally shown at the accelerator KUTI-20 [131, 132].

Results of synchrotron radiation spectroscopy in dependence on the electron number in the ring at the accelerator KUTI-20 are presented in Fig.**3.7** [131]. The vertical figure axis indicates the full width at half maximum of the intensity distribution of the synchrotron radiation $\delta = L\Theta_\lambda$ with $L = 75$ cm as effective measuring basis. On the horizontal axis the electron number N_e in the ring is given in relative units. Simultaneously through corresponding lines the functional dependence $\delta(N_e)$ is presented, how it would receive from the equations for the moments of second order of the electron and ion distribution functions. The derivation of these equations resulted in chapter 2.

The calculations were provided for two models in reference to the ring compression. In the first case (curve a) the reduction of the electron ring cross-section was not considered due to ion focussing and the initial radius of the ring cross-section was assumed to $a_0 = 1.7$ cm. The curve b would receive under consideration of ion focussing at the ion storage and for $a_0 = 1.5$ cm.

Measurements have shown, that at low electron numbers $N_e \leq 1 \cdot 10^{12}$ the ion eigenfields can be neglected and that $\delta = L\Phi$ does not depending on the electron number. At higher electron numbers $N_e > 2 \cdot 10^{12}$ the influence of the ion eigencharges on the focussing becomes visible, the width of the angle distribution of the synchrotron radiation is determined then through the Betatron oscillations and is proportional to the electron number in the ring and it counts $\delta = L\Theta_z$.

3.3   Particularities of the Production of Highly Charged Ions in Electron Rings

3.3.1   Basic Problems

Already with the first works on the collective acceleration method of ions in electron rings it became clear, that the increase of the charge state of ions stored in electron rings is an urgent task. The basic principles of the collective acceleration method as well as equations (3.2) and (3.3) point to it, that the effectiveness of the collective method is determined by the charge to mass relationship of the ions to be accelerated.

Independently by the effort, to receive high ion charges with the goal of an increase of the acceleration velocity, suggestions became known, to use the relativistic electron rings as an efficient source of highly charged ions [20, 21, 128, 133, 135]. Under this point of view relativistic electron rings display actually a series of special important characteristics:

1. The energy of the relativistic electrons exceeds a multiple of the ionization energy of the heaviest ions and makes possible so in principle the production of very highly charged ions up to completely ionized atoms. This characteristic does not point to other known ion sources.
2. Electron rings are an universal ion source. Atoms of arbitrary elements, which arrive in the ring area, are ionized with equal success.
3. The relatively high realized electron densities (10^{12} cm^{-3} and more) make it possible, to receive in the courses of a comparatively short storage time highly charged ions.
4. Electron rings are an ion trap of high capacity (up to 10^{12} ion charges in a ring).
5. The electrons of the electron ring ionize in the sense effectively, that at the ionization processes one and the same electron entirety participates, without introducing in the ion trap new electrons.
6. Through the possibility, to observe the emitted infrared, visible, ultraviolet and X-ray radiation during the entire process of the ring formation and during the ion storage, it makes possible, to pursue the dynamic of the ion assembly and to determine with spectroscopic methods characteristics of electron levels of multiple charged ions, without eliminating the ions from the electron ring.

For the characterization of possibilities of electron-ion-rings as ion traps for storage and ionization of ions became accomplished calculations of the time development of the ionization in electron rings with high density at constant in time geometric dimensions [20, 21, 51, 135].

In Fig.**3.8** an example for the calculation of the development of the charge distribution of Xenon ions in an electron ring with the parameters $N_e = 1 \cdot 10^{13}$, R=4 cm, a=0.2 cm and E=20 MeV is shown [51].

In the Tables **3.2** and **3.3** are indicated cross-sections for Coulomb ionization [22], ionization factors $j\tau_i = 1/\sigma_i$, which are necessary for the production of a charge state i from the preceded one and corresponding ion storage times τ_i for an electron density of $n_e = 3.2 \cdot 10^{12}$ cm^{-3}. The calculations were accomplished for the supreme ten charge states of Krypton, Xenon and Uranium at electron energies of 1 MeV and 20 MeV.

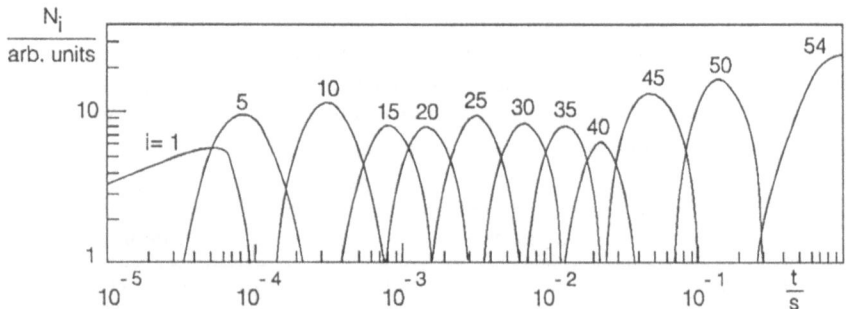

Figure 3.8   Time development of Xenon ions in electron rings with an electron density of $n_e = 3.2 \cdot 10^{12}$ cm^{-3} and a Xenon gas impulse of a duration time of 70 μs. As parameter the ionization degree is indicated [51].

The ion collecting times $j\tau_i$ and storage times τ_i determine the conditions for the transition of an ion of a charge state to another. The ionization factor, which is necessary for the production of an ion of a given charge state from a neutral particle and the corresponding time τ_i can be determined from (1.49) or estimated from relations (1.50) and (1.51)

$$\tau_i \approx \sum_{k=1}^{i} \tau_k$$

The results, which are shown in Fig.3.8 and in the Tables 3.2 and 3.3, are given without consideration of Auger ionization.

The consideration of Auger processes leads for low and medium charge states in Fig.3.8 to a shift on the left, but does not influence practically the characteristic times, which are necessary for the receipt of very highly charged ions [20, 21]. This depends from the fact, already noticed in the first chapter, that Auger ionization processes bring only a contribution, if in the atom or in the ion more than only one atomic subshell is occupied. However, if several subshells are occupied, Auger processes deliver an important contribution to the total process. Auger ionization processes can accelerate the production of highly charged ions with more than 10 electrons up to 2 ... 3 times, without having influence on the ionization times for the highest possible charge states.

Numerical calculations of the ion storage in multicomponent electron-ion rings at changeable ring dimensions yield on the basis of the equations, which would receive with the momentum method (see Chapter 2.) total values $(j\tau_i)_{\text{tot}}$, which are necessary for the production of ions of a certain charge state

$$
\begin{aligned}
\text{Kr}^{26+} &: \quad (j\tau_{26})_{\text{tot}} = \quad 0.05 \cdot 10^{22} \text{ cm}^{-2} \quad (\tau_{26})_{\text{tot}} = 0.05 \text{ s} \\
\text{Xe}^{44+} &: \quad (j\tau_{44})_{\text{tot}} = \quad 0.2 \cdot 10^{22} \text{ cm}^{-2} \quad (\tau_{44})_{\text{tot}} = 0.02 \text{ s} \qquad (3.18) \\
\text{U}^{82+} &: \quad (j\tau_{82})_{\text{tot}} = \quad 1.0 \cdot 10^{22} \text{ cm}^{-2} \quad (\tau_{82})_{\text{tot}} = 0.1 \text{ s} .
\end{aligned}
$$

Assumed became thereby an electron energy $E_e \approx 20$ MeV and an electron density $n_e \approx 3.2 \cdot 10^{12}$ cm^{-3}.

For the determination of the total ionization factor $(j\tau_i)_{\text{tot}}$ and the ion storage time $(\tau_i)_{\text{tot}}$ for the production of ions with fewer than ten electrons an addition of the values

Table 3.2 Ionization cross-sections for direct ionization σ_i according to [22], ionization factors $j\tau_i = 1/\sigma_i$, which are necessary for the production of the charge state i from the preceded one and storage times τ_i for an electron density in the ring of $n_e = 3,2 \cdot 10^{12}$ cm^{-3} for Krypton (Z = 36) and Xenon (Z = 54) at electron energies E_e of 1 MeV and 20 MeV. Z gives the atomic number of the considered element.

E_e	i	Krypton			Xenon		
		σ_i $/10^{-22}$cm^{-2}	$j\tau_i$ $/10^{22}$cm^{-2}	τ_i /s	σ_i $/10^{-22}$cm^{-2}	$j\tau_i$ $/10^{22}$cm^{-2}	τ_i /s
1 MeV	Z-9	22.1	0.045	0.005	7.5	0.13	0.015
	Z-8	18.5	0.054	0.006	6.4	0.16	0.018
	Z-7	15.2	0.066	0.007	5.3	0.19	0.021
	Z-6	12.1	0.083	0.009	4.3	0.23	0.026
	Z-5	9.4	0.106	0.012	3.4	0.29	0.033
	Z-4	6.9	0.145	0.016	2.5	0.39	0.044
	Z-3	4.7	0.213	0.024	1.7	0.58	0.065
	Z-2	2.6	0.38	0.043	0.99	1.01	0.113
	Z-1	0.75	1.33	0.15	0.28	3.57	0.40
	Z	0.36	2.78	0.31	0.14	7.4	0.83
20 MeV	Z-9	30	0.033	0.004	10.7	0.093	0.010
	Z-8	25	0.040	0.004	9.1	0.110	0.011
	Z-7	21	0.048	0.005	7.6	0.132	0.013
	Z-6	16.6	0.060	0.006	6.2	0.161	0.017
	Z-5	12.9	0.078	0.008	4.9	0.205	0.022
	Z-4	9.5	0.105	0.011	3.7	0.273	0.029
	Z-3	6.5	0.155	0.016	2.52	0.40	0.042
	Z-2	3.7	0.27	0.028	1.46	0.68	0.072
	Z-1	1.11	0.90	0.095	0.44	2.26	0.24
	Z	0.54	1.87	0.197	0.215	4.7	0.49

from Tables **3.2** and **3.3** as well as from (3.18) is necessary. So for instance for the production of Xe^{48+} ions at an electron energy of 20 MeV yields

$$(j\tau_{48})_{\text{tot}} = (j\tau_{44})_{\text{tot}} + j\tau_{45} + j\tau_{46} + j\tau_{47} + j\tau_{48} = 0.7 \cdot 10^{22} \text{ sm}^{-2}$$

and

$$(\tau_{48})_{\text{tot}} = (\tau_{44})_{\text{tot}} + \tau_{45} + \tau_{46} + \tau_{47} + \tau_{48} = 0.07 \text{ s} .$$

From the indicated correlations is obvious, that the values $(j\tau_i)_{\text{tot}}$ and $(\tau_i)_{\text{tot}}$ for highly charged ions and here particularly for completely ionized atoms through the very small values for the direct ionization are determined by the last bound electrons [22]. In

Table 3.3 Ionization cross-sections for direct ionization σ_i after [22], ionization factors $j\tau_i = 1/\sigma_i$, which are necessary for the production of the charge state i from the preceded one and storage times τ_i for an electron density in the ring of $n_e = 3.2 \cdot 10^{12}$ cm^{-3} for Uranium (Z = 92) at electron energies E_e of 1 MeV and 20 MeV. Z indicates the atomic number of Uranium.

i	Uranium 1 MeV			Uranium 20 MeV		
	σ_i $/10^{-22}$cm^{-2}	$j\tau_i$ $/10^{22}$cm^{-2}	τ_i /s	σ_i $/10^{-22}$cm^{-2}	$j\tau_i$ $/10^{22}$cm^{-2}	τ_i /s
Z-9	1.83	0.55	0.061	2.83	0.035	0.037
Z-8	1.56	0.64	0.072	2.42	0.41	0.043
Z-7	1.30	0.77	0.086	2.03	0.49	0.052
Z-6	1.05	0.95	0.107	1.65	0.60	0.064
Z-5	0.81	1.23	0.138	1.29	0.78	0.082
Z-4	0.61	1.63	0.183	0.98	1.02	0.107
Z-3	0.42	2.36	0.265	0.68	1.46	0.154
Z-2	0.241	0.41	0.47	0.40	2.5	0.26
Z-1	0.065	1.55	1.74	0.124	8.1	0.85
Z	0.032	3.2	3.5	0.061	16.4	1.72

this case the summary values $(j\tau_i)_{\text{tot}}$ and $(\tau_i)_{\text{tot}}$ are not influenced by Auger processes practically.

In (3.18) the values $(\tau_i)_{tot}$ are about two times lower than the values for Xe^{44+}, which are shown in Fig.**3.8** and in comparison with the values for Kr^{26+} and U^{82+} from reference [51], although at equation (3.18) is proceeded on the assumption that the ions of the working gas are produced permanently from a residual gas atmosphere. The observed difference can be explained with it, that in [51] Auger processes had not considered and that the values $j\tau_i$ and τ_i are 2...3 fold excessive for low and medium charge states.

It shows itself, that highly charged ions can be received from heavy elements, if these can be held in the relativistic electron rings about a time $\tau_i \sim 1$ s . With it however immediately the following questions are connected

- Can a ring of relativistic electrons in a magnetic field configuration of the KUTI-type exist up to a second or longer?
- Can processes exist, which limit the ionization process and the ion charge in electron rings of the KUTI?

Not considered here are technical problems, which appear at the increase of the life time of electron rings in the milliseconds to the second region. For instance, it is especially difficult to develop a magnetic system, which maintains a magnetic induction of about 2 T in the volume included by the electron ring in a stationary or quasistationary regime.

3.3.2   Physical Problems

The basic physical problems of the production of highly charged ions in electrons rings can be formulated in the following way [8, 73, 136]

a) energy losses of the relativistic ring electrons by synchrotron radiation;

b) electron scattering at in the ring by stored ions and with it joined growth of the radius of the ring cross-section;

c) neutralization of the eigencharge of the electron ring by ion production from the residual gas;

d) limitation of the maximum achievable charge state of the stored ions by charge exchange at neutral atoms of the residual gas.

The last two points present a limit for the permissible residual gas pressure in the compressor chamber.

The other problems become treated in more detail in Ref.[73].

a) The temporal change of the ring radius R in a time variable magnetic field yields under consideration of the synchrotron radiation loses to

$$\frac{dR}{dt} = -\frac{R}{(1-n)B_z}\left(\frac{dB_z}{dt} - \frac{1}{2}\frac{d\bar{B}_z}{dt}\right) - \frac{R}{1-n}\frac{P}{E} \ . \tag{3.19}$$

Thereby B_z describes the axial component of the magnetic induction, $\bar{B}_z$ the mean value of B_z within the electron ring and P the power loss by synchrotron radiation of the electrons. Further yields

$$n = -\frac{R}{B_z}\frac{\partial B_z}{\partial R} \qquad \text{(magnetic field index)} \tag{3.20}$$

and

$$E = \gamma mc^2 \qquad \text{(electron energy)} \ . \tag{3.21}$$

The possibility, to use synchrotron radiation and radial inhomogeneties of the magnetic field to the formation and long stabilization of electron rings in collective accelerators, was proposed at first in Ref.[137].

In a magnetic field with the axial component B_z the electrons move on equilibrium paths with the radius R, which is determined by (3.4)

$$R = -\frac{mc\beta\gamma}{eB_z} = \frac{\beta E}{ecB_z} \ . \tag{3.22}$$

Electron energy losses through synchrotron radiation (3.17)

$$\frac{dE}{dt} = P$$

reduce according to (3.22) the electron ring radius and the generalized momentum M_Θ.

For $\gamma = 30 \ldots 40$ and $R \approx 4$ cm the characteristic time for the reduction of the ring dimensions totals in a time-stable homogeneous magnetic field of about $30 \ldots 60$ ms. In the axialy symmetric, weakly focussing magnetic field of the ADGESATOR (compressor) evidently no possibility exists of a compensation of energy losses and of the generalized momentum of the ring electrons. This leads to, that the synchrotron radiation becomes the primary cause for the temporal limit of the life time of relativistic electron rings on about 100 ms.

b) In the electron ring the electrons are scattered at the ions and atoms of the residual gas elastically in small angles. This leads to a successive growth of the amplitude of the Betatron oscillations and enlarges the dimensions of the ring cross-section. Simultaneously the synchrotron radiation reduces, as we have seen already from equations (2.70) to (2.72), the dimensions of the ring cross-section. This phenomena was marked as *radiation friction*. If (2.71) is rewritten for the effective phase space volume of the ring cross-section $\varepsilon = \gamma \omega a^2$, follows

$$\frac{d\varepsilon}{dt} = \frac{\gamma S}{\omega} - \frac{P\varepsilon}{E} \tag{3.23}$$

with ω as frequency the the Betatron oscillations and S as collision integral (2.66).

Thus two competing processes exist, whereby one process reduces the phase space volume of the ring cross-section and the dimensions of the ring cross-section and the other one enlarges the named quantities. For heavy ions with high atomic numbers the electron scattering dominates over the radiation friction. For example in the compressed electron ring there are 10^{11} Xenon ions, so for $\gamma = 30 \ldots 40$ the first term in the right part of (3.23) exceeds the second one by the factor two and the dimensions of the ring cross-section increase. Thereby the radiation friction (3.17) decrease and the collision integral and scattering probability (2.66) increase as a result of the reduction of the electron energy by synchrotron radiation loses. Characteristic times for the enlargement of the ring cross-section dimensions due to Coulomb scattering amount in dependence of the ion number and of the atomic number of the considered element to about $30 \ldots 100$ ms.

c) Additional production of light ions from the residual gas occurs during the entire electron ring confinement time and does not depend on the way of electron ring loading with ions of working matter. If we start from (3.14) an estimation, follows, that at a residual gas pressure of 10^{-9} Torr the electric charge of the electron ring is neutralized approximately after about 50 ms by light ions. For instance, this happens in the essential by N^{7+}. From this circumstance follows high demands on the vacuum in the compression chamber.

d) Charge exchange processes of the residual gas atoms with in the electron ring stored ions decrease their charge states and present thus a process working in competition to the ionization. On the basis of the in paragraph 1.6 and in Fig.1.11 presented results can be closed, that for the production of Xe^{52+} or U^{86+} ions a residual gas pressure by better than $1 \cdot 10^{-9}$ Torr must be demanded.

The listed circumstances limit the life-time of the electron-ion rings on a time limit of $50 \ldots 100$ ms and put simultaneously high demands on the vacuum to be realized in the compressor chamber. Under these conditions the maximum achievable ion charge can be estimated at constant ring dimensions for Xenon to $i = 44 \ldots 46$ and for Uranium to $i = 70 \ldots 76$. A further increase of the ion charge would require an considerable increase of the electron density.

The final radius of the electron ring and accordingly also the electron ring density depends from the maximum reached magnetic induction of the compressing field. From the equations (3.3) and (3.4) follows

$$M_\Theta = -\frac{eB_z R^2}{2} = \text{const} \,.$$

Corresponding counts

$$R \sim \frac{1}{\sqrt{B_z}} \,.$$

From this it becomes obvious, that for a reduction of R around the factor two B_z must be increased around the factor four! Up to now the in experiments at the KUTI preserved induction totaled $B_z = 2.5$ T. Corresponding to this value a ring compression to an radius $R = 3.2$ cm was received [122].

Another way, using the reduction of the electron energy at the electron injection in the compressor chamber to the reduction of the generalized momentum of the electrons is not likewise possible, because such a procedure leads to a reduced effectiveness for the capture of the electron current on an initial equilibrium radius in the compressor chamber and with it to a decrease of the electron number in the formed ring.

In [138, 139] it was proposed, to use the synchrotron radiation of the electrons at the end of the compression, to reduce the initial moment of the electrons and to reach so an additional ring compression. Through synchrotron radiation losses the generalized electron moment M_Θ in a time constant magnetic field is reduced according to

$$\frac{dM_\Theta}{dt} = \frac{1}{1-n} \left(2M_\Theta - nmcR\gamma\right) \frac{P}{E} \,. \tag{3.24}$$

Simultaneously the radius of the electron ring R changes accordingly

$$\frac{dR}{dt} = \frac{R}{1-n} \frac{P}{E} \,. \tag{3.25}$$

Through the generation of an inhomogenuos magnetic field with a field index n closely to one at the end of the ring compression can be reduced at comparatively insignificant electron energy losses by the emission of synchrotron radiation with a sicnificant reduction of the generalized moment M_Θ and the final ring radius R considerably. An additional ring compression results with the reduction of the electron energy. This decreases the necessary maximum value of the magnetic induction B in the final compression phase (3.22).

On the basis of the presented idea a mathematical modelling was undertaken a system was proposed for additional compression and long ring storage [139]. After a reduction of the ring radius by the factor two (to $R \approx 2$ cm) in a magnetic field with a field index $n \leq 1$ arrives the electron ring in a slowly according to

$$B_z = B_{z_0} \, e^{-t/\tau}$$

falling magnetic field with $\tau \approx 50$ ms. As result of the decrease of the magnetic field the reduction of the electron ring radius was compensated by synchrotron radiation losses and the ring dimensions remained in a period of $40\ldots50$ ms roughly constant. The realization of such a compression diagram makes it possible, to receive ionization factors of $j\tau \approx 10^{22}$ cm^2 and so with it Kr^{34+}, Xe^{50+} and U^{82+}.

3.4 The Electron Ring Ion Source ERIS

The creation of an electron-ring ion source (ERIS) in the JINR Dubna goes back to a suggestion from the year 1975 [133] and became begun with the realization in 1991 [73, 135, 136, 138, 139, 141]. The goal of the development of an electron-ring ion source are experiments to study the physics of electronic shells and correlated problems in highly ionized atoms.

Present a large amount of calculated data for ionized atoms is available, which is determined on the basis of different physical models (see e.g. Refs.[10, 7, 11]). As a rule, with increasing nuclear charge the precision of the available data decrease clearly. The consideration of relativistic effects and of contributions from Auger- and Coster-Kronig processes in the different models leads to divergenting results.

The employment of the ionizer ERIS for investigations of excited or ionized atoms, for the determination of ionization cross-sections as well as for the determination of different dynamical and structure data of highly charged atoms makes it possible, to receive a multitude of new information about the electronic shells of multiply charged ions.

Electron rings at the ERIS device should be used as source of highly charged ions with the following parameters:

electron number – $N_e \approx 5 \cdot 10^{12}$;

ring radius – $R \approx 1.8 \ldots 2.2$ cm;

root-mean-square radius of the ring cross-section – $a_{r,z} \approx 1$ mm;

relativistic factor of the azimuthal electron movement – $\gamma \geq 10$;

life time of the electron ring since begin of his formation – $\tau \approx 60 \ldots 100$ ms;

ring life time of the compressed ring – $\tau' \approx 40 \ldots 80$ ms.

The load of the ring volume with dosed clusters of neutral atoms of the element to be ionized makes it possible to produce and store ions with high charge states, so $Xe^{30+} \ldots Xe^{50+}$ and $U^{44+} \ldots U^{82+}$ with up to 10^{10} ions in the ring.

In contrast to the ion acceleration in electron rings the process of production and storage of multiply charged ions is not a threshold process. Here in contrast to the ion acceleration a deviation of the electron ring parameters does not lead to the extinguishing of the process itself, but only to a certain decrease of the achievable ion charge stages.

The ions exited through electron collisions in the ring emmit characteristic X-ray radiation and radiation in the VUV region. The spectrometry of these radiation components and the measurement of the intensity of individual lines make it possible to get information on the ion number in the ring at known other ring parameters. An analysis of the temporal course of emission spectra allowed it, to pursue the dynamics of the ion production and of the ion assembly in the ring. A measurement of the time dependence of transitions of individual ions from a charge state in another allows it to solve the inverse problem for the ionization makes it possible, to determine ionization cross-sections for ionization by relativistic electron impact and charge exchange cross-sections of neutral atoms at ions. The wavelength dispersive X-ray spectroscopy can be determine parameters of the characteristic X-rays of highly charged ions (energy shifts relative to the parent diagram lines, intensity ratios, line widths), radiation corrections (Lamb-Shift for hydrogen-like ions) and other effects of small amplitudes.

An interesting aspect of the ERIS is the high number of stored ions (up to 10^{10} per ring) for practically each element of the periodical system of the elements and the

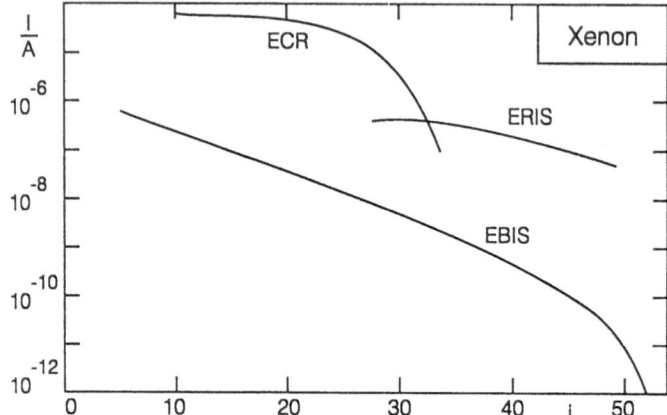

Figure 3.9 Comparison of extracted Xenon ion currents I as a function of the ion charge i for electron-ring ion sources (ERIS), for electron-cyclotron-resonance ion sources (ECR) and for electron beam ion sources (EBIS) [141].

possibility of the analysis of time development of the ionization as a function of the storage time for the ions in the electron ring.

In Fig.**3.9** are shown the calculated ion currents for the extraction of Xenon ions from an electron-ring ion source in comparison with other sources of highly charged ions [141].

The ionizer ERIS is composed of a linear induction accelerator SILUND and of a compressor chamber (ADGESATOR) for production and compression of electron rings. The vacuum chamber and the magnetic system of the ionizer are presented in Fig.**3.10** [142].

For a long confinement of the electron rings on very small radii magnetic fields are necessary, which change only little in the courses of some ten milliseconds. A simple process to achieve such fields is to shunt current coils of the last compression stages at the collective accelerator in the moment, where through the coils the maximum current flows [135]. As result the current falls and so also the magnetic field, which holds the ring, proportionally to e^{-t/τ_c} with the decrement $\tau_c \approx L_c/R_c$ with R_c as active coil resistance and L_c as their inductivity.

A corresponding technical solution was realized at the collective accelerator. The life-time of the electron ring totaled $\tau_c \approx 36$ ms on a radius of $R \approx 4$ cm and at about $30\ldots 40$ ms without a considerable increase of the dimensions of the ring cross-section. Thereby, the ionization factor reached values of about $0.5 \cdot 10^{21}$ cm^{-2} [143, 144].

At the ionizer ERIS is anticipated to reduce the ring radius by a combination of several stages of ring compression and by phases of ring confinement. In Fig.**3.11** is shown the temporal course of the currents in the decisive compression coils of the magnetic system [142]. As basis of a modified variant was used the magnetic system of the ADGESATOR [145], which became completed by additional compression stages [142, 143, 144]. These coils are marked in Fig.**3.10** with the numbers 11 and 12. Reached became ring radii $R \leq 4$ cm at a field index $n \approx 0.5$.

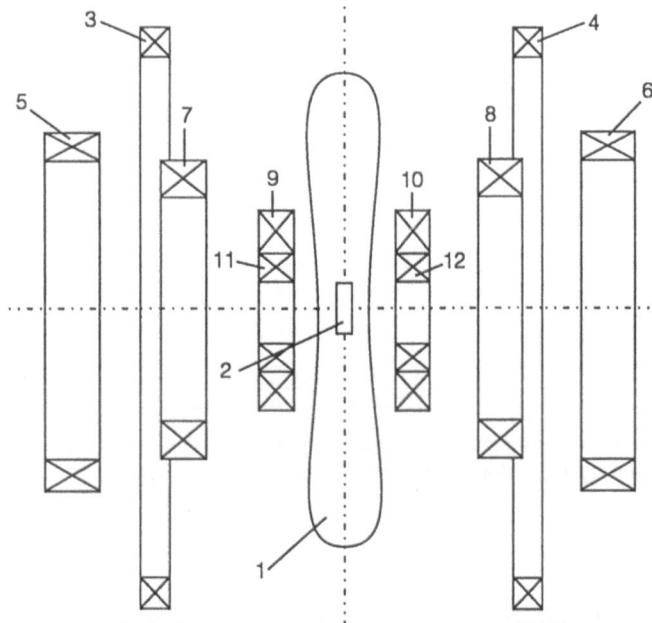

Figure 3.10 Construction of the magnetic system of the compressor chamber of the ionizer ERIS [142]. 1 – vacuum chamber; 2 – electron ring; 3,4 – coils of the zeroth compression stage; 5,6 – coils of the first compression stage; 7,8 – coils of the second compression stage; 9,10 – coils of the third compression level; 11,12 – coils for an additional compression and long storage time of the electron ring.

A standard compression cycle in a magnetic field with $n \approx 0.5$ lasts about 4 ms, whereby a ring radius of 4.5 cm is reached. After reaching the current maximum in the coils the first confinement cycle of the electron ring loaded with ions begins. To it the current sources are shunted and the current in the compression coils falls according to e^{-t/τ_c} ($\tau_c \approx 35$ ms). As result the radius of the synchrotron radiation emitting electron ring remains at 4...4.5 cm and the generalized momentum decreases approximately around the factor 1.7. In the courses of the first confinement cycle the field index has a value of $0.5 \geq n \geq 0.45$. This in accordance with (3.24) enlarges the velocity of decreasing the generalized momentum of the ring about the factor two.

In a second compression cycle the coils 3. and 4. are supplied with a further current impulse, whereby the magnetic field is increased again on up to 3 T. As a result the ring radius decreases up to 2.5 cm and it begins a further ring confinement cycle with a duration time of about 60 ms. During this cycle the ring radius changes in the boundaries 1.8...2.5 cm, the field index n between 0.05...0.1 and the generalized momentum decreases once more around the factor eight.

In Fig.**3.12** is shown the calculated temporal change of the field index and of the electron ring radius in the courses of the entire ring life time [142]. Fig.**3.13** shows results of the calculation of the temporal change of the mean ion charge at the storage of Nitrogen, Argon, Krypton and Xenon ions and of the ionization factor $j\tau$. The ionization factor thereby reaches values up to $0.5 \cdot 10^{22}$ cm^{-2} [142].

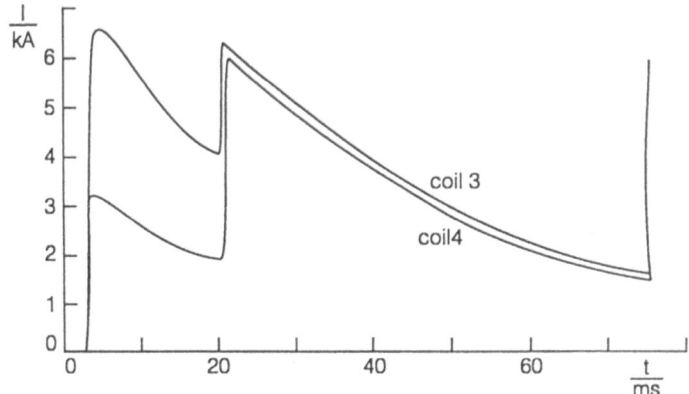

Figure 3.11   Time development of the currents in the 3. and 4. stages of the magnetic system of the ionizer in the regime of long ring confinement [142].

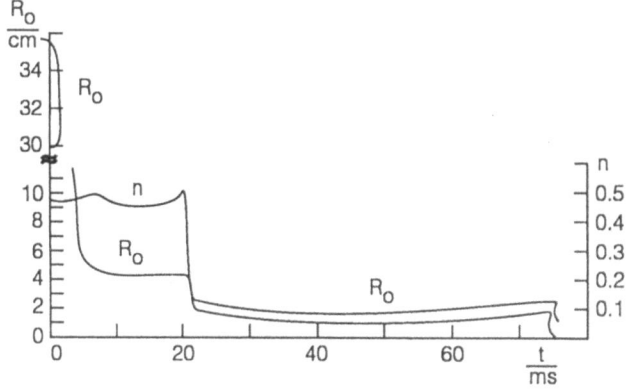

Figure 3.12   Time dependence of the field index n and of the radius of the electron ring R_0 in the regime of long confinement of the electron ring [142].

Further in principle a third compression cycle with a reduction of the generalized momentum around that 10-15 fold facing of the beginning value is possible. The third cycle can begin about $60\ldots70$ ms after the injection of the electron beam in the compressor chamber. In this cycle the radius of the electron ring can be reduced to values of $1.2\ldots1.3$ cm, whereby the entire ring confinement time grows up to 100 ms, which corresponds to an ionization factor of $j\tau$ up to 10^{22} cm^{-2}. According to Tables **3.2** and **3.3** this is sufficient for the receipt of U^{82+}, Xe^{50+} and for Kr^{35+} and possibly still for higher charge states. Equal results can be received also, if the decrement τ_c of the coil currents would be enlarged in the coils up to 50 ms.

The described device is anticipated particularly for spectroscopical investigations at highly charged ions. Thereby spectra, energies, intensities and line widths should be investigated by x-rays appearing in highly charged ions at the filling of inner-shell vacancies. The effectiveness of such studies is determined in decisive degree by the resolution

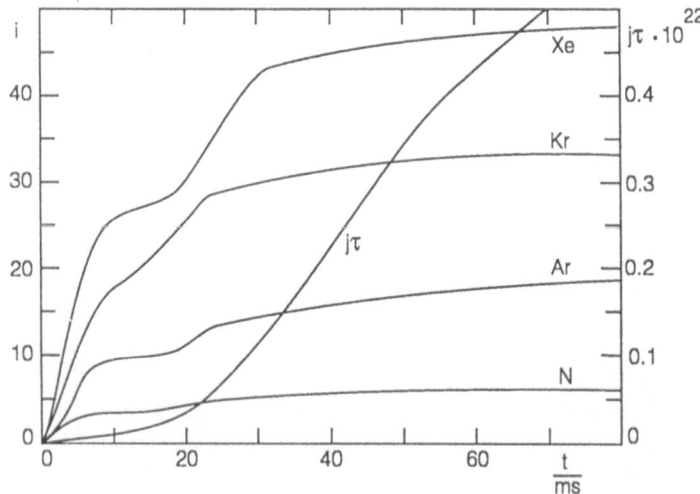

Figure 3.13 Time dependence of the ionization factor $j\tau$ and of the mean charge $\bar{i}$ of Nitrogen, Argon, Krypton and Xenon ions for the regime of long electron ring confinement [142].

of the inserted spectrometer and by the collected statistics at relevant events. Especially usefull for these purposes appears the employment of a crystal diffraction spectrometer for wavelength dispersive X-ray spectroscopy [146, 147].

The intensity of X-ray radiation can determined for a ring formation cycle as follows

$$I_X = \frac{dN_X}{dt} = N_e \frac{N_i}{V} c \sigma_{nlj} \omega_{nlj} \tag{3.26}$$

with σ_{nlj} as ionization cross-section for the subshell (nlj), ω_{nlj} as fluorescence yield for the concerned subshell (nlj), V as volume of the electron ring, N_X as number of the emitted X-ray quanta, N_e as electron number in the ring and N_i as ion number in the ring.

The number of counted by a spectrometer in the time interval Δt registered X-ray quanta totals

$$N = I_X \Delta t \frac{\Delta V}{V} L \varepsilon f_0 \tag{3.27}$$

with $\Delta V/V$ as part of the ring volume, which is seen by the spectrometer from, ε as the registration efficiency of the detector registered quanta with defined energy, L as light power of the spectrometer (respectively solid angle) and f_0 as attenuation coefficient for the weakening of the X-ray radiation on the way from the radiation source to the detector.

With it yields for the number of the registered quanta

$$N = \frac{N_e N_i}{V} \frac{\Delta V}{V} c \Delta t \sigma_{nlj} \omega_{nlj} L \varepsilon f_0 \ . \tag{3.28}$$

Estimations show, that at the employment of a crystal diffraction spectrometer for the registration of K-shell X-rays of Krypton or Xenon per acceleration cycle 1 . . . 3 events in the courses of 10 ms are registered.

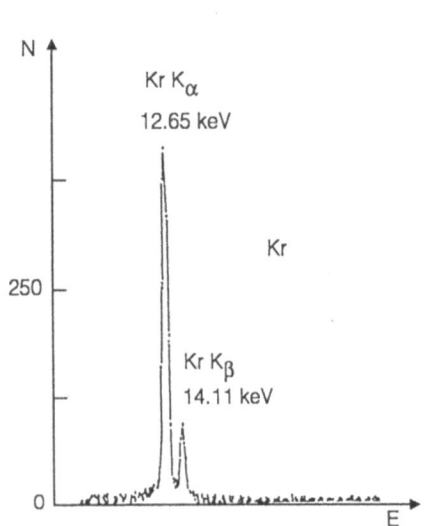

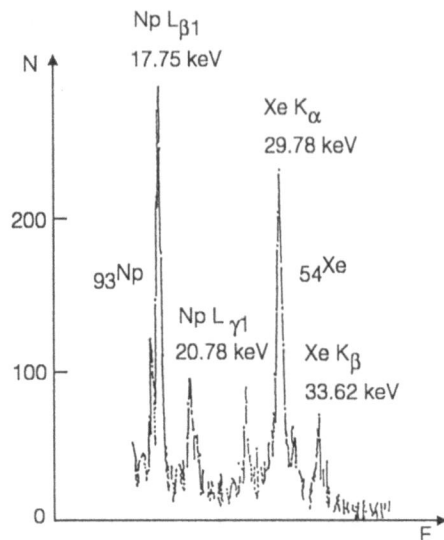

Figure 3.14 Spectrum of the characteristic X-ray K-series of Krypton ions, measured at the electron rings of the ERIS[142].

Figure 3.15 Spectrum of the characteristic X-ray K-series of Xenon ions, measured at the electron rings of the ERIS. For spectrometer calibration additionaly the Neptunium L-series from an $^{241}_{95}$Am source was measured [142].

Beside investigations basing on X-ray spectroscopy the electron rings also emit synchrotron radiation in the region of 1 μm to 100 μm. Thereby the ERIS with their long lived electron rings with 10 ... 15 MeV electron energy can compete with such traditional radiation sources in the infrared region how absolutely black bodies, gas discharge lamps and CO_2 lasers. High pressure Mercury lamps and lasers of high performance own as a rule a discreet spectrum in the region 0.8 ... 3 μm; the far spread CO_2 laser works with a wavelength of 10.6 μm. Synchrotron radiation emitted by the electron rings of the ERIS owns a continuous spectrum with $\lambda_{max} = 1 ... 10$ μm, which falls off slowly in the region of long wavelengths up to 100 μm and further there. The total performance of the synchrotron radiation of the electron ring of the ERIS totals 1 kW and exceeds around two to three orders of magnitude the radiation performance of a black body heated up to the necessary temperature in the long-wave area. Thereby it is possible, to influence the spectral distribution of by the electron rings emitted synchrotron radiation through the change of the radius of the electron rings and through a variation of the electron energy.

In the years 1991/1992 the first stage of reconstruction of the accelerator KUTI-20 to an ionizer ERIS was essentially completed [142]. A new magnetic system for the ring compression was installed (see Fig.**3.10**) and there were begun first experiments to detect of from the ring emitted characteristic X-ray radiation under using a semiconductor detector. In the Figs.**3.14** and **3.15** are presented corresponding spectra of the X-ray K-series of Krypton and Xenon ions, which are registered at electron rings with a radius of $R = 4.5 ... 5$ cm in the course of a ring life time of 30 ... 40 ms.

The completion of the ERIS device bordered as a novel ion source presents a further step in the development of sources of highly charged ions. Here the high charge states achievable in an EBIS are joined with the high ion numbers for the corresponding charge states (compare Fig.**3.9**). Of it independent are theoretical and experimental investigations accomplished together with this device to the production of ions in relativistic electron beams doubtless of fundamental interest and of practical meaning for the development of the physics of ion sources on the basis of electron collision processes and for the physics of the electron shells of highly charged ions.

Chapter 4

Electron Beam Ion Sources

4.1 The Electron Beam Method for the Production of Highly Charged Ions

4.1.1 History and Operation Principles

The electron beam method was proposed 1967 by E.D.Donets [149]. It was the basis of a new ion source, which received in the subsequent time the designation EBIS (Electron Beam Ion Source), and was conceived originally as source of highly charged ions for high energy ion accelerators.

Of course in ion accelerators and in ion storage rings the use of beams of completely ionized atoms is the most optimal case. Of particular interest is the acceleration of heavy ions, at which the relationship of nuclear charge to mass is smaller than 0.5. Since the sixties heavy ion accelerators used Penning and Duoplasmatron sources which could not deliver high charge to mass ratios. Thus ions were shot at relatively low charges through stripping folies or gas targets to increase the ion charge state. The ion charge states obtained are determined by the corresponding ionization processes and by the effective charge exchange processes, whereby their cross-sections are dependent on the energy of the colliding particles. This is shown for Hydrogen in Fig.1.1. For the production of highly charged ions by the interaction with gas targets or stripping foils ion energies of $10 \ldots 100$ MeV/nucleon or more are necessary.

To increase the ion charge states in the ion sources an increase of the electron energy and a larger life time of the atoms in the source are necessary. The idea of an EBIS emerged in accelerator laboratories and was realized with elements of the accelerator technology. The basis for this source is a continuously or quasicontinuously working electron linear accelerator.

An essential advantage of an EBIS is that the electron beam fulfills simultaneously two essential tasks. On the one hand the electrons serve as well as at the majority of the other ion sources as the ionizer for the atoms and ions in the source. On the other hand the space charge produced by the electrons works as ion trap, which stores the ions produced in the electron beam. The principle of the EBIS allows it, to control the injection of neutral particles or of lowly charged ions in the source, the life time of the ions in the electron beam and the moment of ion extraction from the EBIS. Thus it becomes possible, to control the ionization stage of the produced ions and to vary it in

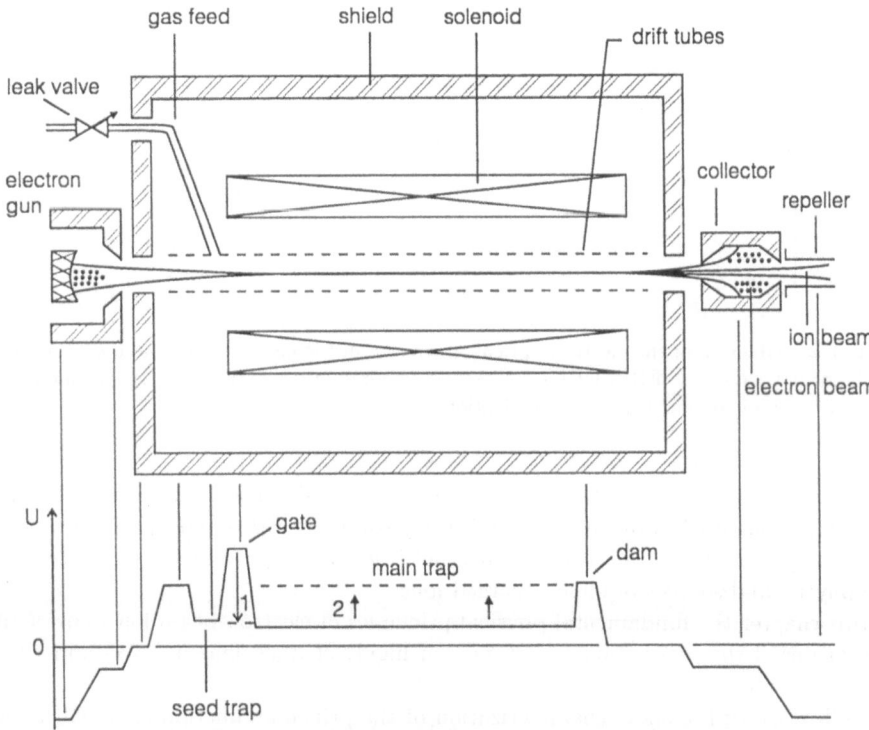

Figure 4.1 Function diagram and distribution of the electrostatic potential along the electron beam of the electron beam ion source CRYEBIS of the Kansas State University [152].

a wide range. This is not only a considerable advantage of the EBIS as an injector for heavy ion accelerators, but also for providing experiments with highly charged ions.

In the course of the last twenty years the EBIS principle has been developed dynamically. So EBIS sources are inserted for example in Orsay and Stockholm at synchrotrons located there and in Dubna at the synchrophasotron as heavy ion injectors. In the source KRION-3 in Dubna for the first time completely ionized Xenon atoms were produced [71].

The production of high ion charge stages requires however ion confinement times in the electron beam of some ten seconds. Until today it has not been successful to avoid ion losses from the electron beam of the EBIS source completely. So only a small part of the ions injected primarily in the EBIS remains up to the end of the ion collecting time in the electron beam. In comparison with other ion sources the main defect of the EBIS are ion losses and the relatively small intensity of the preserved ion beams.

4.1.2 The Construction of an EBIS

Presently there exists in the world about 20 plants of the EBIS type. Sources with electron beams of a length less than 10 cm were developed as an independent direction, the so-called EBIT (Electron Beam Ion Trap) [150]. Sources of this type are well developed at the Lawrence Livermore National Laboratory (see also section 4.4.).

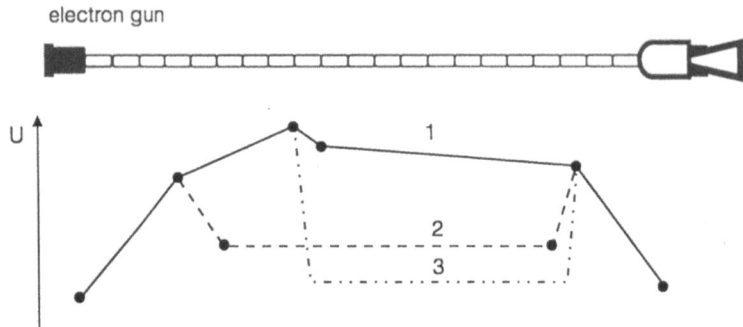

Figure 4.2 Diagram of the electron-optical sytem and of the electric potential distribution U along the drift tube of the EBIS KRION-2 [70]. 1 – regime of ion extraction; 2 – regime of ion injection; 3 – regime of ion storage and ionization.

It is not the object of the existing book, to discuss all construction variants of the above named sources. Details of the technical realization of EBIS and EBIT sources are described in [6, 15, 70, 71, 150, 151] and in a multitude of articles and conference lectures on the physics and sources of highly charged ions.

In this chapter the fundamental physical principles of electron beam ion sources and particularities of the production and storage of highly charged ions in EBIS and EBIT sources are treated.

We will start with a short characterization of the principle function of an EBIS. An EBIS represents an electron linear accelerator with a drift tube. Thereby the electron beam becomes focussed by an axial magnetic field. Dependent of the manner of the production of the necessary magnetic field the EBIS are divided into two fundamental kinds of sources:

- "warm" EBIS, at which the magnetic field is produced with magnetic field coils at room temperature;
- "cold" or cryogenic EBIS, at which the magnetic field is produced with superconducting coils [70].

The function diagram of an EBIS is presented in Fig.4.1 [152]. An electron gun, which exists of an anode and a cathode, produces an electron beam, which is localized in the interiors of a solenoid. The solenoid coil is surrounded by a soft Iron yoke. At superconducting temperatures all systems are housed in cryostats for liquid Helium and liquid Nitrogen. The electron beam meets with the electron collector and the ion beam with the extraction electrode.

As an example the construction and the function principle of the EBIS KRION-2 [70] are shown in Fig.4.2.

The cryogenic magnet system of the EBIS KRION-2 produce a maximum magnetic induction of the electron beam focussing field of $B = 2.25$ T on a length of 1.2 m. The cathode of the electron gun is localized at a diameter of 0.8 mm at a place of growing induction of the solenoid field at $B = B_{max}/6$. This guarantees a compression of the electron beam up to an equilibrium radius $a = 0.15$ mm. The structure of the drift tube is composed of 25 isolated sections. The working gas, from which the ions are formed, is injected in the third section. The temperature of 4.2 K maintained in the working regime in the drift tube guarantees a rest gas pressure of less than 10^{-12} Torr.

The pulsed gas admission in the electron beam, the ion storage and the ion extraction is guaranteed by a successive change of the potential V along the axis of the EBIS: $2 \Rightarrow 3 \Rightarrow 1$.

Results of experimental investigations on the storage of highly charged ions are presented in [70, 71]. At the EBIS KRION-2 the following parameters are achieved:

- electron energy up to 120 keV;
- electron current density up to 1 kA;
- ionization time up to some ten seconds.

4.2   Production and Storage of Multiply Charged Ions in an EBIS

The self-charge of the electron beam of an EBIS serves in the transversal direction as a trap for positively charged ions. For ion storage in the axial direction usually such a potential distribution is produced along the beam, that displays potential barriers at the ends of the beam. In sources of the KRION type the drift tube is constructed in the form of individual sections (Fig.**4.2**). At each of these sections an independent potential can be realized. The potential distribution can be controlled along the beam and different injection regimes can achieved as well as the storage or the ion extraction in different variants (for example the regime 1, 2 and 3 in Fig.**4.2**).

The depth of the potential gap is determined by the potential distribution in the transversal ring cross-section. The cathode construction of the electron gun in the KRION source and the strong longitudinal magnetic field guarantees an electron density in the beam cross-section, which approaches the equilibrium density. The potential distribution of the electric field in a beam cross-section with constant density is determined by the following dependence

$$U(r) = \begin{cases} eN_e \left[2\ln\dfrac{R}{a} + \left(1 - \dfrac{r^2}{a^2}\right) \right] & r \leq a \\[3mm] 2eN_e \ln\dfrac{R}{r} & r > a . \end{cases} \tag{4.1}$$

with N_e – linear electron density in the beam, a – beam radius and R – radius of the drift tube.

The potential $U(r)$ can be expressed also by the electron current I_e and the electron energy E_e

$$U(r) = \begin{cases} \dfrac{I_e}{c}\sqrt{\dfrac{mc^2}{2E_e}} \left[2\ln\dfrac{R}{a} + \left(1 - \dfrac{r^2}{a^2}\right) \right] & r \leq a \\[3mm] \dfrac{2I_e}{c}\sqrt{\dfrac{mc^2}{2E_e}} \ln\dfrac{R}{a} & r > a \end{cases} \tag{4.2}$$

For example the depth of the potential gap for $I_e = 1$ A and $E_e = 10$ keV is about $U = 150$ V. Simultaneously the potential on the surface of the tube ($r = R$) can be a multiple higher for $R \gg a$.

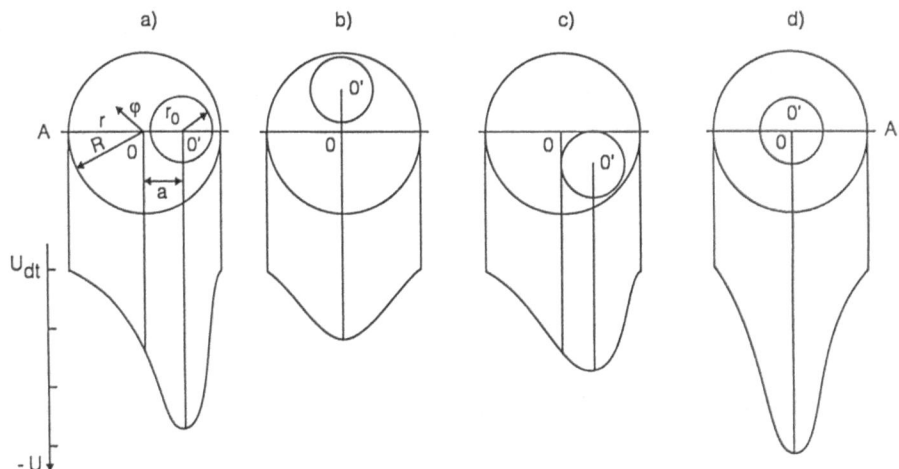

Figure 4.3 Potential distribution of the electric field U along the radius of the electron beam in dependence of the beam position relatively to the axis of the drift tube [70].

As in [70] is shown, a deviation of the beam axis facing the center of the drift tube cross-section leads to a disturbance of the potential profile and to a reduction of the depth of the potential well (compare Fig.**4.3**). This leads to very high demands on the precision of the production and adjustment of the magnetic system of the source.

From (4.2) it is obvious, that the depth of the potential well in the beam center is determined relatively to this boundary ($r = a$) only by the linear density of the beam or by the total current of the electrons and that they does not depend on the beam diameter.

Positive ions, which move in the potential (4.2) and display within the electron beam a square dependence of the radius, execute harmonious oscillations with a frequency (2.18)

$$\omega_i^2 = \frac{iN_e r_e mc^2}{AMa^2} \ .$$

Ion distribution functions and their first moments in electron beams with constant and Gaussian distributed densities in the beam cross-section were considered already in the second chapter.

An important characterization of an arbitrary ion source is the number of ions, extracted from the source. The amount of ions preserved in an EBIS depends on the capacity of the effective ion trap, which is limited by the electric charge Q of the electron beam

$$Q_e = N_e\,l \tag{4.3}$$

with l as the length of the effective volume of the electron beam.

The corresponding maximum ion number N, which can be stored in the electron beam, totals

$$N = N_i\,l = \frac{Q_e}{\bar{i}} \tag{4.4}$$

with N_i as linear ion density and $\bar{i}$ as mean ion charge in the beam.

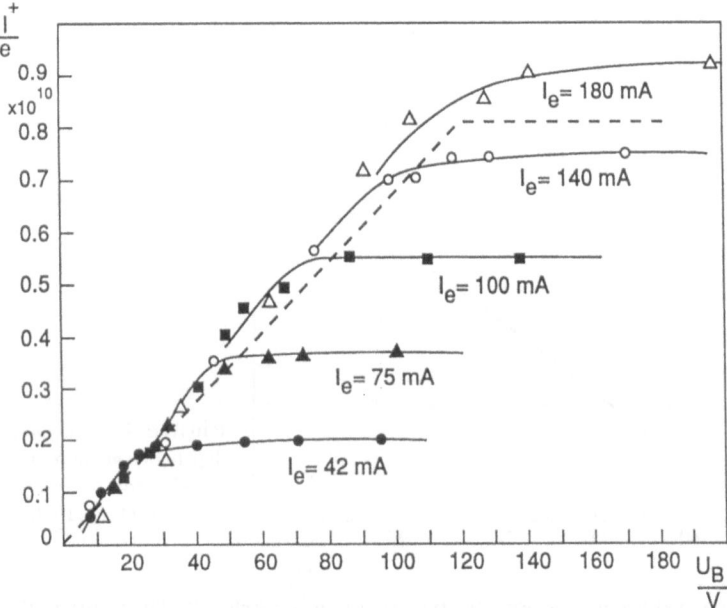

Figure 4.4   Dependence of the total ion charge I in the trap from the amount of the inclusion potential U_B [153].

Under actual working conditions the ion number is a multiple smaller than obtained according to (4.4), since otherwise the space charge of the electrons is compensated by the ion charges leading to a compensation the Coulomb potential of the electron beam.

The ability of the electron beam to store ions was discussed in Ref.[153]. In this work the ion number was calculated, which could be achieved in the KRION source in dependence of the value of the potential barrier at the ends of the beam. Through a variation of the potential at the ends of the drift tubes the depth of the potential well can be changed in axial direction and with it also the capacity of the ion trap.

Results from measurements for different electron energies are presented in Fig.4.4. It is obvious, that an enlargement of the inclusion potential U_t at the ends of the beam enlarges the capacity of the trap linearly. When reaching a certain amount of the potential wall the growth of the collected charge however ends at this point. This saturation effects enters, if the amount of the longitudional and transversal potentials and a further increase of the inclusion potential leads no more to a growth of the depth and capacity of the potential well.

In Fig.4.5 for the saturation case the dependence of the wall potential U_t from the current of the electron beam is presented. The linear law preserved at these measurements for the change of the wall potential is found in full agreement with (4.3), i.e. the electron number in the beam is determined by this current.

One of the most important goals in the development the EBIS was the receipt of highly charged ions. This required the highest possible ionization factors $j\tau$ with especially high electron current densities j.

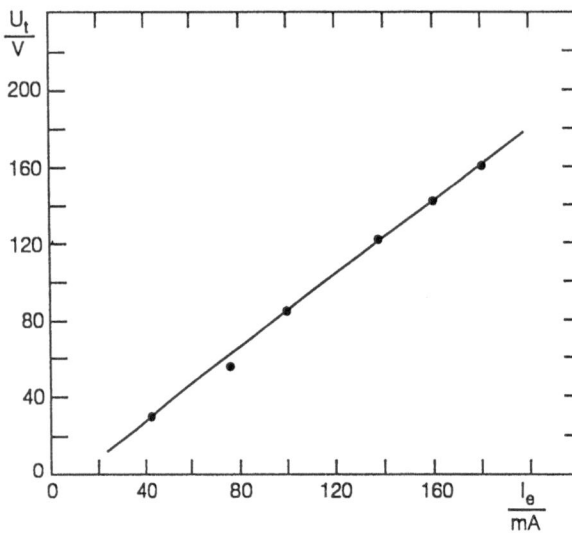

Figure 4.5 Dependence of the depth of the potential well U_t in the electron beam as a function of the current I_e [153].

Necessary values for $j\tau$ for the receipt of noble gas and Uranium ions are already shown in Fig.**1.10**. Thus, for Xe^{54+} ionization factors $j\tau > 1 \cdot 10^{23}$ cm^{-2} are necessary at an electron energy of more than 50 keV. On the other hand, for U^{92+} already values of $j\tau > 1 \cdot 10^{24}$ cm^{-2} at an electron energy of more than 200 keV are necessary. The optimal energy for ionization of electrons from the K-shell from uranium ions is about 300 keV.

The production of precise, strong focussing continuous electron beams with an energy of several hundred keV is a difficult technical problem. Interestingly it is to be mentioned here also, that the accomplishment of $j\tau \cong (1 \ldots 2) \cdot 10^{24}$ cm^{-2}, which is necessary for the production of U^{92+}, is comparable with the problems of the thermonuclear synthesis. For this case $n_e\tau$ reaches 10^{14} cm^{-3} s, which is equivalent for reaching of the Lawson criterion for thermonuclear reactions in a Deuterium-Tritium plasma.

In the first chapter of our book we already pointed out, that the maximum achievable degree of ionization in electron impact ion sources is determined not only by the electron energy and the ionization factor, but also by the density of the neutral particles and by the rest gas pressure in the working area of the ion source because of decreasing ionization cross-sections for highly charged ions and growing charge exchange cross-sections at increasing ion charges. As result the increase of the ionization degree of ions in the beam is limited and the ionization degree assumes a stationary value. This is seen in Fig.**1.12**, where the influence of the presence of neutral Helium on the mean charge of ions in the source KRION-2 was shown.

From (1.54) can be estimated the necessary density of neutral particles in the electron beam. So for an electron energy of $E_e = 100$ keV and a current density in the beam $j = 1000$ A cm^{-2} for the receipt of completely ionized Xenon ions a rest gas pressure of $P < 10^{-12}$ Torr is necessary. Presently such a pressure can be guaranteed only with cryogenic systems.

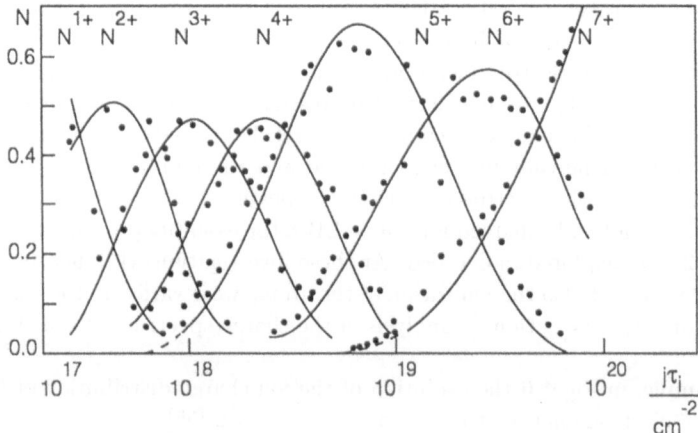

Figure 4.6 Evolution of the ion charge distribution of Nitrogen atoms in an electron beam ion source [55].

As a general tendency in the development of the EBIS is the effort to increase the electron density in the source. For electron densities $n_e > 1 \cdot 10^{12}$ cm^{-3}, which are characteristic for an EBIT, the recombination of electrons from the beam at highly charged ions becomes to the most essential process, which limits the growth of the ion charge [154].

Of the same importance as the electron energy is the electron current. This quantity determines the capacity of an EBIS as ion trap and the current density in the beam. For continuous electron beams, which find application in the EBIS, the current is limited to some Ampere. Higher currents can not be reached because of different technical problems and of the appearance of different beam instabilities. An enlargement of the length of the electron beam increases the capacity of the ion trap, but simultaneously increases also the probability for the appearance of instabilities [155, 156].

Instabilities increase the transversal movement of the electron beam. This lead to a heating of the stored ions and so also to increased ion losses. Conditions for the appearance of instabilities in an EBIS in dependence of the beam parameters and of the induction of the outer longitudinal magnetic field were explored in the Lawrence Berkeley Laboratory [156].

A sufficiently strong magnetic field is able to suppress the development of beam instabilities. So it could be shown, that for the stabilization of beam oscillations at resonant harmonics for the magnetic induction of the longitudinal field holds [157]

$$B > \frac{jmcf\lambda}{2E_e} \ . \tag{4.5}$$

Thereby f describes the neutralization factor of the beam and

$$\lambda = 2\sqrt{\frac{\pi E}{r n_e f m c^2}}$$

the wavelength of the resonance.

At a current density of j =1 kA cm^{-2} and at a beam energy of E_e =10 keV the necessary value of the magnetic induction reaches about 10 kG.

A high density of the electron beam, needed for the production of highly charged ions in an EBIS, is reached by the magnetic compression of the electron beam. In sources of the KRION type the magnetic compression reaches values around the factor ten. Thereby, the strong longitudional magnetic field and the high electron energy increase the achievable limit of the electron current in the beam.

For the storage of highly charged ions in an EBIS the essential processes at the KRION and KRION-2 were explored in detailed. At these investigations the electron energy, the ion collecting time and also the ionization of the atoms were varied in the electron beam. In addition, after ion extraction an analysis of the charge spectrum of the extracted ions was done.

As an example, in Fig.**4.6** the evolution of the ion charge distribution of Nitrogen in an EBIS at an electron energy of $E_e = 5{,}45$ keV is shown [55].

The results allow it by solving of the inverse problem for the charge balance equations, to determine ionization cross-sections from the known distributions at different points in time [55]. With this method for the first time ionization cross-sections were determined for all charge states of Carbon, Nitrogen, Oxygen, Neon and Argon in the electron energy region of several keV [16]. Ionization cross-sections are presented in Fig.**4.7** together with cross-sections calculated according to the Lotz formula (see equation (1.8) and [12, 13]).

In Fig.**4.6** inked lines characterize the calculated evolution of the charge distribution of Nitrogen ions. For these calculations experimental values of electron impact ionization cross-sections are shown in Fig.**4.7**.

From indisputable interest for the physics of multiply charged ions were also measurements of ionization cross-sections of hydrogen-like ions of all gaseous elements up to Argon (Z=18). These cross-sections are shown in comparison with calculated values already in the first chapter (Fig.**1.2**).

Later measurements of the energy dependence of effective electron capture cross-sections at collisions of Carbon, Nitrogen, Oxygen and Neon nucleii with Hydrogen molecules at the ionizer KRION-2 are given in [157].

Receiving beams of slow highly charged ions at the EBIS new possibilities emerged for investigations of the interaction of almost fully ionized atoms with solid state surfaces. In the result of such interactions occurs a nearly instantaneous neutralization of the incoming ion, whereby a large amount of electrons is trapped in highly excited atomic states. It emerges a highly excited atom with a large number of vacancies in inner shells [158, 159]. The deexcitation results by cascades of Auger and X-ray processes, whereby in comparison with neutral atoms energetic shifted X-ray quanta and Auger electrons are observed. In experiments at the source KRION-2 the interaction of Ar^{17+} and Kr^{36+} with solid state surfaces was investigated [71, 160, 161]. In Fig.**4.8** an X-ray spectrum, measured with a semiconductor detector from the interaction of Kr^{35+}-ions with a solid state target is shown.

Analyzing the up to now accomplished experiments and measurements at EBIS sources, we see, that such ion sources are effective plants for providing investigations to the physics of electron-ion beams and to the physics of highly charged ions.

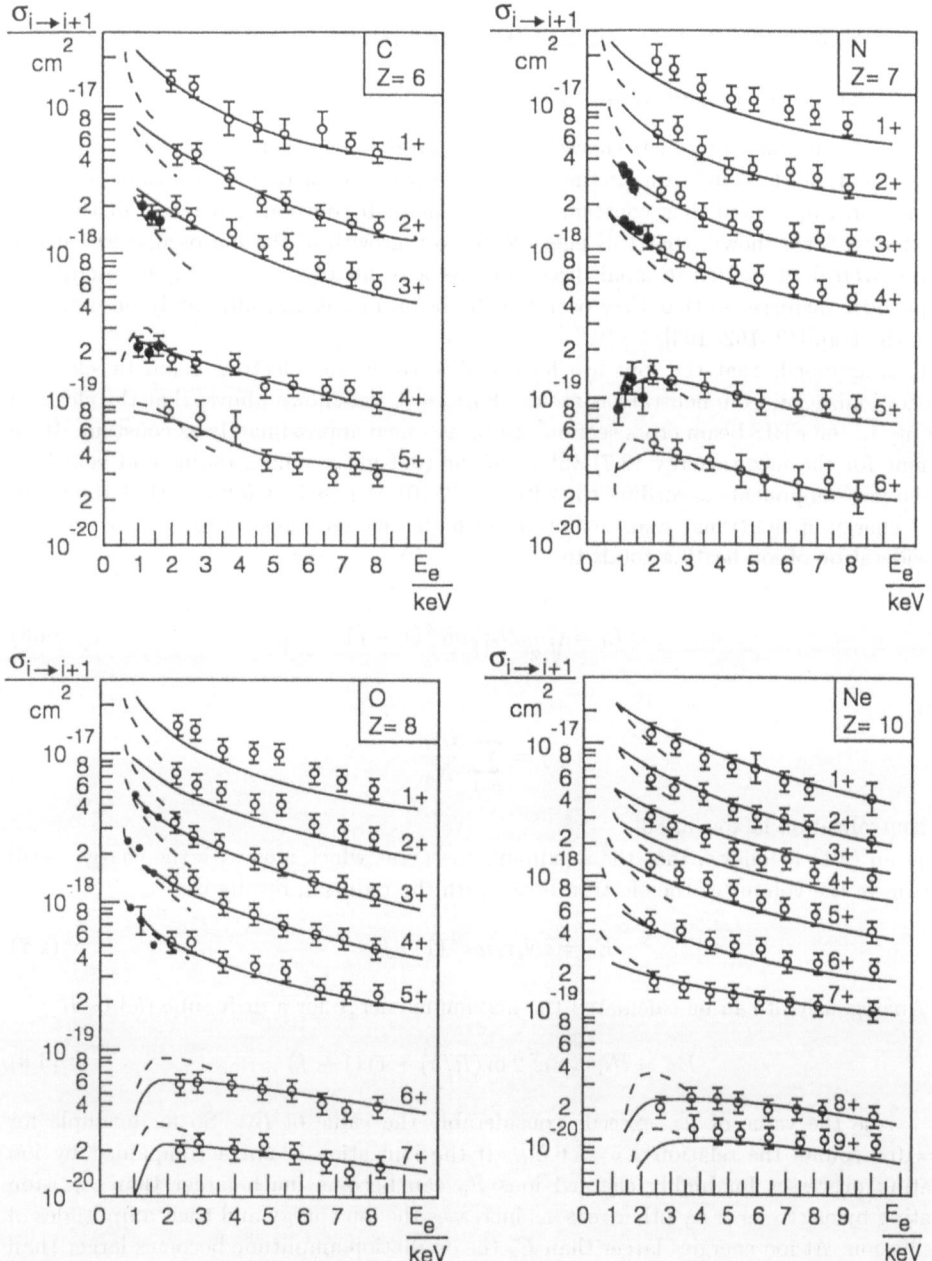

Figure 4.7 Experimentally determined ionization cross-sections σ for ions of different charge states of Carbon, Nitrogen, Oxygen and Neon at different electron energies E_e. Points indicate experimental values from Ref.[16] and solid lines calculated values according to the Lotz formula [13].

4.3   Heating and Cooling of Ions in an EBIS

4.3.1   Electron Losses from the Electron Beam

Electron beam ion sources are characterized by large ion storage times in the Coulomb potential of the electron beam. In the course of several seconds beside ionization processes the in the trap stored ions constantly suffer Coulomb collisions. Already in the first chapter we have shown, that this effect leads to a growth of the ion oscillatory energy in transversal as well as in axial direction. As a result of this process the oscillation amplitudes increase, so that they overstep the beam radius and ultimately become lost from the trap [42, 162, 163].

It is assumed, that the ions are formed directly in the electron beam by electron impact ionization from neutral gas atoms. Further it was shown above, that the electron density in the EBIS beam cross-section can be assumed approximately as constant. If we assume for the mean energy (2.7) values of the root-mean-square radius and velocities of the ion components according to (2.9) and (2.10) at $i \gg 1$, it follows, that the mean total energy of the transversal oscillations of highly charged ions in the source without consideration of ion heating totals to

$$\bar{E}_i = \sqrt{\frac{i}{\pi}}\, N_e r_e mc^2 \left(1 - f\right) \tag{4.6}$$

with

$$f = \sum_{i=1}^{Z} \frac{i N_i}{N_e}$$

as neutralization factor [8, 80].

From (4.1) it follows, that the maximum energy at which ions with the charge i still remain in the volume of the electron beam with the radius a, results in

$$E_a = i N_e r_e mc^2 \left(1 - f\right) . \tag{4.7}$$

Analogously, it can be calculated the maximum energy for a drift tube radius R

$$E_R = i N_e r_e mc^2 \, 2 \ln\left(R/a\right) + 1) \left(1 - f\right) . \tag{4.8}$$

As a rule the value of E_R exceeds considerably the value of E_a. So for example for $R = 10\,a$ counts the relation $E_R = 5.6\,E_a$. If the ionization is not accompanied by ion heating processes for highly charged ions E_a can become much larger than $\bar{E}_i$. Ion heating by electrons or by other reasons increases the ion energy and their amplitudes of oscillation. At ion energies larger than E_a the oscillation amplitude becomes larger than the beam radius and the ion is found a certain time outside the electron beam. If the ion reaches the energy E_R, it hits the wall of the drift tube and gets lost from the source.

According to the dependence of the electric field potential from the radius (4.1) the ions leaving the beam as the power F decreases by $1/r$ (within the beam counts $F = 2ieN_e r/a^2$). So the ion velocity outside the beam is considerably smaller than in the beam and thus the ion ejected from the beam spends a considerable part of his time outside the beam.

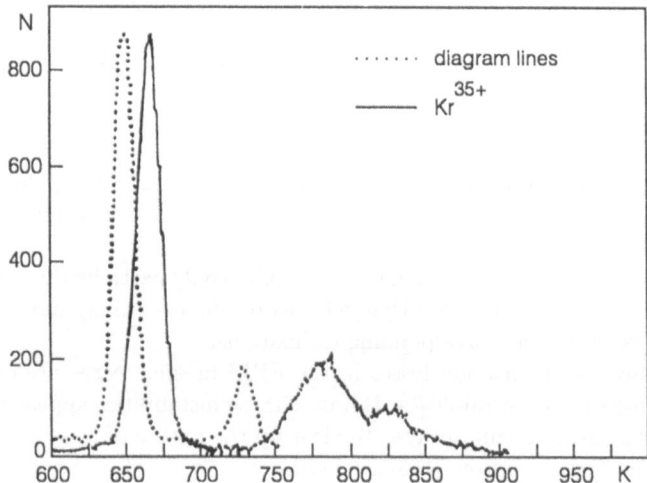

Figure 4.8 K_α-X-ray spectrum from the interaction of Kr^{35+} ions with a solid state target [160, 161].

By the integration of the law of ion motion in the field (4.1) it can be shown that, for example, at a total ion energy of $E = 2E_a$ the ion is found at about two thirds of his time outside the beam. Counts $E = 5E_A$, than the ion is found only 5% of the time in the beam. In addition, the strong magnetic direct-axis field, which is always available in electron beam sources, further reduces the storage time of the ions in the beam.

Ions leaving from the electron beam suffer no further ionization and heating outside of the beam volume. Accordingly to an increase of the energy on values above E_a the ions are not effective trapped in the beam and their heating decreases quickly.

The accomplished considerations lead to the conclusion, that a complete loss of an ion is less probable from the source for the case, if its energy reaches E_R at $E_R \gg E_a$ and if its oscillation amplitude enlarges itself to the size of the radius R of the drift tube. On the other hand an ion emerging from the beam does not influence the electrons and other ions of the beam and thus participates no more in the collective motion of the charged particles in the beam. An analysis of ion losses from linear ion beams leads to the conclusion, that ions of an energy larger than E_a must be considered as lost from the beam [164].

Further, we assume above the energy E_m the ions are lost from the beam, as the end-point energy of the ions and in this sense $E_m = E_a$.

Intense ion losses are one of the fundamental problems in the operation of an EBIS. An analysis of the corresponding literature leads to the conclusion, that the end of the ion confinement time, which totals as a rule several seconds, only a slight part of the initially stored ions remains in the source . Unfortunately, precise statements of velocities of ion losses are not given in the literature.

According to the results from paragraph 1.4 and from Table **1.2** the characteristic time for ion heating in the electron beam as a result of elastic Coulomb collisions can be described by the relation

$$\tau_{ie}^{+} = \frac{A_i M T_i}{2\pi i^2 m^2 c^4 r_e^2 n_e L} \sqrt{\frac{E_e}{2m}} \; . \tag{4.9}$$

The designations used here are the same as used above.

An estimation of the ion heating according to (4.9) with parameters characteristic for the EBIS (electron energy E_e=100 keV; beam current I=0.1 A; beam radius a=0.015 cm) yields for τ_{ie}^{+} for Xenon ions to about 1 s [42].

The preserved estimations for the ion heating explain only partially the actual observed ion losses, because the loss of highly charged ions results essentially more quickly, as it would be to be expected by the corresponding estimations.

Unquestionably the intense ion losses in the EBIS in some cases are caused by the appearance of electron-ion instabilities. However these instabilities appear under certain conditions, but the losses result always. To clear up the nature of these losses, results of paragraph 1.4 and that ones from [163] are used.

In electron-ion beams not only do elastic electron-ion collisions result, but also collisions between the ions. Thereby the collision frequency for collisions between ions is essentially higher than the frequency for collisions between electrons and ions. The characteristic collision time between two different ion components and the formation of an equilibrium between them is determined by (1.36). If this equation is rewritten for two identical ions and if we assume, that all ions own the equal temperature T_i, it follows that

$$\tau_{ii} = \frac{3}{4} \sqrt{\frac{A_i M}{\pi}} \frac{T_i^{3/2}}{i^4 m^2 c^4 r_e^2 L n_i} \; . \tag{4.10}$$

This equation can serve as the estimation of the collision frequency between identical ions. Simultaneously with this equation the time for the formation of an equilibrium distribution within an ion component can be determined. Within a small factor the results agree with the values from (1.39). It is known, that the equilibrium distribution originates as a result of elastic Coulomb collisions of charged particles and that it suffices a Boltzmann distribution function for the particle energy

$$f_i(E_i) = \frac{2}{\sqrt{\pi}} \frac{1}{T_i} \sqrt{\frac{E_i}{T_i}} \exp\left(-\frac{E_i}{T_i}\right) \; . \tag{4.11}$$

Due to the small ion energy in the electron beam particularly for highly charged ions the time for the formation of an equipartition for the ions proves relatively small. For instance in the electron beam results with above indicated beam parameters and at ion densities of $n_i = 3 \cdot 10^8$ cm^{-3} for Xe^{50+}-ions a value of $\tau_{ii} \cong 10^{-6}$ s. Thus the balance distribution develops almost momentarily.

The Boltzmann distribution (4.11) owns the characteristic particularity, that even at a mean ion energy or temperature $T_i \ll E_m$ always a slight amount of ions ΔN_i exists with energies larger than E_m.

This statement is demonstrated in Fig.4.9, where the Boltzmann distribution for the ion temperature T_i is presented. Likewise presented are values of the end-point energy $E_m = ieU$ for ions of different charge states. Thereby the potential of the electric field at the beam boundary has the value $U = eN_e(1 - f)$.

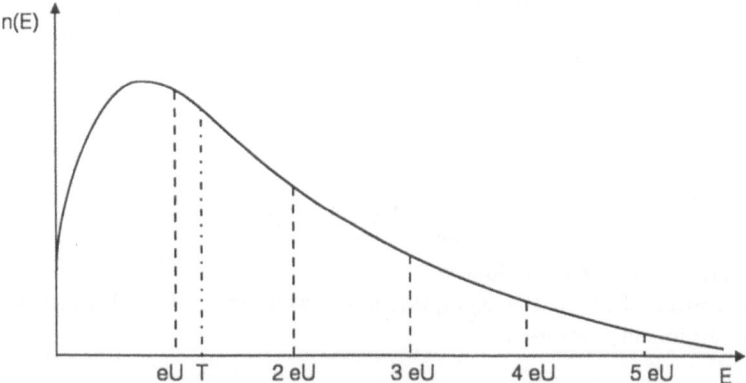

Figure 4.9   Boltzmann distribution of the energy for ions of the temperature T and values of the ion end-point energies for ion charges $i = 1, 2, 3, 4, 5$ in the electric field with a potential barrier U.

The size ΔN_i can be found by integration of the distribution function (4.11) [163]

$$\Delta N_i = N_i \int_{E_m}^{\infty} f_i(E)\,dE = \frac{2N_i}{\sqrt{\pi}}\,\Gamma\left(\frac{3}{2}, \frac{E_m}{T_i}\right) \tag{4.12}$$

with $\Gamma(\alpha, x)$ as incomplete Gamma function.

As was already stated, ions with an energy larger than E_m are lost from the beam. In this case the distribution function differs from a Boltzmann distribution and is no longer stationary. However in a time $\Delta t \cong \tau_{ii}$ collisions between the particles lead to a redistribution of the energy and a stationary ion distribution function is formed again in the beam. Again ions also reappear with energies larger than E_m, which get lost from the beam, etc.

This kind of ion loss appears constantly at arbitrary mean energies. These losses result even without an additional heating of the ion component. Thereby the ion loss velocity is determined by the relationship of the ion temperature to the ion end-point energy in the beam, in our case by the relationship T_i/E_a. Counts $T_i/E_a \ll 1$, follows $\Delta N_i/N_i \ll 1$ and the velocity of the relative ion losses becomes small, so that to each point in time the ion distribution function differs only little from the stationary Boltzmann distribution (4.11).

This process can be described by the kinetic ion equations under consideration of elastic Coulomb collisions and of the balance equation for the ion energy. An exact solution requires the application of numerical methods. Quantitative estimations can however result on a much simpler method [163].

In each time interval Δt there are produced ΔN_i ions with an energy larger than E_m and which are also lost again. Thus the ion loss velocity can be described by

$$\frac{dN_i}{dt} = -\frac{\Delta N_i}{\Delta t}. \tag{4.13}$$

By integration and under use of the values of ΔN_i and Δt from equations (4.10) and (4.12) the time dependence of the ion number in the beam can be determined

$$N_i = \frac{N_{i0}}{(1 + t/\tau_i)} \qquad (4.14)$$

with

$$\tau_i = \frac{1}{2} \sqrt{\frac{\pi A_i M N_e}{2 i m r_e}} \frac{a^2 N_e}{i^2 N_{i0} c L} \left(\frac{T_i}{E_m}\right)^2 \exp\left(\frac{E_m}{T_i}\right) (1 - f)^{3/2} \qquad (4.15)$$

as characteristic time for the ion losses.

In the derivation of (4.15) the asymptotical characteristics of the Γ-function at large values of the argument were used

$$\Gamma\left(\frac{3}{2}, x\right) = \sqrt{x} \exp\left(-x\right) \qquad x \gg 1 \; .$$

The quantity τ_i can be considered also as characteristic life time of the ions in the electron beam. In this time about the half of the stored ions are lost .

In Table **4.1** [163] values for τ_i in dependence in the relationship T_i/E_m for Argon, Krypton, Xenon and Uranium ions are indicated.

The calculations were accomplished for an electron beam with the parameters: linear electron density $N_e = 3 \cdot 10^8$ cm^{-1}; beam radius $a = 0.015$ cm; neutralization factor $f = 0.5$. The results in Table **4.1** show, that even at an ion temperature of up to 5-10 times smaller than the end-point energy of the ions in the beam during the ion storage a large part of ions is lost from the beam in the course of any few seconds. An additional ion heating by electrons still enlarges this effect. A consideration of the actual exponential ion energy distribution in the beam can explain the ion losses observed in electron beam ion sources. The characteristic times τ_i are in essential determined by the exponential dependence of the relationship T_i/E_m in (4.15) and thus are only slightly sensitive to the change of other sizes, for example, of the beam parameters.

Table 4.1   Dependence of the characteristic ion life-time τ_i (in seconds) in the electron beam from the relationship between ion temperature T_i and ion energy E_m at the beam border for Argon, Krypton, Xenon and Uranium.

T_i/E_m	Argon	Krypton	Xenon	Uranium
0.5	$4.7 \cdot 10^{-5}$	$2.4 \cdot 10^{-5}$	$1.6 \cdot 10^{-5}$	$1.0 \cdot 10^{-5}$
0.2	$1.5 \cdot 10^{-4}$	$7.7 \cdot 10^{-5}$	$5.3 \cdot 10^{-5}$	$3.2 \cdot 10^{-5}$
0.1	$5.6 \cdot 10^{-3}$	$2.9 \cdot 10^{-3}$	$1.9 \cdot 10^{-3}$	$1.2 \cdot 10^{-3}$
0.07	0.37	0.19	0.13	0.078
0.05	31	16	11	6.5
0.04	2900	1500	1000	620

It should be pointed out here, that in the derivation of equations (4.14) and (4.15) the gradual reduction of the ion temperature at the escaping of the hottest ions was neclegted. The cooling due to the ion evaporation of the ion component should reduce ion losses in the course of time. Independent of it the quantity τ_i can serve as criterion for appearing of ion losses and for the determination of the moment of appearance of the most intense ion losses.

Independently from the presented ideas and from the still more comprehensive treatment of the problem in [164] it can be examined as questionable, to consider the ions as lost from the beam, if their energy oversteps the maximum energy at the beam border in radial direction. Results do not change, if as boundary is not chosen $E_m = E_a$, but any other energy, for example the value of the electrostatical potential at the beam ends for the inclusion of the ions in axial direction or the energy E_R at the surface of the drift tube (4.8). For this case the ion temperature must be set in relationship to the corresponding end-point energy.

4.3.2 Cooling of Multiply Charged Ions

In Ref.[41] it was proposed to use ions of low charge states for the cooling of multiply charged heavy ions in order to avoid losses from the beam of the EBIS. An analogous suggestion was made and successfully applicated for the same reason indepently for the EBIT [67, 68, 165, 166]. The application of ion cooling gave positive results in EBIS later [71, 167].

The nature of the method of the ion cooling [41, 42], or the so-called "evaporation cooling", was in the literature connected with EBIT plants [165, 166], whereby ions of low charge states were injected in an electron beam with trapped highly charged ions.

Elastic collisions of ions with different charge states and masses lead to equilibrium distributions of the energy for each ion component and to an equilibrium between all individual ion components. According to the thermodynamic laws the energy distribution of all ions corresponds to a Boltzmann distribution with an ion temperature common for all ion sorts. Simultaneously, the distribution within each ion component also follows a Boltzmann distribution with a common temperature for all ions of this component. The characteristic time for the adjusting of the balance distribution follows the expression (1.36) and has a microsecond scale for the electron beam.

High charged and low charged ions have at equal temperatures different end-point energies. The ion end-point energies in the potential gap of the electron beam are dependend on the ion charge (4.7). Low charged ions have small end-point energies, high charged ions have high end-point energies in the beam. Accordingly the loss rates and the ion life-times differ also in the electron beam. Light low-charged ions live in the beam shorter and are lost more quickly. If the initial energy of the low-charged ions was smaller than the initial energy of the highly charged ions, so the low-charged ions are heated by collisions and lost from the beam, more sharply formulated, evaporated from the electron beam. Through the loss of low-charged ions from the source energy is also extracted from the working volume of the source, and thus the highly charged ions are cooled. Therefore, the ion life time increases in the electron beam.

For the simplification of further descriptions a simplifying symbolism should be introduced. Heavy multiply charged ions are marked in the following with the index "1" and low-charged ions with the index "2".

Let the mean initial energy (or temperature) of heavy multiply charged ions equals to E_1. This energy E_1 is produced by the electrostatic potential of the electron beam. Thereby, the expressions "mean energy and "temperature are physically equivalent and differ only by a factor of 1.5 for the particle equilibrium distribution. Further it is assumed, that in the beam arrive low-charged ions with a medium energy E_2 (by injection from outside or by electron impact ionization in the beam). After the levelling of the available energies adjusts a general mean ion energy

$$E = \frac{E_1 N_1 + E_2 N_2}{N_1 + N_2} \ . \tag{4.16}$$

It is evident, that light ions are used preferably as cooling ions, because their maximum charge and end-point energy assume only small values. With the application of low-charged medium and heavy ions for cooling purposes it may be, that ions, which are not lost from the beam, are ionized up to such a level, that their end-point energy becomes comparable with the end-point energy of the heavy ions of the element to be ionized and that they lose so their ability to cool.

To hold the mean energy of heavy ions on a low level and to extract energy, which the ions have received by their heating by electrons, a constant current of light ions in the working volume of the source is necessary. For this case the source develops an equilibrium, which depends on the charge and the current of the injected light ions. The adjusting parameters are determined by the balance of the number and the total energy of all ions in the beam.

For instance, to cool well highly charged Xenon ions in a KRION source and to extract the energy transferred through heating processes, it is sufficient, to inject about $3 \cdot 10^5$ C^{3+} ions per centimetre of the electron beam length [42] in 10 ms.

As already shown, it is however not sufficient for the production of highly charged ions, only to extract the energy of the heated ions. For the storage of ions in the electron beam for seconds or even for some ten seconds it is necessary, to hold the ion energy on a very low level. From Table **4.1** it is obvious, that the energy of highly charged ions must be around 20...25 times smaller than their end-point energy. Simultaneously the cooling ions must be chosen so, that they quickly leave the beam and can effectively transport energy from the beam. This means, that their mean energy should differ only little from their end-point energy. From this result two relations allow us to determine the permissible relationship for the mean ion charges of the working gas and the cooling ions

$$E = (0.04 \ldots 0.05)\, E_{m1} = (0.04 \ldots 0.05)\, Z_1 N_e r_e mc^2$$

$$E = (0.2 \ldots 0.5)\, E_{m2} = (0.2 \ldots 0.5)\, Z_2 N_e r_e mc^2$$

with Z_1, Z_2, E_{m1} and E_{m2} as ion charges and corresponding end-point energies for ions of the working gas and of the cooling ions. Thus, it follows [42]

$$Z_2 = (0.1 \ldots 0.2)\, Z_1 \tag{4.17}$$

and in some cases even

$$Z_2 = 0.05\, Z_1 \tag{4.18}$$

in dependence of the cooling conditions and the charge of the stored ions.

So, for an effective cooling two contradictory demands are needed to be true

- according to (1.36) the cooling ions N_2 must have the highest possible charge, to be able to rapidly exchange energy with ions N_1 of the working gas;
- according to (4.17) the cooling ions N_2 must have the smallest possible charge $Z_2 \ll Z_1$, to be able to quickly escape from the beam, carrying away the energy of the heavier ions.

Summing up the conditions for an effective ion cooling, they can be indicated under application of the already introduced designations in the following form

$$\tau_{11}, \tau_{22}, \tau_{12} < \tau_2 \ll \tau_1 \cong \tau_c \qquad (4.19)$$

with $\tau_{11}, \tau_{22}, \tau_{12}$ as characteristic times for the adjustment of equilibrium distributions between heavy and light ions and for the energy exchange between the different ion species. The quantities τ_1 and τ_2 describe the characteristic life times of heavy and light ions and τ_c the storage time necessary for highly charged ions.

To satisfy the listed conditions, very light ions must be used for the cooling [163].

As a convincing example the ion cooling experiments at the EBIS of the University of Frankfurt/Main [167] can be used. The ion accumulation resulting here from the residual gas at a pressure of $5 \cdot 10^{-10}$ mbar and the partial pressure of the used Argon gas totaled to $2 \cdot 10^{-10}$ mbar. According to the estimations, the electron beam should have been neutralized after about 0.2 s by the accumualtion of ions.

In Fig.**4.10** the temporal development of the ion charge distribution in the electron beam is presented. At the beginning low-charged ions dominate the residual gas in the beam. This is particularly true for ions of the elements Nitrogen, Oxygen and Carbon. After about 100 ms storage time in the electron beam multiply charged Argon ions begin to dominate and gradually replace the light ions. Therefore, this initiates a process in which the light ions become heated and escape from the beam and where Argon ions occupy the place of the escaping ions.

The results of work [167] beside the experimental confirmation the ion heating process and the ion cooling by elastic Coulomb collisions also allow the conclusion, that during the ion storage in electron beams at bad vacuum conditions and for the case of complete charge neutralization of the electron beam multiply charged ions in the beam are remained better than low charged species.

4.3.3 Ion Cooling at Continuous Injection of Light Ions

One of the variants of ion cooling is the application of continuous beams of light ions [163]. Thereby low-charged ions of low energy become injected in the electron beam from one of the source front ends with an energy lying lower as the value of the potential barrier, which limits the ion trap in axial direction. Thereby the conditions for the cooling process are chosen so, that during the ion crossing time through the electron beam in axial direction the light ions are heated by the heavier ions, cooling thereby the heavier ions and escape then from the beam.

The advantages of this method are

- reduction of the number of the cooling ions, which neutralize the ion charge and reduce the depth of the electron beam potential well;

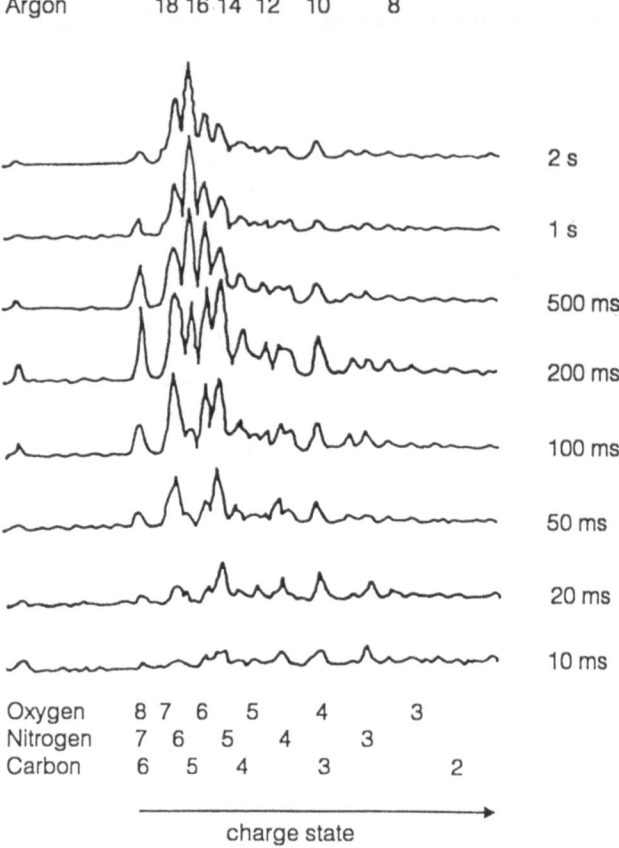

Figure 4.10 Time development of the ion charge distribution in the electron beam of the EBIS of the University of Frankfurt/Main at the accumulation of Argon ions from a mixture of Argon gas and residual gas components [167].

- avoidance of an additional injection of neutral atoms in the working area of the source, which by charge exchange processes can limit the maximum achievable ion charge.

Thus, for the cooling of heavy ions by a continuous current of low-charged light ions the following conditions result:

1. The cooling rate of heavy ions must be higher than the heating rate by the electrons

$$\frac{dE_1^-}{dt} \geq \frac{dE_1^+}{dt} \tag{4.20}$$

where the heating rate is determined as follows

$$\frac{dE_1^+}{dt} = \sqrt{\frac{2m}{E_e}} \; \frac{r_e^2 Z_1^2 N_e m^2 c^4 L}{a^2 A_1 M} \tag{4.21}$$

and the cooling rate yields to

$$\frac{dE_1^-}{dt} = \frac{4N_2^2 r_e^2 m^2 c^4 Z_1^2 Z_2^2 L}{a^2 A M_1} \sqrt{\frac{A_2}{M}} \frac{(E_1 - E_2)}{E_2^{2/3}} . \tag{4.22}$$

2. During the transit time through the source the light ions must get sufficient energy from the heavy ions to be able to overcome the potential barrier at the end of the beam

$$\frac{dE_2^+}{dt} l \geq (E_{2m} - E_{20}) v_2 \tag{4.23}$$

with E_{20} as energy, with which the light ions are injected in the beam and

$$v_2 = \sqrt{\frac{E_{20} + E_{2m}}{A_2 M}}$$

as mean velocity of the light ions during the transit through a beam of the length l.

Further it must be considered, that the total energy, which is lost by heavy ions, must be equal the energy, which is transfered to light ions

$$N_2 \frac{dE_2^+}{dt} = N_1 \frac{dE_1^-}{dt} .$$

Further, we proceed on the assumption, that the relation between the charges of the heavy and the light ions satisfies the condition (4.18), e.g., the ion loss velocity for the heavy ions oversteps the permissible boundaries.

Knowing the current in the electron beam

$$I_e = e v_e N_e$$

the size of the current of light ions to be injected can be determined by

$$I_2 = Z_2 e v_2 N_2$$

$$\frac{I_2}{I_e} \geq \frac{m^2 c^2 N_e r_e}{4\sqrt{2} A_2 M E_e} \frac{(1 - f - \varepsilon_0)^2}{(2\varepsilon_1 - \varepsilon_0 - 1 + f)} .$$

where we introduce the designations

$$\varepsilon_0 = \frac{E_{20}}{Z_2 N_e r_e mc^2} \quad \text{and} \quad \varepsilon_1 = \frac{E_1}{Z_2 N_e r_e mc^2} .$$

According to the assumptions $E_1 = 0.05\, E_{1m}$, it follows that

$$\varepsilon_1 = \frac{z_1 (1 - f)}{20\, Z_2}$$

and ultimately

$$\frac{I_2}{I_1} \geq 1.4 \cdot 10^{-10} \frac{N_e I_2}{E_e A_2} \frac{(1 - f + \varepsilon_o)^2}{((1 - f)(Z_1 - 10\, Z_2) - 10\, Z_2 \varepsilon_0)} . \tag{4.24}$$

Simultaneously results from the relationship (4.21)

$$\frac{(1 - f - \varepsilon_o)\,(1 - f + \varepsilon_0)^2}{(1 - f)\,(Z_1 - 10\,Z_2) - 10\,Z_2\varepsilon_0} \geq \alpha f \qquad (4.25)$$

with

$$\alpha = \frac{2\sqrt{2}Z_1 A_2 L l}{5\,Z_2 A_1 N_e a^2}\;.$$

The relationships (4.24) and (4.25) make it possible, to determine the necessary current and the initial energy of the beam of light ions at permissible values of the ionization factor.

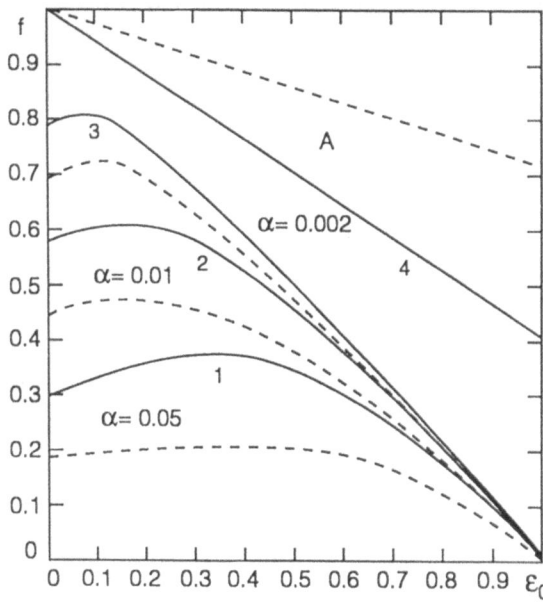

Figure 4.11 Dependence of the permissible values of the neutralization factor f in the electron beam of an initial energy ε_0 of the light ions at the cooling of heavy ions at a single transit of the ions through the source [163]. The field of work is found for the parameters between the curves 4 and in dependence of the concrete values of the parameter α between one of the curves 1, 2 and 3. The inked lines refer to Xenon ions and the dotted lines to Uranium ions.

Permissible values of the ionization factor f depending on the injection energy ε_0 of the light ions are indicated in Fig.**4.11**. The work area for the individual parameters is found between curve 4 and one of the curves 1, 2 and 3.

In Fig.**4.11** curve 1 corresponds the equality sign in (4.25) for $\alpha = 0.05$, curve 2 for $\alpha = 0.01$ and curve 3 for $\alpha = 0.002$. The curve 4 corresponds to the condition $E_1 = E_2$ in (4.22) and to the condition, that in the equations (4.24) and (4.25) the denominators are equal to zero. For this case the energies of the light and heavy ions are equal and cooling becomes impossible. The inked lines describe the storage of Xenon ions ($Z_1 = 54$) and the dotted lines the storage of Uranium ions ($Z_1 = 92$).

For the source KRION-2 ($E_e = 100$ keV, $l = 100$ cm, $E_{20} = 60$ eV) results $\alpha = 0.01$, $\varepsilon_0 = 0.7$ and thus from Fig.**4.11** for Xenon ions

$$I_2 > 4.2 \cdot 10^{-9}\, I_e = 4.2 \cdot 10^{-8}\ \text{A} \quad \text{and} \quad 0.6 > f > 0.3\;.$$

Accordingly yields for Uranium

$$I_2 > 5 \cdot 10^{-9}\ \text{A} \quad \text{and} \quad 0.8 > f > 0.25\;.$$

At the end of this paragraph the results are summed up once more:

- Elastic Coulomb collisions lead to the formation of a Boltzmann distribution of the ion energies in the beam. This is the reason for ion losses even if the mean ion energy is below the end-point energy.

- To avoid ion losses from the electron beam during the storage of ions in the course of some seconds the ion temperature should be not higher than 1/20 of the ion end-point energy in the beam.

- For the cooling of highly charged heavy ions ions of the lightest elements should be used.

4.4   The Electron Beam Ion Trap

Initially developed as an ion source for heavy ion accelerators the EBIS has been used up to now most successfully for investigations in atomic physics of highly charged ions. However, some constructive particularities of the EBIS complicate the use for atomic physics experiments. For the production of the transversal magnetic field along the beam axis the closed solenoid magnet complicates the observation of ions stored in the source considerably. The large length chosen for increasing the ion storage capacity of the electron beam leads to difficulties of the production of the axial magnetic field, at the beam transport and at the beam focussing and increases the probability for the appearance of beam instabilities.

In 1986 the Lawrence Livermore National Laboratory began the development of a source, whose design was layed out specially to the particularities of investigations of highly charged ions and their interactions with the electrons of the beam [67, 68]. The most important particularity of this new source was their short electron beam with a length of 10 cm and the very small ion trap with an axial extension of 2 cm. Thus it became possible, to renounce the solenoids along the entire beam. The magnetic focussing of the electron beam was realized with two superconducting coils. Between the coils in the middle of the drift tube a special window is placed for the observation of characteristic X-ray radiation emitted in the source by the stored ions. Through the small dimensions of the electron optical system the compression of the electron beam could increased considerably and the measurements of the plant are reduced. In consideration of the constructive particularities the developed source of highly charged ions was called EBIT (Electron Beam Ion Trap).

In the past years the EBIT proved itself as a very successful instrument and there were accomplished a series of important experiments, for example

- measurement of cross-sections for electron collisions, excitation and ionization [168, 169, 170];
- determination of energy levels and of the Lamb shift in selected ions [171, 172, 173];
- study of the interaction of highly charged ions with solid state surfaces [174, 175];
- study of the ion trap under the conditions of ion production and ion storage [176].

Special investigations were accomplished on the problem of ion cooling to increase the ion charge state and to minimize ion losses [165, 166, 177]. Altogether there are up to now a lot of works published on experiments at the EBIT; a considerable part of these works was summed up in [150].

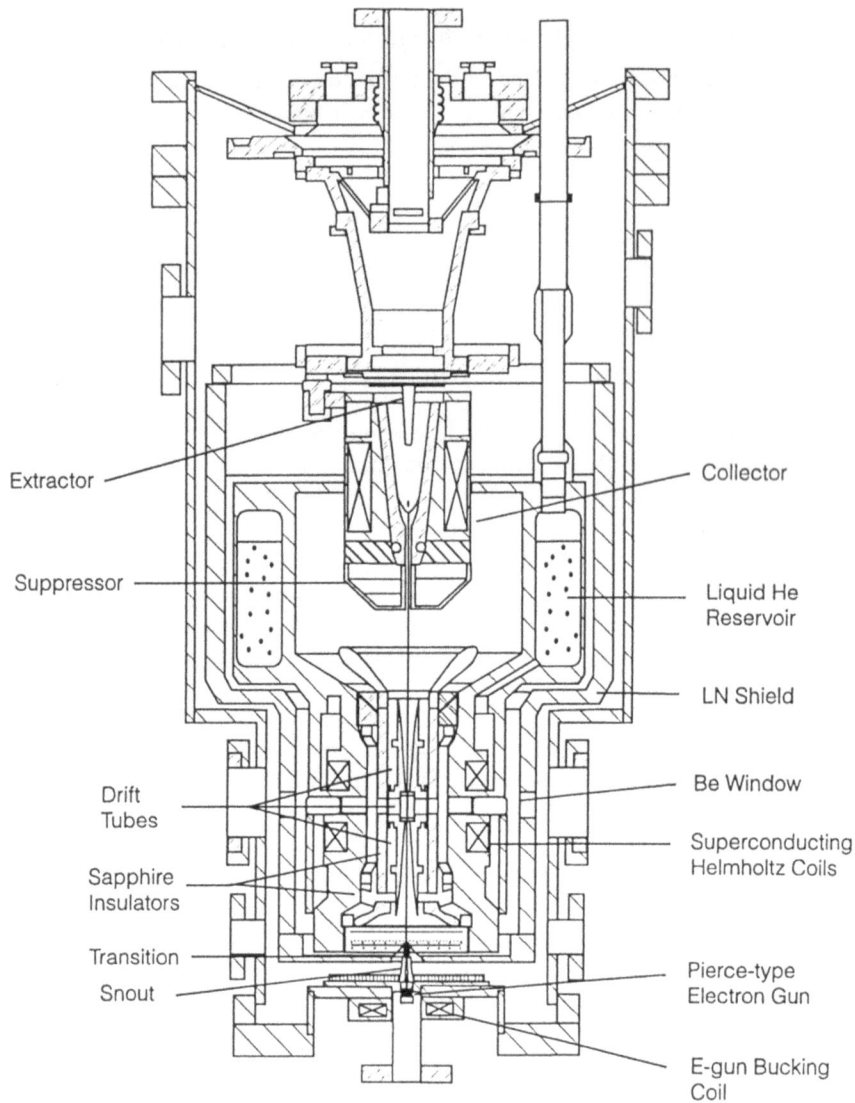

Extractor

Suppressor

Drift
Tubes

Sapphire
Insulators

Transition

Snout

Collector

Liquid He
Reservoir

LN Shield

Be Window

Superconducting
Helmholtz Coils

Pierce-type
Electron Gun

E-gun Bucking
Coil

Figure 4.12   The function diagram of the electron beam ion source EBIT [176].

Today, it is evident that the EBIT represents an independent direction in the development of sources of highly charged ions and therefore it should be treated in detail.

4.4.1   The Construction of an EBIT

The construction of an EBIT makes it possible to measure characteristic X-ray radiation from the working area of the source in the course of the entire ion storage time.

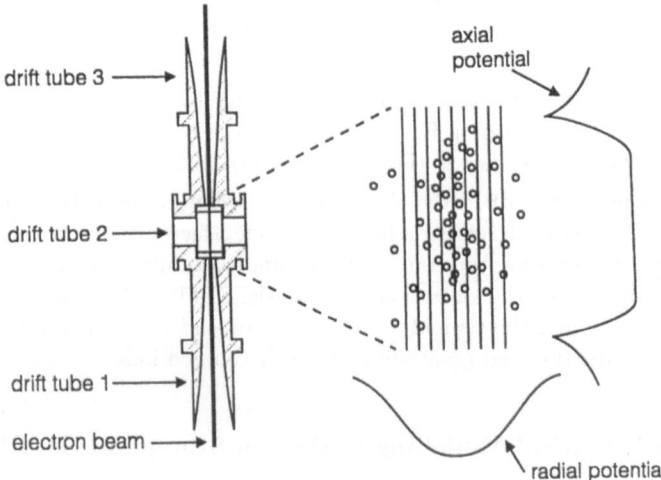

Figure 4.13   Sketch of the electron beam, the drift tubes, the potential distribution of the electric field over the radius and along the axis [177].

Furthermore, beside the installation of semiconductor spectrometers, crystal diffraction and VUV spectrometers for the registration of the quantum radiation from the source also an ion extraction system and a mass spectrometer became installed. This essentially expanded the possibilities for investigations of ion-atom interactions and made it possible to measure ion charge distributions and interaction processes of highly charged ions with solid state surfaces directly [174].

The function diagram of an EBIT is presented in Fig.**4.12**. An electron gun produces an electron beam with a current of 200 mA, which is accelerated in the drift tubes in an electric potential. The axial magnetic field with a magnetic induction of 3 T compresses the electron beam up to a radius of $a = 30\mu$m. This guarantees a current density in the compressed beam of up to 6000 A/cm^2. The focussing longitudinal field is produced by two superconducting solenoid coils and the drift tube for the electron beam is composed of three sections (Figs.**4.12** and **4.13** [177]). In the mean section between the two producing the magnetic field solenoid coils a Beryllium window is mounted, through which the registration of emitted from the source X-ray radiation is possible. The temperature of the drift tubes is held on the level from liquid Helium. This is a necessary condition for maintaining a pressure of 10^{-12} Torr or less in the source.

Ions of gaseous elements are produced in the trap by ionization of the residual gas, whereby the composition of the gas is controled and which can contain also heavy elements. Metal ions from a special ion source MEVVA (Metallic Vapor Vacuum Arc) can become injected in the source.

In radial direction the ion trap is closed by the electric charge of the electron beam. The axial inclusion of the ions results from potentials (Fig.**4.13**) existing between the individual sections of the drift tube. The depth of the potential well U_a between the beam axis and the limiting surface totals 10 ... 20 V per ion charge. The diameter of the drift tube in the central part of the tube is 10 mm and decreases at the edge to 3 mm. Thus the potential at the drift tube $U_R \cong 200$ V oversteps the potential U_a around tenfold.

The axial dimensions of the ion source are determined by the length of the central drift tube section and total altogether only 2 cm. Thereby the electron beam has a length of 10 cm. This is the condition for an increase of the beam compression and increases the beam stability in reference to the origin of electron-ion instabilities.

In comparison to the EBIS the short electron beam of the EBIT was successful in increasing the density of the electron beam by about a factor of ten. Under consideration of the large ion life times in the trap, which is practically unlimited and was determined to up to five hours [178], the ionization factor in the EBIT could be increased in contrast to the EBIS by around the tenfold. All these facts lead to the EBIT being an unique plant for the production and for studies on highly charged ions.

4.4.2   Mathematical Modelling of the Ion Storage in the EBIT

All works in the Lawrence Livermore National Laboratory are marked by a careful mathematical modelling and by consideration of very detailed particularities of the processes in the electron-ion beam.

In 1988 in Livermore the model of the evaporative cooling [165] was worked out analogously to that in 1984 Dubna created model for the cooling of heavy ions in electron beams [41, 42]. Evaporative cooling first was realized at the LLNL [166] and there also resulted corresponding computer simulations of this method [165].

Modelling of the ionization processes and of the ion storage in the electron beam were accomplished on the basis of balance equations for the density and the energy of the ions in the beam. The balance equation for the electron density was chosen thereby in conventionally usual form. The processes considered were single ionization and single charge exchange, ion losses in radial and axial directions and the possibility of an external ion injection in the source. The ionization cross-sections were calculated on the basis of the Lotz formula [12, 13] and the charge exchange cross-sections on the basis of the formula of Salzborn and Mueller [35].

A particularity of modelling ionization processes in the EBIT is that in the balance equations radiative recombination was considered. If the energy of the free electron decreases or becomes comparably with the ionization energy, so the electron can be captured in an excited state of the ion under X-ray emission. This is marked as radiatiative recombination.

The construction of an EBIS allows it to tune the electron energy very exactly and fix it so the value of the achievable charge states. In case that the energy of the electrons is very near to the ionization potential of the ions of a given charge state, the confinement time necessary for ions of a given charge state can be increased essentially by radiative recombination.

The cross-section for radiative recombination was determined by Kim and Pratt [296]

$$\sigma_i^{rr} = \frac{8\pi}{3\sqrt{3}} \, \alpha \lambda_e^2 \chi \ln \left(1 + \frac{\chi}{2n_{\text{eff}}^2} \right) \tag{4.26}$$

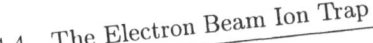

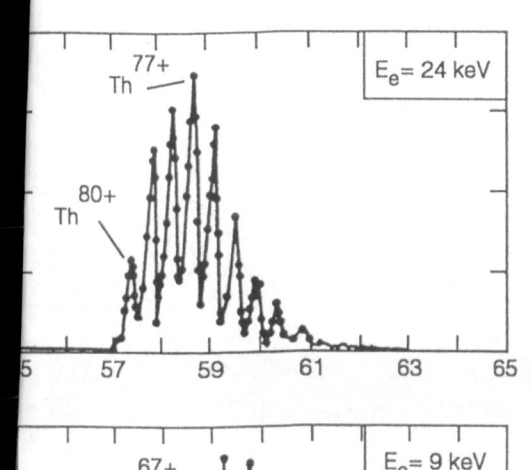

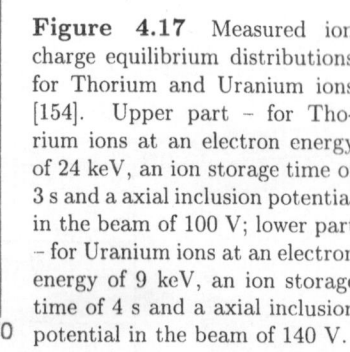

Figure 4.17 Measured ion charge equilibrium distributions for Thorium and Uranium ions [154]. Upper part – for Thorium ions at an electron energy of 24 keV, an ion storage time of 3 s and a axial inclusion potential in the beam of 100 V; lower part – for Uranium ions at an electron energy of 9 keV, an ion storage time of 4 s and a axial inclusion potential in the beam of 140 V.

; electron energy at the ionization of Uranium – 9 keV; residual gas pressure in Fig.**4.16**) at the ionization of Thorium – $7 \cdot 10^{-11}$ Torr; residual gas pressure in Fig.**4.16**) at the ionization of Uranium – $4 \cdot 10^{-11}$ Torr.

.**4.14** the calculated development of the ion charge distribution of Thorium nium ions in the electron beam is presented. From Fig.**4.14** is obvious, that , necessary for the receipt of Thorium and Uranium ions with ionized three st principal shells totals about 1...2 seconds. Thereby the ionization factor $j\tau$ values of about 10^{23} cm^{-2}. The relatively low ion charges for such an ionization an be explained by the small energy of the electron beam (24 keV for Thorium eV for Uranium), which is insufficient for the ionization of atomic inner-shells the low ionization efficiency. This can be caused by the circumstance, that the -point energy E_m, which is equal to the amount of the including potential at the nd, is much higher than the energy E_a at the lateral beam boundaries, so that s are found a large time outside the electron beam in the area between the beam drift tube.

ig.**4.15** the ionization rates and the characteristic times for recombination pro- and radiative recombination for different ion charges in Thorium and Uranium icated. According to the accomplished calculations for highly charged ions the of radiative recombination processes in the trap is higher than the rate of charge

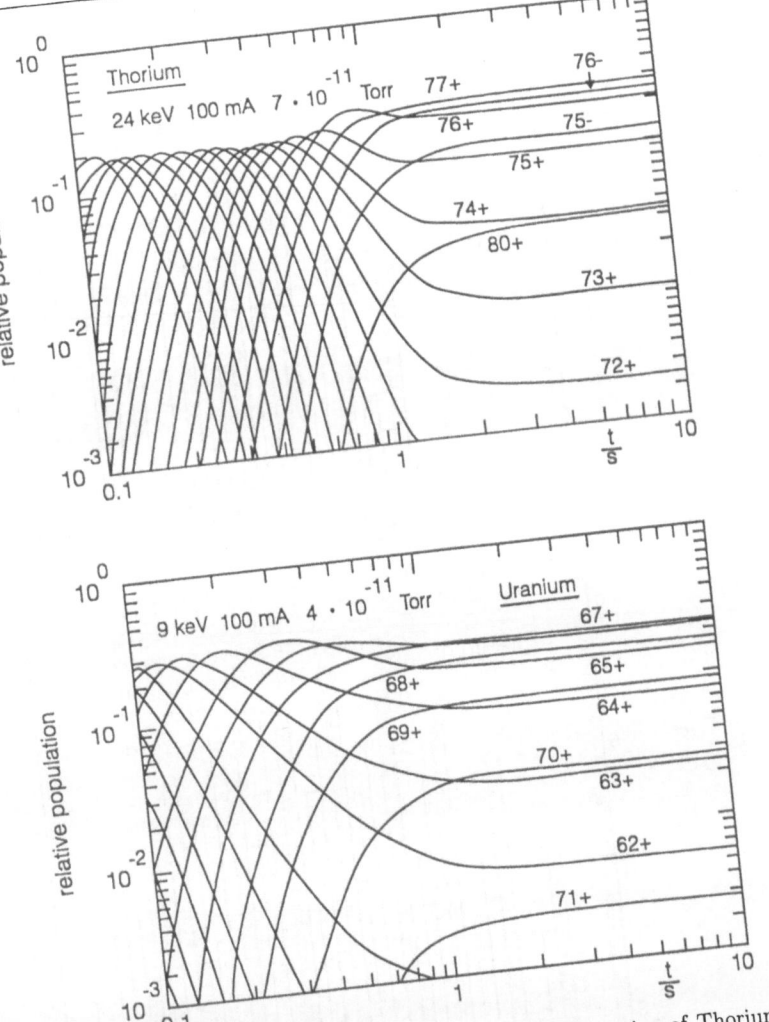

Figure 4.14 Calculated time development for the charge distribution of Thorium and Uranium ions in the electron beam of the EBIT [154].

with $\alpha = 1/137$ as fine-structure constant,

$$\chi = \frac{2Z_{\mathrm{eff}}^2 I_H}{E_e}$$

with $I_H = 13.6$ eV as the ionization potential of the Hydrogen atom, Z_{eff} as effective charge of an ion with the charge i and nuclear charge Z and $n_{\mathrm{eff}} = n_0 + (1 - w) - 0.3$ as effective quantum number of an ion with an outer non-filled principal shell n_o and the ratio w between the vacancy number and the electron number of the possible electron occupation in this shell.

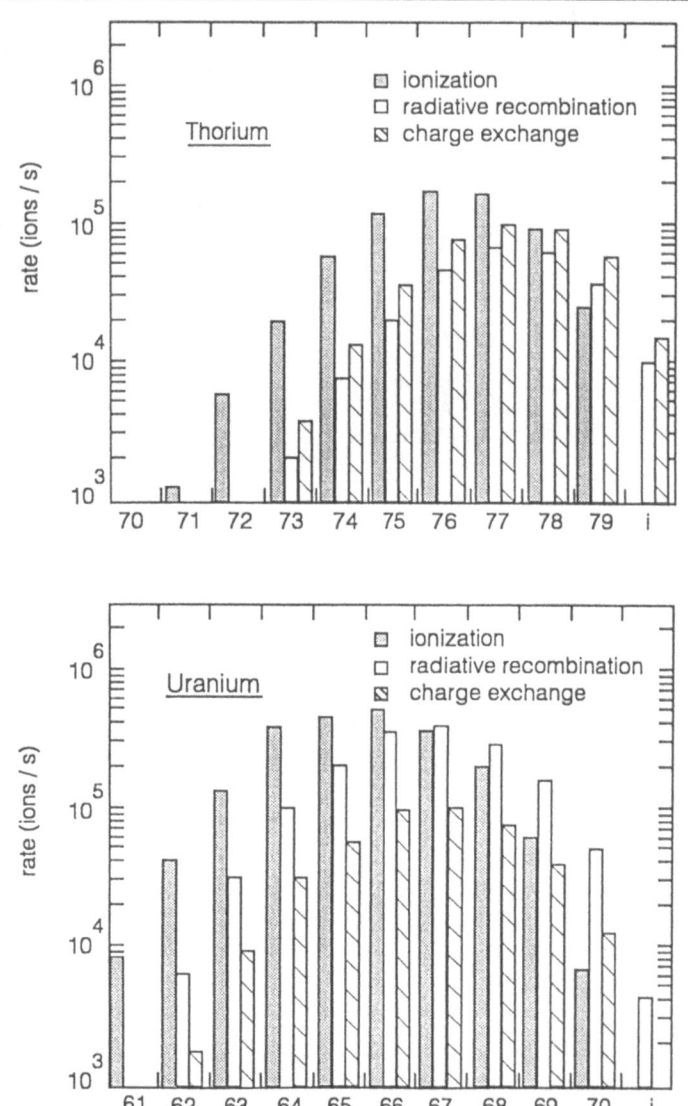

Figure 4.15 illustration (Thorium and Uranium rate plots)

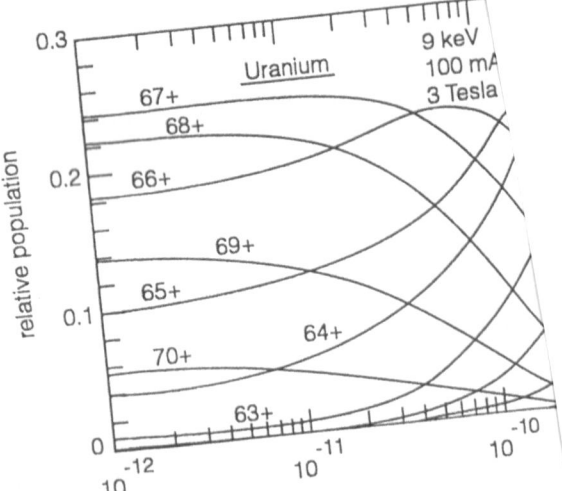

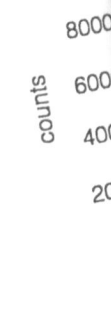

Figure 4.15 Calculated characteristic reaction rates for ionization processes, radiative recombination and charge exchange for different charge states of Thorium and Uranium ions in the electron beam of the EBIT [154]. Electron energy, electron current and residual gas pressure in the drift tube are thereby assumed as already indicated in Fig.**4.14**.

For the calculation of the ion loss velocity results from the theory of Pastukhov [180, 181] to the storage and losses of charged particles in open magnetic traps are used. The effective potential, which holds the ions in radial direction, considers the axial magnetic field, although, as estimations show, the influence of the magnetic field on the ion movement in the beam is neglegibly small.

Figure 4.16   Calculated equilibrium charge distributions of Th
dependence of the residual gas pressure in the electron beam of the

Computer modelling preceded practically all investigation
merical modellings the influence of a large amount of beam
experimental conditions on the ion production and the ion sto
zed. Results of these abundant calculations are indicated for e
Here only some characteristic results should be indicated in th

The presented results are calculated with the following para
tion – 3 T, electron current – 100 mA; electron energy at th

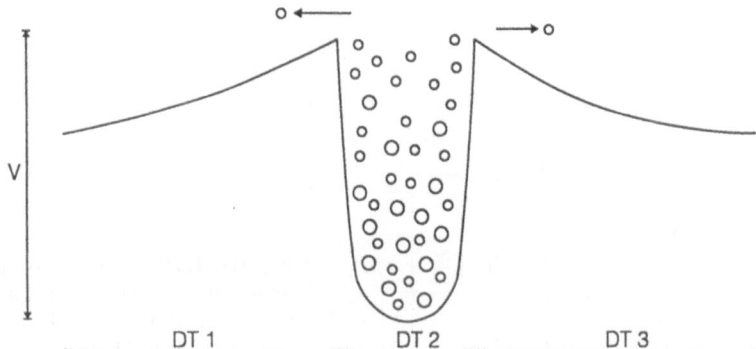

Figure 4.18   General diagram of the ion heating of light ions at collisions with heavy ions and of the evaporation of light ions in the axial potential well of the EBIT beam [175].

exchange processes. With it the radiative recombination becomes the limiting element for the achievable ion charge at the considered electron energies in the beam. This is conditioned by the high electron density in the beam in relation to the density of the neutral particles under the assumed conditions for the modelling.

In Fig.**4.16** the dependence of the equilibrium charge distribution of Thorium and Uranium ions with the residual gas pressure in the drift tube of the EBIT is presented.

For comparison in Fig.**4.17** the changes of the charge spectra of Thorium and Uranium ions are presented. The storage time for Thorium ions at an axial inclusion potential of $U_m =$100 V totaled 3 s and the storage time for Uranium ions at $U_m =$140 V was 4 s.

In publications of the modelling of basic processes in the EBIT the process of ion or evaporative cooling was also considered [165, 166, 176, 177, 182]. In these works all possible cooling conditions and the parameter of the cooled ions were varied.

In Fig.**4.18** the cooling of heavy ions and the evaporation of light ions in the axial potential well of the EBIT is presented schematically [175].

In Fig.**4.19** the calculated dependence of the equilibrium temperature of Dy $^{66+}$ ions on the kind of the cooling ions is presented. The calculations were accomplished for the most current variant of an EBIT construction, the so-called SUPER-EBIT [177]. This variant have the following parameters: ion density in the beam $n_i = 10^9$ cm^{-3}; density of the neutral particles in the beam $n_0 = 10^4$ cm^{-3}; electron energy in the beam $E_e = 150$ keV; electron current in the beam $I_e = 150$ mA; axial inclusion potential $E_m = 300$ V; magnetic induction $B = 3$ T.

Results from modellings are doubtless from a large practical interest and of meaning for the understanding of the processes taking place in the source, although the theoretical results were not always in full agreement with the experiment. To be able to reproduce in modelings the experimental charge distributions, it was necessary, to analyze a series of parameters detailed. As variable quantities the pressure of the neutral gas, the beam current and the inclusion potential at the ends of the trap were analyzed. This procedure

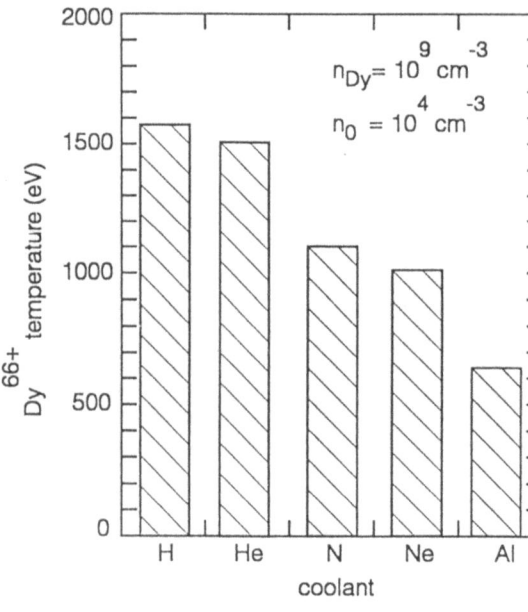

Figure 4.19 Calculated equilibrium temperature of Dy^{66+} ions at ion cooling with Hydrogen, Helium, Nitrogen, Neon and Aluminum in the electron beam of the EBIT at the following beam parameters: $E_e = 150$ keV, $I_e = 150$ mA, $U = 300$ V and $B = 3$ T.

made it possible, to receive a good agreement of the calculations with experimental values of the ion charge distributions [154, 176]. The results indicate for an right choice of the calculation methodology and for the consideration of the fundamental physical processes, which takes place in the EBIT.

The still remaining differences between modelling results and measurements can be caused by inaccuracies in the measurements. The authors of corresponding works point to it, that not all parameters could be determined precise. For example, the density of the residual gas could be determined only with an error of 20%. However this does not explain the necessary reduction of the inclusion potential around the factor two in the calculations to reach an agreement with measured results [176]. In the preceding paragraph we have already seen, that a change of the inclusion potential around a factor of two can change the rate of ion losses around some orders of magnitude [163]. Thus the size of the inclusion potential is one of the fundamental parameters, which can influence the equilibrium charge state.

In calculations the balance equation was used for modellings of the densities of each ion component with defined charge. Thereby it was assumed, that the ions own a stationary Maxwell distribution of the velocities and have a constant density over the source cross-section. Deviations in the spatial distribution of the ion and electron components are considered by the introduction of a so-called overlap factor, which going out from a constant density in the beam cross-section considering the overlap of different components.

The conditions, under which the calculations and experiments were done, correspond to a value of the inclusion potential U_m of 5...10 times higher than the potential U_a at the lateral beam boundaries. Therefore, the ion temperature was comparable or even larger than U_a. Under this conditions the largest ion number was found in the area between the beam and the drift tube wall. Under these conditions the cooling light ions suffered immediately intense losses. On the basis of these facts the following conclusions can be made about the character of the storage and movement of ions in the EBIT:

1. The ion components are rarefied strongly and the rate of the energy exchange between them is very small. The effectiveness of ion cooling is small and the lightest low-charged ions can not dissipate energy of heavy ions from the trap. This can explain, that neon was found as the most effective cooling medium [176]. Neon can be up to tenfold charged and exceeds in reference to the rate of the energy exchange with heavy ions the lightest ions considerably, without holding simultaneously the temperature of the heavy ions on the necessary low level. On the other hand, Neon has a relatively high ionization potential and therefore Neon neutrals have a low probability to charge exchange with highly charged ions and thus reduce achieved charge states.

2. The ions display no homogeneous density over the cross-section of the electron beam. It already was shown in chapter 2, the density of the ions formed in the electron beam increases in direction to the beam center and can reach there relatively high values. Elastic ion collisions lead to the formation of a Gaussian distribution along the radius of the beam cross-section. Under these conditions it is meaningful, to use at the calculation of charge distributions as balance equations the equations for the total ion number in the trap or for the ion density per length unit, as this already was done in the third chapter for calculations of the ion storage in electron rings. Such equations can be determined by integration of equations of the type (1.42), if the electron and ion density distributions are known.

It appears evident, that the in paragraph 2.2. we described the momentum method for the distribution function which can be applied very effectively for the mathematical modelling of ion production and storage processes in the electron beam of the EBIT.

3. Ions do not display a stationary distribution function. This coheres with it, that in the EBIT an ion temperature $T_i \ll iU_a$ can be maintained and thus the ions essentially do not move in the beam surrounding area. Thus, ions with low kinetic energy are heated by electrons, which is the essential process in the beam. On the other hand, ions with high energy can overcome the potential barrier and escape from the trap. Especially high are the losses for light, low-charged ions. In the result the stationary ion distribution is disturbed. Outside the electron beam the ion density is small and the energy redistribution in the resulting elastic collisions can not restore the initially stationary distribution. Consequently the ion distribution function does not satisfy a Boltzmann equilibrium distribution in reference to the energies and a Maxwell distribution in reference to the velocities.

This leads to the result that the mean rates of the electron-ion and ion-ion interactions can differ even under consideration of the overlap factor essentially from that in [174] obtained quantities and present an essential reason for the insufficient agreement between calculations and experiments.

4. In accordance with calculations and the experiment (Fig.**4.14** and Fig.**4.17**) for the receipt of $U^{66+}\ldots U^{68+}$ and $Th^{76+}\ldots Th^{78+}$ ions a ionization factor of $j\tau \cong 10^{23}$ cm^{-2} is demanded [182]. That is approximately thirty times more than at optimal ionization conditions (Fig.**1.10**), if all ions are found in the beam. The slowdown of the ionization processes in the trap is evidently in reference to the high ion temperature $T_i > iU_a$ of one of the fundamental defects at the operation of the EBIT.

4.4.3   Experimental Investigations of Highly Charged Ions

The advantages of the EBIT construction made it possible, to accomplish a series of unique experiments and investigations of properties of highly charged ions. The number of experiments at the EBIT is so high and the results are so interesting and important for the physics of highly charged ions, that this theme would have to be treated in a separate monography. However, here only some characteristic examples should be given for the illustration of the accomplished works (for more details see chapter 7.).

All investigations at the EBIT can be subdivided in two groups.

To the first group count measurements, which are accomplished during the storage of the ions in the trap. The ions have here relatively small kinetic energies, so that the Doppler effect only little disturbs the X-ray spectroscopy of excited ions. This makes it possible, to accomplish spectroscopic precision measurements at highly charged ions during the whole storage time of the ions in the trap (see for example [183, 184]).

To the second group belong experiments, at which the beam is extracted from the trap as ion beam. Thereby, it becomes possible, to measure directly the charge and the charge distributions of ions produced in the trap and to accomplish experiments to ion-atom, ion-ion and ion-solid state interactions.

Very recently experimental investigations at the plant began at the SUPEREBIT with electron energies in the beam of up to 150 keV [185]. This ion source is anticipated for the receipt of completely ionized atoms themselves of the heaviest elements. The choice of the inclusion potential in the ion trap on the level of the potential at the margin of the beam made it possible to increase the ionization efficiency and to improve the ion cooling and with it to reduce the time necessary for the complete ionization of individual atoms.

Chapter 5

Electron-Cyclotron Resonance Ion Sources

5.1 Operation Principles of Electron-Cyclotron Resonance Ion Sources

In the preceding chapters we considered ion sources where the ions are produced by electron impact and are stored inside of the potential well of the electron self-charge (EBIS and EBIT) or of a relativistic electron beam (ERIS). The working area of an electron-cyclotron resonance (ECR) ion source is a plasma, which is held by an outer magnetic field.

ECR-sources have their historic starting point in devices for plasma confinement for the thermonuclear synthesis. First 1969 Postma [186] proposed to utilize the plasma heating by the electron-cyclotron resonance for producing multiply charged ions. The first ECR sources were created as open magnetic traps for the thermonuclear synthesis in France by the group of Geller [187] and in Germany by Wiesemann [188] at the beginning of the seventies. These plants were very large, expensive and energy consuming.

The first successes were reported by Geller and co-workers with the source SUPER-MAFIOS [189]. This source used an azimuthal magnetic hexapole structure for the suppression of magneto-hydrodynamical instabilities and was composed of two sections for the plasma storage. In the first stage the initial ionization of the plasma is realized at a relatively high pressure of 10^{-3} Torr. The second stage was designed for the production of highly charged ions and the rest gas pressure totaled here 10^{-6} Torr. In the source SUPERMAFIOS Ar^{12+} and Xe^{26+} ions were produced [190]. However, the SUPERMA-FIOS source also has all defects of their predecessors. The longitudional dimensions of the source working area totaled 120 cm at a performance of 3 MW.

The further development of ECR ion sources was effectively influenced by the development of the MICROMAFIOS source, which became renamed in the courses of further modifications in MINIMAFIOS [191]. The working volume of this source totaled only at about a length of 30 cm at a diameter of 7 cm. Thereby, the source length was in the order of one meter. The small plasma volume made it possible, to insert permanent magnets for the production of the longitudinal magnetic field and to reduce the power consumed by the source around a factor of ten.

In the course of the eighties ECR sources became the most widely used sources of multiply charged ions. At this time these sources find applications in a multitude of accelerator centers as injectors of heavy ions and in many institutes and universities in connection with investigations in atomic physics. The main advantages of ECR ion sources are the high achievable beam intensity for multiply charged ions, their compactness, reliability and stability. The source does not own a cathode for the production of an electron or ion component in the plasma that requires a regular exchange of sub-assemblies of the source.

Table 5.1 Ion currents (in μA) for different elements from the CAPRICE ECR ion source at a microwave frequency of 14.5 GHz [196]. i – ion charge.

i	Ar	Kr	Xe	Ta	Pb	U
8	500					
9	300					
11	80					
12	36					
13	13	115				
14	5					
15		90				
16	0.5					
17		50	70			
19		20	55			
20		8.5	41			
21			30			
22			30	21		
23			30		25	21
24				21	30	22
25			17	18	30	18
26			14	16		14
27				12	25	10
28				7.5	20	
29			3	4		
31				2	8	

As example in Table **5.1** ion currents for different elements from the CAPRICE ECR source (GANIL/France) at a microwave frequency of 14.5 GHz [196] are indicated.

The principle functions, different construction variants and research results at ECR ion sources are comprehensively presented in [191, 194, 195, 197], likewise also in the proceedings of traditional international workshops on ECR sources and in conferences on ion sources, which are accomplished every two years.

5.1.1   Principle Function of an ECR Ion Source

An ECR source is based on a plasma, which is held in an open magnetic trap. The plasma forming electrons and ions are created by electron impact ionization processes from neutral atoms or molecules. The emitted electrons are heated by the microwave field and participate in the further ionization of atoms or molecules. Ions can be ionized up to the moment, until they are lost from the working area of the source.

The ion charge state in the ECR ion source is determined by the density of the electron component and by the life time of the ions in the plasma. Further, the maximum achievable charge state of the ions is determined by the electron energy. In ECR sources the conditions for the production of very highly charged ions do not apply like in EBIS sources. Thus charge exchange processes as a rule have no determining influence on the production of highly charged ions. However, these processes must be considered in the study of processes participating in the production of highly charged ions.

In Fig.5.1 a cross-section through the ECR ion source MINIMAFIOS [198, 199] and the distributions of the magnetic field B_z along the axis and the radial distribution B_r are presented.

The MINIMAFIOS source is composed of two sections, which are indicated with the numbers 1 and 2 in Fig.5.1 [198]. The magnetic field in the source results from the superposition of the hexapole field and the field, which is produced by the coils S1 ... S7. The resulting field displays an azimuthal varying hexapole structure and increases with the radius and along the source axis in the working area of the source. In some designs of ECR sources superconducting coils are inserted for the generation of the magnetic field.

For the heating of the electron component the microwave is injected in the source by the wave guide 3. Thereby, the microwave frequency determines the achievable electron density and with it also the charge distribution of the ions in the source.

The injection of neutral particles occurs by the valve 4. The ion extraction is realized with the electrode system 6, at which a negative potential is laid out. Usually this voltage totals some kV.

The first known ECR-sources all worked continuously. Later it was noticed, that a switching off of the microwaves leads to a strong short-term growing of the output of highly charged ions instead of evident reduction of the ion current due to the cooling of the electrons. This abnormal ion output received the designation pulse regime or *afterglow mode* and is presently applied for ECR sources as injectors for pulsed accelerators of heavy ions.

The further improvement of ECR ion sources results from the increase of the microwave frequency for the electron heating up to 20 ... 30 GHz, magnetic excitation conditions, the reduction of the source measurements and the increase of the reliability of the source. For increasing of the output of highly charged ions ion mixing, pulsed regime of source operation, special biased electrodes for additional electron production and other techniques are used.

Simultaneously with the development of ECR sources of highly charged ions an independent direction for the operation of ECR-sources with low microwave frequencies in the range of 2 ... 3 GHz was developed. These sources are used for special applications both for the receipt of intense currents of light, low charged ions as well as for industrial

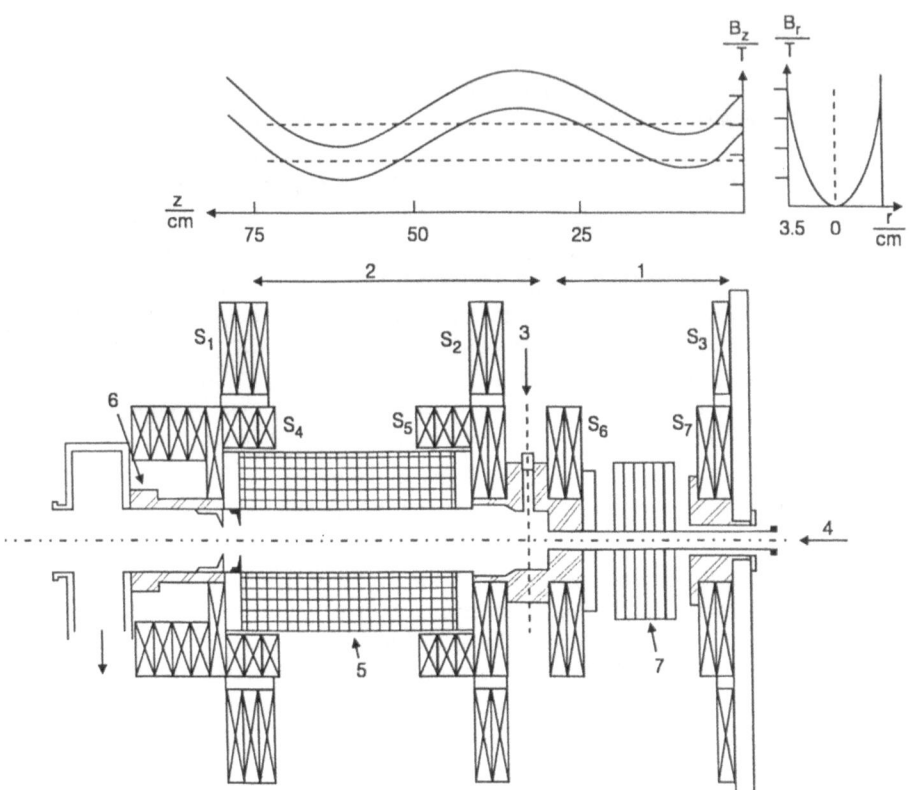

Figure 5.1 Sketch of the ECR ion source MINIMAFIOS and magnetic field distribution B_z along the axis and the field B_r along the radius in the central part of the source [198]. In the upper part the inked line corresponds to the operation of the source at 10 GHz and the dotted line to the work of the source at 16.6 GHz. In the lower part the following designations are introduced: S_1, S_2, ... S_7 – coils for the production of the axial magnetic field configuration, 1 – first stage of the source, 2 – second stage of the source, 3 – wave guides for the injection of the electromagnetic microwave field, 4 – gas injection, 5 – hexapole with permanent magnets for the production of the radial B-field, 6 – electrode system for the ion extraction, 7 – radiator.

applications. In the present book these sources of low-charged ions are not treated as well as a multitude of other corresponding sources of low charged ions.

5.1.2 High Frequency Heating of Electrons in the Plasma of an ECR Ion Source

Particles of the charge ie (for electrons counts $i = 1$) move in the magnetic field B on spiral orbits with the Lamour radius r_L

$$r_L = \frac{P \sin \Theta}{ieB} \tag{5.1}$$

with P as particle impulse and Θ as the angle between the directions of the magnetic field and the particle movement. In ECR sources the Lamour radius of electrons and ions usually totals fractions of a millimeter.

The progressive movement of the charge and the step of the spiral is determined by the velocity component parallel to the direction of the magnetic field.

The rotation frequency ω_{ce} of the electrons in the magnetic field, the so-called cyclotron frequency is

$$\omega_{ce} = \frac{eB}{m} \; . \tag{5.2}$$

Excitation processes are spread in the plasma with the eigenfrequency of the plasma ω_{pe}, which is determined by the plasma density. The eigenfrequency of the electrons in the plasma is determined by

$$\omega_{pe} = \sqrt{\frac{4\pi n_e e^2}{m}} \; . \tag{5.3}$$

An electromagnetic wave, which becomes fed into the plasma in resonance with one of the frequencies (5.2) or (5.3) or with one of their combinations

$$\omega = \sqrt{\omega_{ce}^2 + \omega_{pe}^2} \tag{5.4}$$

can exchange energy with the electron component of the plasma [200]. Usually the cyclotron resonance is used for the plasma heating at the frequency (5.2) or in some cases at the hybrid resonance (5.4).

So that the electromagnetic wave can expand in the plasma and effectively transmit energy to the electrons, the frequency must be higher than the eigenfrequency of the plasma ω_{pe}. This condition limits the electron density in the plasma

$$n_e \le \frac{\omega_{ce}^2}{4\pi r_e c^2} = 1.24 \cdot 10^{-8} \, f^2 \; . \tag{5.5}$$

The quantity f describes the plasma heating frequency in Hertz.

At a given microwave frequency the resonance condition is fulfilled only at certain values of the magnetic field B_{res}, which are determined from (5.2). For the design of the magnetic field of an ECR source the value of B_{res} is determined from the condition $B_{\text{min}} < B_{\text{res}} < B_{\text{max}}$, whereby B_{min} indicates the minimal value of the magnetic induction in the center of the magnetic trap and B_{max} the maximum value of the field along the source axis.

In Fig.5.2 the spatial distribution of the magnetic field for the second stage of the 16.6 GHz MINIMAFIOS source is shown [201]. The resonance surface for a microwave frequency of 16.6 GHz is presented here as a dotted line.

As a consequence only electrons which are found on the resonance surface are heated, the electron energy distribution is sufficiently complicated and as a rule determined only with difficulty. The part of the electrons, which do not arrive in resonance conditions, keeps this primary energy. Electrons, which are produced by the ionization of atoms and ions, have energies on the order of the ionization energy, i.e. some ten to hundred electron volts. After the microwave heating the electron energy can total some ten, in some cases even hundreds of keV [202].

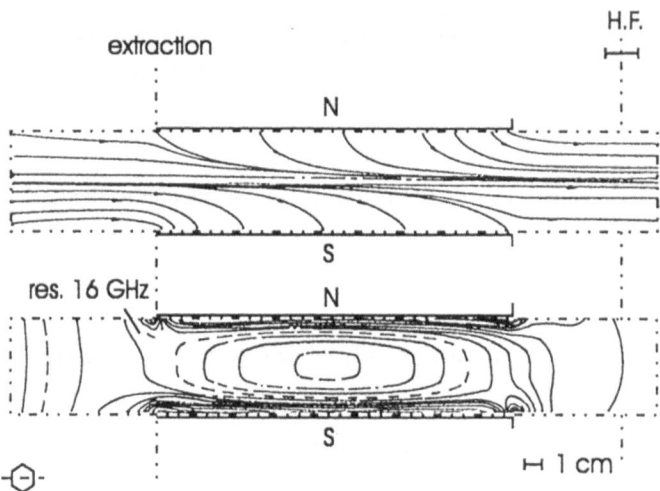

Figure 5.2 Spatial distribution of the magnetic field in reference to two each to other vertical planes for the second stage of the 16.6 GHz Minimafios source. The solid lines indicate equipotential lines for the field of the magnetic induction between 0.44 T and 1.04 T. The dotted line describes the resonance surface for a frequency of 16.6 GHz [201].

The relationship (5.5) shows, that an increase of the heating frequency leads also to an increase of the maximum possible electron density. This is one of the ways, to perfect ECR sources of highly charged ions.

5.1.3   Storage of Charged Particles in a Magnetic Trap

An increase of the magnetic field in both radial directions works as magnetic plugs and the axial distribution of the magnetic field acts as a magnetic trap for the charged particles moving in the source.

Let us consider the movement of a charged particle in the inhomogeneous magnetic field. Thereby it is assumed, that the particle moves under an impulse P, which is directed under an angle Θ to the field direction in the central source plane where the induction of the axial magnetic field ($B = B_{\min}$) is minimal. The charge moves in the magnetic field on a spiral, whose radius is determined from (5.1).

In a static magnetic field the total impulse of the charged particle is conserved

$$P^2 = P_r^2 + P_z^2 = P_0^2 \ . \tag{5.6}$$

If the spatial change of the magnetic field is relatively small and the value of the magnetic field within a trajectory orbit is constant, then the magnetic flux through the plain of an orbit of the particle is conserved

$$r_L^2\, B = r_{L0}^2\, B_{\min} = \text{const.} \tag{5.7}$$

with r_{L0} as the Larmour radius in the central plane.

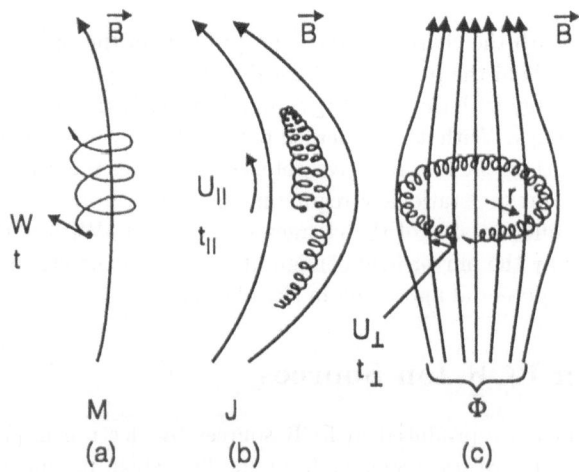

Figure 5.3 Movement of charged particles in the magnetic field [203]. a – movement along a field line in the axial magnetic field; b – reflection at the magnetic mirror; c – azimuthal particle drift.

From the combination of equations (5.1), (5.6) and (5.7) follows

$$\frac{P_r^2}{B} = \frac{P_{r0}^2}{B_{min}}$$

and so

$$P_z(z) = P_0 \left(1 - \frac{B(z) \sin^2 \Theta}{B_{min}} \right)^{1/2} . \tag{5.8}$$

From (5.8) it is understood, that at $B(z) \sin^2 \Theta = B_{min}$ charged particles discontinue their movement away from the center of the trap, linger and then turn back their movement. The particle already initially in reference to the source axis has a small movement angle $\Theta \leq \Theta_{min}$ with

$$\Theta_{min} = \arcsin \sqrt{\frac{B_{min}}{B_{max}}} \tag{5.9}$$

so it is not reflected at $B(z) = B_{max}$ and escapes from the source.

The relationship B_{max}/B_{min} will marked as *mirror ratio* of the magnetic trap and the cone, which emerges through the solid angle Θ_{min} around the source axis as the *loss cone*.

The radial magnetic field gradient can cause an azimuthal drift of the charged particles. In Fig.**5.3** the Larmour motion of particles around the magnetic field lines (a), the reflection of particles at the magnetic mirror (b) and the azimuthal drift (c), which is caused by the radial gradient of the magnetic field are shown.

Charged particles, which move under an angle smaller than Θ_{min} to the axis of the magnetic field, escape from the working volume of the source. As a result of elastic Coulomb collisions new particles emerge with small angles of the velocity vector to the axis, which are lost again. In this way electron and ion losses in ECR ion sources appear continuously.

The loss velocity is determined by the mirror ratio and by the frequency of the particle collisions. The ions have an energy small in comparison with the electrons and thus mutually collide essentially more often. The ion losses appearing at the beginning of the source operation conditionally disturb the charge balance and produce in the plasma a negative potential, which begins to hold the positively charged ions in the plasma. The strength of this potential is comparably with the ion energy. Simultaneously the electrons have a considerably high kinetic energy. With it the emerged potential does not influence the electron movement. The strength of the new negative potential regulates the velocity of the ion losses in the sense, that in the stationary regime in the source working area the balance between particles of different sign of the charges is conserved. With it the electrons are held in ECR sources by the mirror field configuration of the magnetic field and the ions by the small negative potential appearing in the plasma.

5.2   Storage of Ions in ECR Ion Sources

As well as in other ion sources ions are produced in ECR sources by electron impact ionization of atoms or molecules. In two-stage sources from the first stage pre-ionized ions are injected in the second stage, where the main ionization process results. In the course of the ion life time in the source the ions are successive ionized and so gradually increase their charge.

The adjusting ion charge is determined by the electron density n_e, the ion life time τ_i and by the ion confinement factor $n_e \tau_i$ (see section 1.6.). In contrast to other sources of highly charged ions an ECR source does not have a directed electron current and it is difficult in this connection, to speak of an electron current density j_e, which is applicable to this kind of ion source. Thus as a criterion for the ionization efficiency of an ECR source instead of the ionization factor $j_e \tau_i$ usually the *ion confinement factor* $n_e \tau_i$ is used. As more precise parameters for the determination of possible ion charge stages can serve the product $n_e < v_e > \tau_i$, whereby for $< v_e >$ the mean electron velocity is used. The size $n_e < v_e > \tau_i$ serves here as analogue to the ionization factor.

5.2.1   Ion Confinement Time

As a rule in the previously considered ion sources ion losses appear. These prove to be an undesirable element and there must be additional efforts, to reduce these losses or to remove them. Under ideal conditions the ion life time in the trap is identical to the life time in the electron beam or the electron ring sources.

A particularity of the work of an ECR source is their continuous working regime (even in the pulsed working regime all processes up to switching off of the microwave seems to be stationary). Thus the ion confinement time τ_i is determined only by the rate of the ion losses. The extracted ion current is also dependent on the ion loss rate. Thus the problem of ion storage and of ion losses becomes one of the most essential problems for the work of ECR ion sources.

ECR ion sources were created on the basis of open magnetic traps for the thermonuclear synthesis. In this connection the efforts of many scientists were concentrated on the development of this direction of plasma physics and the theory of the ion storage and the ion losses in open magnetic traps was worked out [180, 181, 203]. Independently from differences at different parameters between ECR ion sources and plants for the thermonuclear synthesis results of these works can be aquired well for the study of processes in ECR ion sources.

Two fundamental kinds of physical processes determine the loss velocity of charged particles in a plasma, which are located in outer electromagnetic fields. On the one hand classic processes contribute, which are established in the movement of particles in electromagnetic fields and in elastic Coulomb collisions. These processes always appear in the plasma.

On the other hand processes appear, which are connected with turbulences and different kinds of instabilities in the plasma. As a rule these processes are considerably stronger than classic processes and lead to current changes of plasma parameters and with it also to particle losses. Thus turbulences and plasma instabilities cause important changes in the ion charge distribution and with it also in the parameters of the extracted ion currents from the source.

In ECR sources the plasma is found in a strong axially, azimuthal varying magnetic field. Such a configuration of the magnetic field is chosen specially for the suppression of turbulencies and instabilities. Here a careful optimization of the spatial distribution of the magnetic field is necessary, to guarantee a successful working source. One can proceed on the assumption, that a careful source optimization can be found, which excludes the appearance of turbulences and instabilities in the plasma. Contrary to this classic ion loss processes always exist and can not be excluded. Thus it follows, that in a well constructed and optimized ECR source ion storage and ion losses are mainly determined by classic processes.

In accordance with the theory of Pastukhov [180, 181] we will consider only classic processes, since this makes it possible, to explain many important processes and effects in ECR ion sources [42, 204, 205].

Neclegting turbulences and instabilities the main mechanism for ion heating is the appearance of elastic Coulomb collisions of ions with hot electrons. According to the results from paragraph 1.4 the velocity of the ion heating is determined by

$$\frac{dT_i}{dt} = \frac{4\sqrt{2\pi}\, n_e i^2 r_e^2 m^2 \sqrt{mc^4}}{AM\sqrt{T_e}}\, L \; . \tag{5.10}$$

The ions own a comparatively low temperature and provide intense elastic interactions. The collision frequency ν_i for elastic collisions for ions of different masses and charges yields

$$\nu_i = \frac{4\pi\, m^2 c^4 r_e^2 i^2}{\sqrt{M}} \sum_{k=1}^{n} \frac{k^2 Ln_k \sqrt{A_i A_k}}{(A_k T_i + A_i T_k)^{3/2}} \; . \tag{5.11}$$

If for the estimation equation (5.5) is used, for modern ECR sources plasma densities of $n_e \simeq 10^{12}$ cm^{-3} and more can be reached. Thereby the preserved ion charges are as a rule considerably larger than unity and thus the density of the corresponding ion component is low. Independently of this at these parameters according to (1.36) the time characteristic for the tuning of a balance distribution is due to elastic Coulomb collisions particularly very small for highly charged ions and displays a time scale in the order of microseconds. This time can be estimated by (5.11)

$$\Delta t \simeq \frac{1}{\nu_i} \; .$$

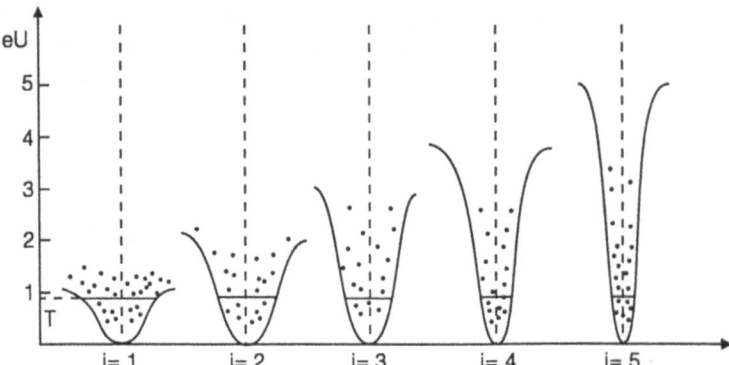

Figure 5.4   Relative positions of ions of the charges $i = 1,2,3,4,5$ at equal temperature T_i in a potential well of the depth U.

The characteristic ion life times and heating times lie in the region upto of milliseconds and are considerably larger than Δt. Thus the ion energy distribution can be examined as a stationary Boltzmann distribution (4.11).

Losses of energetic ions reduce the temperature of the ion component altogether. For the existence of a complete ion charge spectrum a general ion temperature can be found from the condition for the energetic balance at the ion heating and the ion cooling [204, 205]

$$T_i = \frac{dT_i}{dt} \frac{\sum\limits_{i=1}^{Z} n_i}{\sum\limits_{i=1}^{Z} n_i/\tau_i} \tag{5.12}$$

with τ_i as life time of different ion components.

Here the evident circumstance was used, that also in ECR sources all ion components resulting from elastic collisions come to the same general ion temperature.

As we have seen already in the preceding paragraph the plasma emerges in an equilibrium between electron and ion losses from a negative potential U. As well as in the electron beam of the EBIS also in ECR ion sources ions of different charges i have different amounts of the potential barrier ieU and have so also different probabilities to be lost from the source. This was already shown in Fig.**4.9** for an equilibrium Boltzmann distribution of the ion energy which was found in an electrostatic potential well.

Ions of high charge states are found at the mean energy at the bottom of the potential well and only few ions own an energy, which essentially exceeds the mean energy. These ions can overcome the potential barrier and leave the source. Ions of low charge states, which own a potential barrier, which is comparable with their mean energy or temperature, leave the trap with high probability. This situation is presented schematically in Fig.**5.4** where potential barriers are presented for different ion charges with equal temperature.

In papers on the ion storage in open magnetic traps it was shown, that in a wide area of plasma parameters the ion life time becomes sufficiently well approximated by a simple interpolation formula [180, 181]

$$\tau_i = \tau_{i1} + \tau_{i2} \tag{5.13}$$

with τ_{i1} and τ_{i2} as characteristic ion life times for close and distant collisions.

In case of close collisions the ion losses by the gas dynamical penetration of particles through the potential barrier in the trap are described by

$$\tau_{i1} = Rl \sqrt{\frac{\pi AM}{2T_i}} \exp\left(\frac{ieU}{T_i}\right) . \tag{5.14}$$

Here $R = B_{max}/B_{min}$ refer to the mirror ratio of the magnetic trap and l to the effective source length.

For the ion life time at distant collisions counts [180, 181]

$$\tau_{i2} = \frac{G}{(1 + T_i/2ieU)(\nu_{ii} + \nu_{i0})} \frac{ieU}{T_i} \exp\left(\frac{ieU}{T_i}\right) \tag{5.15}$$

with

$$G = \sqrt{\pi} \frac{R+1}{2R/\ln(2R+2)} .$$

Thereby the quantities ν_{ii} and ν_{i0} describe the collision frequencies of the ions mutually and with neutral particles. The value of ν_{ii} is determined by (5.11). For the determination of ν_{i0} an interpolation formula approach can be used [61, 64]

$$\nu_{i0} = 1.5 \cdot 10^{-9} \, i n_0 \, A_i^{-1/4} \tag{5.16}$$

with n_0 as density of the neutral particles in the plasma.

For typical ECR plasma parameters the characteristic time τ_{i1} for close Coulomb collisions exceeds the time τ_{i2} for distant collisions considerably.

If ions have a Boltzmann distribution in reference to the energy from the expressions (5.14) and (5.15) follows that their life time in the potential trap is mainly determined by the relationship between the height of the potential barrier and the ion temperature. Thereby highly charged ions live longer in the trap and are lost less frequently. Thus a reduction of the ion temperature leads to an increase of the ion life time in the trap.

In the strong axial magnetic field with magnetic mirrors at the ends of the source the ion losses result at the limiting side plains of the source. Knowing the ion life time the ion current density through the source cross-section can be estimated [60]

$$j_i = \frac{V i e n_i}{2S\tau_i} . \tag{5.17}$$

The quantities V and S describe the working volume and the sectional area of the source.

5.2.2   Balance Equations for the Ion Components

For the calculation of the ion charge state distributions in ECR ion sources a system of balance equations for all ion components can be used, whose form is similar to that for the usual balance equations (1.42)

$$\frac{dn_0}{dt} = \frac{S}{V} v_0 \left(n - n_0\right) - n_0 \left[\sum_{i=2}^{Z} \sigma_i^{ex} n_i v_i + \sum_{i=3}^{Z} \sigma_i^{2ex} n_i v_i + \left(\sigma_1^i + \sigma_1^{2i}\right) n_e v_e\right]$$

$$\frac{dn_1}{dt} = n_0 \left(\sigma_1^i v_e n_e + \sigma_2^{ex} n_2 v_2 + \sigma_3^{2ex} n_3 v_3 + \sum_{i=2}^{Z} \sigma_i^{ex} n_i v_i\right)$$

$$\qquad -n_1 \left(\sigma_2^{2i} v_e n_e + \sigma_2^{2i} v_e n_e + \frac{1}{\tau_1}\right)$$

$$\frac{dn_2}{dt} = n_0 \left(\sigma_1^{2i} v_e n_e + \sum_{i=3}^{Z} \sigma_i^{2ex} n_i v_i\right) + n_i \sigma_2^i v_e n_e + \left(\sigma_3^{ex} n_3 v_3 + \sigma_4^{2ex} n_4 v_4\right) n_0$$

$$\qquad -n_2 \left[\left(\sigma_3^i + \sigma_3^{2i}\right) v_e n_e + \left(\sigma_2^{ex} + \sigma_2^{2ex}\right) v_2 n_0 + \frac{1}{\tau_2}\right]$$

$$\vdots \quad \vdots \qquad\qquad \vdots$$

$$\frac{dn_i}{dt} = \sigma_i^i v_e n_e n_{i-1} + \sigma_{i-1}^{2i} v_e n_e n_{i-2} + \left(\sigma_{i+1}^{ex} n_{i+1} v_{i+1} + \sigma_{i+2}^{2ex} n_{i+2} v_{i+2}\right) n_0$$

$$\qquad -n_i \left[\left(\sigma_i^{ex} + \sigma_i^{2ex}\right) v_i n_0 - \left(\sigma_{i+1}^i + \sigma_{i+2}^{2i}\right) v_e n_e + \frac{1}{\tau_i}\right] \quad 3 \le i \le Z - 2$$

$$\vdots \quad \vdots \qquad\qquad \vdots \qquad\qquad\qquad\qquad (5.18)$$

$$\frac{dn_{Z-1}}{dt} = \left(\sigma_{Z-1}^i n_{Z-2} + \sigma_{Z-2}1^{2i} n_{Z-3}\right) v_e n_e + \sigma_Z^{ex} n_Z v_Z n_0$$

$$\qquad -n_{Z-1} \left[\sigma_Z^i v_e n_e + \left(\sigma_{Z-1}^{ex} + \sigma_{Z-1}^{2ex}\right) v_{Z-1} n_0 + \frac{1}{\tau_{Z-1}}\right]$$

$$\frac{dn_Z}{dt} = \left(\sigma_Z^i n_{Z-1} + \sigma_{Z-1}^{2i} n_{Z-2}\right) v_e n_e - n_Z \left[\left(\sigma_Z^{ex} + \sigma_Z^{2ex}\right) v_Z n_0 + \frac{1}{\tau_Z}\right]$$

with

S, V	–	plasma surface and plasma volume;
n, n_0	–	density of neutral particels outside and within the plasma volume;
v_0, v_i, v_e	–	mean velocities of the neutral particles, ions and electrons;
$\sigma_i^i, \sigma_i^{2i}$	–	cross-sections for single and double ionization;
$\sigma_i^{ex}, \sigma_i^{2ex}$	–	cross-sections for single and double charge exchange processes.

In the equation system (5.18) single and double ionization processes and single and double charge exchange processes are considered. Terms which can be inserted, describe other processes in the equation system (5.18), which lead to the change of the ion number at each ionization stage.

Equation (5.18) is a system of nonlinear differential equations of first order. For the solution of the equation system numerical methods are used.

In earlier works the plasma potential was determined for the case of an ion component from the balance condition for the ion and electron currents

$$\sum_{i=1}^{Z} \frac{i\, n_i}{\tau_i} = \frac{n_e}{\tau_e} \qquad\qquad (5.19)$$

whereby for the determination of the ion life time τ_i the theory of storage of charged particles in open magnetic traps [42, 181, 204, 205] was used. In the stationary case condition (5.19) leads to an equal number of electron and ion charges in the plasma, e.g., to the condition of plasma neutrality in the source, which corresponds for a homogeneous particle density to the condition

$$\sum_{i=1}^{Z} i\, n_i = n_e \ .$$

A negative potential in the center of the neutral plasma can only exist at differences in the spatial distribution of the electron and the ion component. For a negative value of the potential in the center of the source working area a concentration of electrons in the center and of ions at the periphery of the source is necessary. From the appearance of a negative potential follows that the ions will move to the center of the source and their density will then be high there, which contradicts the above assumptions. Thus the plasma can not be completely neutral at the existence of a negative potential and at a prevailing ion movement in the center of the source but must display a small negative charge. In this case the potential is determined by the charge or by the neutralization factor

$$f = \frac{\sum\limits_{i=1}^{Z} i\, n_i - n_e}{n_e} \ .$$

For a cylindrical source working area of the length l and of a radius R the depth of the potential well at the source axis can be described by

$$U = \frac{\pi}{2} mc^2 r_e n_e R^2 f \left(1 + 2\ln\frac{l}{R}\right) \ . \tag{5.20}$$

It must be noticed, that despite the elemination of the contradictions of the preceding models here also the assumption remains still valid, that the densities of the electrons and ions are distributed independently of each other evenly over the plasma volume.

If the condition (5.19) is excluded from the considerations the equation system must be added to an equation for the determination of the electron density. Than the balance equation for the electron component has the form

$$\frac{dn_e}{dt} = \left(\sum_{1}^{Z-1} \sigma_i^i n_{i-1} + 2 \sum_{1}^{Z-2} \sigma_i^{2i} n_{i-1}\right) v_e n_e - \frac{n_e}{\tau_e} \ . \tag{5.21}$$

The life times τ_e of the electrons are the plasma is determined by the collision rates ν_e for elastic electron scattering, whereby result electrons with velocities directed into the loss cone leaving the magnetic trap. Thereby, the plasma potential is a multiple smaller than the electron energy and does not influence the electron movement. The scattering rates consists as a sum from contributions of the mutual scattering of electrons, from the scattering of electrons at ions and at neutral particles

$$\nu_e = (\nu_{ee} + \nu_{ei} + \nu_{e0}) \ .$$

For the determination of ν_{ee} and ν_{ei} results from paragraph 1.4 and from [61, 64] can be used

$$\nu_{ee} = \frac{4\pi r_e^2 m^2 c^4 n_e L}{(2T_e)^{3/2}\sqrt{m}} \tag{5.22}$$

$$\nu_{ei} = \frac{4\pi r_e^2 m^2 c^4 L}{T_e^{3/2}\sqrt{m}} \sum_{i=1}^{Z} n_i i^2 \;.$$

(5.23)

The collision rate of electrons with neutral particles can be approximated by [61, 64]

$$\nu_{e0} \cong 4.2 \cdot 10^{-7} Z^{3/2} \frac{n_0}{T_e}$$

(5.24)

with Z as the atomic number of the neutral atoms in the plasma.

5.2.3   Calculations of the Ion Charge State Distribution

In first calculations of the ion charge distribution in ECR sources only the stationary solution of the equation system (5.22) and (5.23) were considered. For the stationary case all physical quantities in the source were constant and the left sides of equation system (5.22) become equal to zero and the system itself is a system of nonlinear algebraic equations. The solution of a nonlinear algebraic equation system is however a difficult problem.

First calculations of ion charge state distributions were accomplished by Jongen [58]. For simplicity in these calculations first both equations of the system were not considered. Likewise double ionization and charge exchange processes remained unconsidered. The density of the neutral particles n_0, the electron density n_e in the plasma and the temperatures T_i of the ions and of the electrons T_e were considered as input parameters. Such a presentation made it possible to represent the system of algebraic equations in matrix form and to use a simple algorithm for the solution of the problem. Under variation of the given start parameters (electron and ion temperature, densities of the electrons and of the neutral particles etc.) which values as a rule are not known, in Ref.[58] charge distributions were derived relatively close to experimental results at MINIMAFIOS type sources. Unfortunately, the first two neglected equations led to a violation of the correlation between the densities of the neutral particles and of the electrons as well as the ion density in the plasma. This caused a solution of the problem that did not result from a completely correct physical point of view.

Further progress in the calculation of the ion charge distributions was achieved by West [60]. In the model of West two electron components of different energy were considered – cold electrons, which are produced in the result of ionization processes and hot electrons, which become heated by the microwaves. The consideration of the cold electron component made it necessary to include processes of radiative recombination. For the determination of the plasma potential elements of the theory of Pastukhov [180, 181] were used. The describing the problem algebraic equation system was solved with an algorithm corresponding to that in [58].

In the work of West [60] a multitude of calculations were presented. Numerical results of the influence of plasma parameters on the ion charge distribution and on the from the source extracted ion currents for different kinds of ECR sources, such as for the SUPERMAFIOS and the MICROMAFIOS sources were presented. The results show a good agreement with experimentally measured quantities.

As an example of somewhat comprehensive calculations of charge state distributions in Refs.[61, 63] are considered results for the source DECRIS-14 (Dubna ECR Ion Source with 14.5 GHz) [206, 207]. The parameters of this source correspond to these of the MINIMAFIOS source.

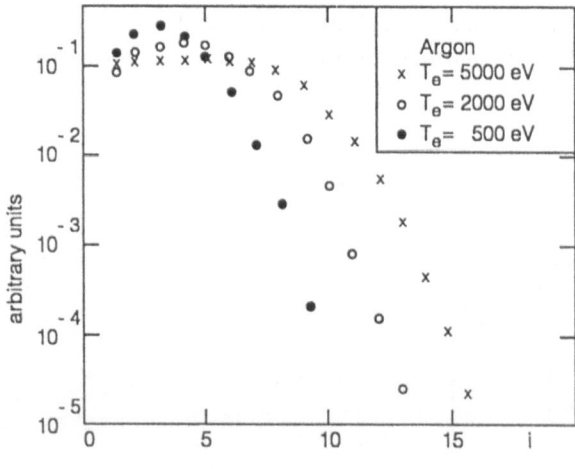

Figure 5.5 Calculated ion charge distributions of Argon ions for different values of the electron temperature T_e at an electron density $n_e = 2 \cdot 10^{12}$ cm^{-3} [61].

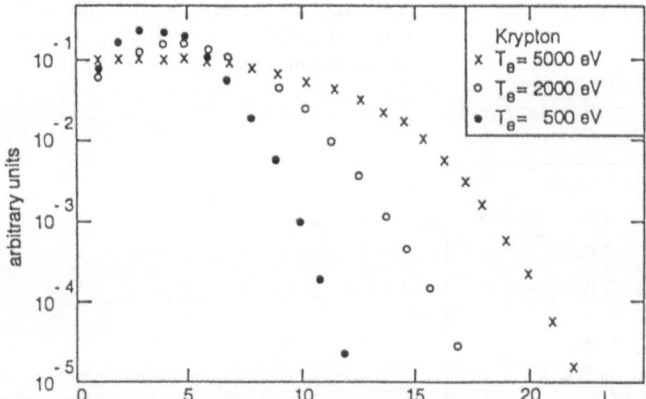

Figure 5.6 Calculated ion charge distributions of Krypton ions for different values the electron temperature T_e at an electron density $n_e = 2 \cdot 10^{12}$ cm^{-3} [61].

For the determination of the ion charge distribution in the DECRIS-14 source a system of nonlinear algebraic equations was used, which was taken from the right sides of equations (5.18) for the stationary case under consideration of the plasma neutrality condition (5.19).

For the calculation of the cross-sections for single ionization processes the Lotz formula (1.8) [12, 13] and for double ionization processes the formula of Mueller and Frodl (1.10) [19] were used. For charge exchange cross-sections the formulas given by Mueller and Salzborn (1.12) [35] were used, which were determined on the basis of extensive experimental results. The ion temperature was determined as a result of the heating processes by electron collisions (5.12).

The dimensions of the working area of the source and the electron temperature were given as outer parameters. The equation system (5.18) together with condition (5.19) or the balance equations for the electron component (5.21) are composed of $Z+2$ equations. The number of the unknowns quantities (n, n_0, n_i, n_e) with $i = 1, 2, \ldots, Z$ however totals $Z + 3$.

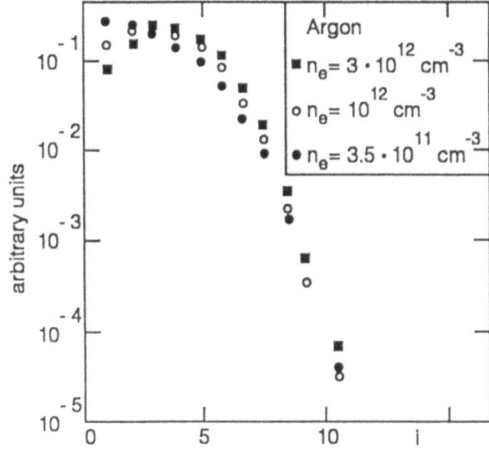

Figure 5.7 Calculated ion charge distributions for Argon ions for differently values of the electron density n_e at an electron temperature $T_e = 1000$ eV [61].

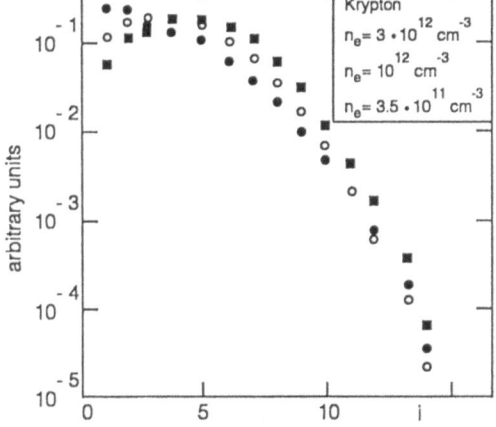

Figure 5.8 Ion charge distributions of Krypton ions for different electron densities n_e at an electron temperature $T_e = 1000$ eV [61].

In Figs.5.5 to 5.8 ion charge distributions of Argon and Krypton ions at the exit of the source as a function of the electron temperature T_e and of the electron density n_e are shown. The ion output is indicated in relative units and each value is normalized to the total current of all ion charge states. Equation (5.17) was used for the determination of the ion current at the exit of the source.

For comparison in Fig.5.9 the charge distribution of Argon ions in the 16.6 GHz ECR source MINIMAFIOS [208] is presented. Fig.5.10 shows the charge distribution of Argon ions as the microwave frequency in the MINIMAFIOS source is increased from 10 GHz to 16.6 GHz [201]. The increase of the microwave frequency leads according to (5.5) to

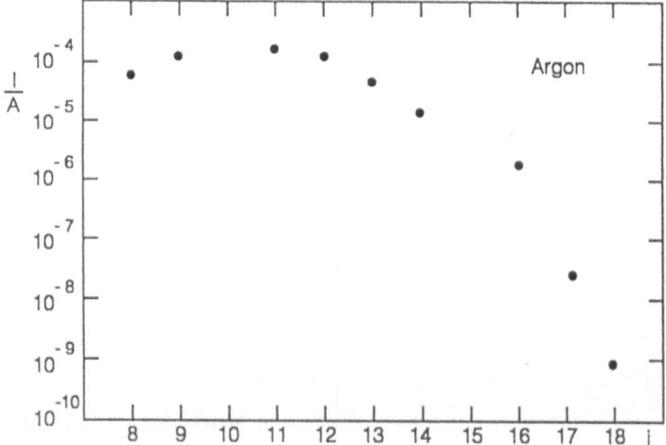

Figure 5.9   Charge state distribution of Argon ions at the exit of the 16.6 GHz MINIMAFIOS source [208].

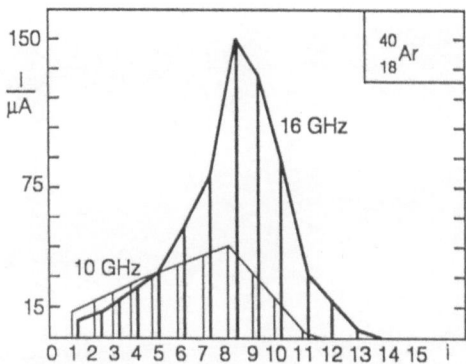

Figure 5.10   Charge distribution of Argon ions at the exit of the MINIMAFIOS source at microwave frequencies of 10 GHz and 16.6 GHz [201].

an increase of the electron density and with it probably also to an increase of the electron temperature in the source.

In Figs. **5.11** and **5.12** the dependence of the ion temperature T_i of Neon, Argon and Krypton on the electron temperature T_e and on the electron density n_e are presented.

Equation (5.17) makes it possible to estimate ion losses. The total ion current I from the source is presented as a function of T_e and n_e in Figs. **5.13** and **5.14**.

The results show that the ion charge distributions and the output of highly charged ions from the ECR source strongly depend on the electron temperature. At an increase of the electron temperature increases the mean ion charge in the plasma and with it also the output of highly charged ions. This fact is found also in accordance with measured results and also with the results of earlier calculations. Simultaneously the total ion current (Fig. **5.13**) decreases. This is in addition to that due to the plasma neutrality condition the net ion losses are equal to the electron losses, which depends again from the collision rate in the plasma. The collision probability with increasing energy decreases

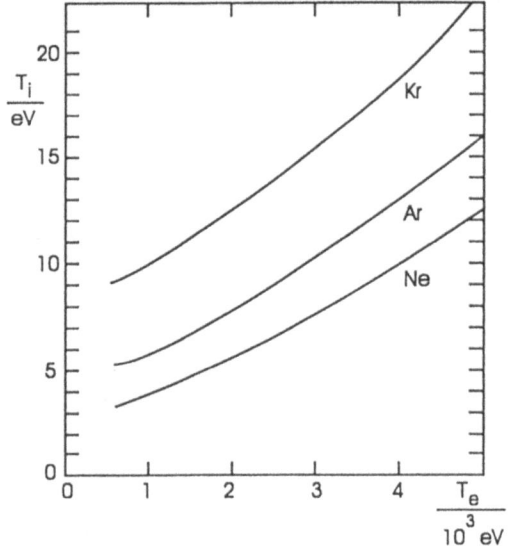

Figure 5.11 Dependence of the ion temperature T_i of Neon, Argon and Krypton from the electron temperature T_e at an electron density $n_e = 2 \cdot 10^{12}$ cm^{-3} [61].

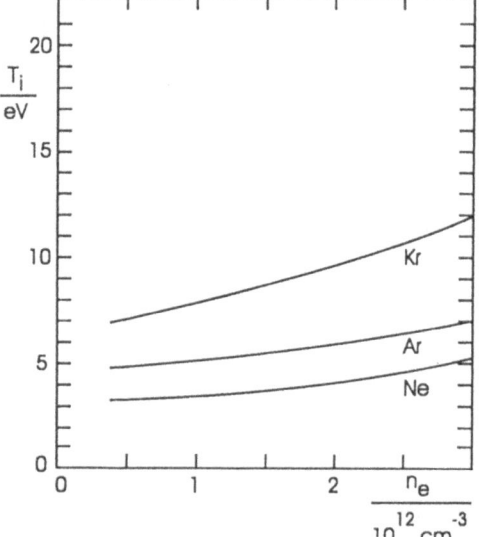

Figure 5.12 Dependence of the ion temperature T_i of Neon, Argon and Krypton from the electron density n_e at an electron temperature $T_e = 1000$ eV [61].

which leads to a reduction of electron losses and to a smaller ion output from the source, as it is seen from Fig.**5.13**. An increase of the electron density n_e causes a strong growth in the ion output from the source (see Fig.**5.14**).

A result of the calculations is that the charge distribution practically does not depend on the electron density (see Figs.**5.7** and **5.8**). This is found for calculations using the equation system (5.18). It is solved and so a connection between the density of the neutral particles, the ion number and the electron number in the plasma exists. So an increase of the electron density requires an increase of the neutral particle density in the plasma

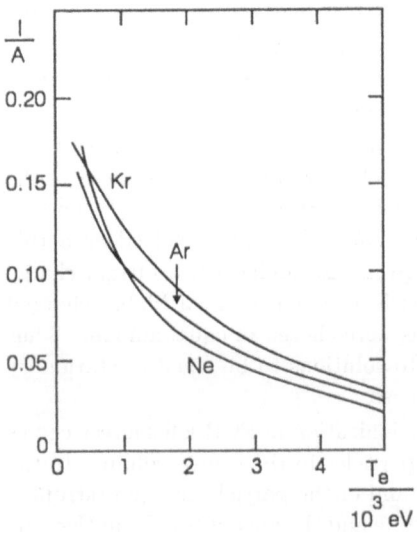

Figure 5.13   Dependence of the total ion current I for Neon, Argon and Krypton from the electron temperature T_e at an electron density $n_e = 2 \cdot 10^{12}$ cm^{-3} [61].

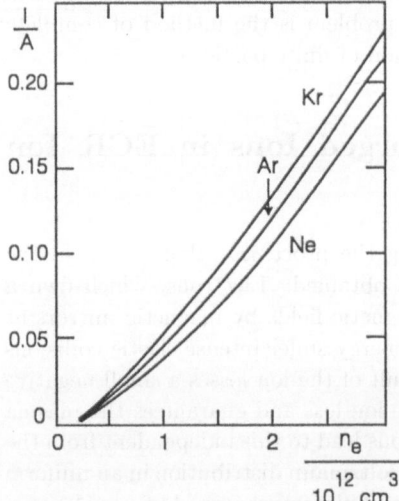

Figure 5.14   Dependence of the total ion current I of Neon, Argon and Krypton from the electron density n_e at an electron temperature $T_e = 1000$ eV [61].

and conversely. On the other hand, the high neutral particle density is a factor which reduces the ion charge. This leads to a simultaneous increase of the electron density and the normalized charge distribution remains in the average unchanged. Simultaneously an increase in the ion density results in an increase of the ion output from the source (see Fig.**5.14**).

A numerical solution of (5.18) is as a rule connected with significant numerical difficulties. In the right sides of the equations the terms with τ_i give a considerable contribution. The ion life time τ_i is determined by the ion density n_i. For this reason the equation system is nonlinear and is only poorly solved numerically.

In the above calculations [61, 63] the iterative Seidel method was used. For the convergence of the iteration process the right choice of the initial conditions for the charge distribution was of decisive meaning. For some starting parameters the iteration process does not converge and it is difficult to find a solution.

To facilitate the solution for ion charge state distributions in ECR ion sources it was proposed not to solve the algebraic equations but to solve the stationary solutions of the complete differential equations (5.22) at given initial conditions [62, 66]. It is known, that with numerical methods nonlinear differential equations can be solved as a rule effectively without especially high expense. In the given case sufficiently arbitrary charge distributions were chosen as initial conditions which do not contained highly charged ions. For instance single charged ions and electrons were chosen in equal amount. Thus the integration of the equation system (5.22) led to solutions to be equal to stationary values more quickly and with fewer numerical expense.

A general defect of all calculations of the plasma ionization in ECR ion sources comes from assuming an equidistribution of the charged particles in the source volume. In the actual case the distribution of the particle density and of the particle energy is strongly inhomogenous, which influences the charge distribution and the ion output. Doubtless the solution of the self-consistent problem for the description of the plasma in ECR sources is not a simple task. Thereby the fundamental physical processes can be understood in a simpler way. As effective start to the described problem is the method of complete moments of the distribution functions and the method of finite particles.

5.3 Production of Multiply Charged Ions in ECR Ion Sources

In both the preceding paragraphs by considerating the processes taking place in the plasma of an ECR source the following picture is obtained: Electrons, which own a relatively high energy, are held by strong outer magnetic fields by magnetic mirrors at the ends of the source. Ions with considerably low energy suffer intense elastic collisions and are weakly held in the magnetic field. As a result of the ion losses a small negative potential emerges in the plasma which regulates the ion loss and guarantees the plasma neutrality. Intense elastic interactions between the ions lead to this independent from the charge and the mass of the ions which approaches a Boltzmann distribution in an uniform temperature and energy distribution. Thus ions of different charge, but equal mean energy have different amounts of the potential barrier (Fig.**5.4**) and different probabilities to become lost from the potential well.

Thus, three fundamental principles can be formulated for the storage of ions in the potential trap of ECR ion sources [204]:

1. Ions with high charge states have larger life times and it is difficult to them to leave the working area of the source.
2. The ion charge distribution displays at the ion extraction a mean charge which is more lower than the mean charge in the interior of the ion trap.
3. A reduction of the ion temperature or of the mean ion energy in the source increases the ion life time and leads with it to an increase of the mean ion charge.

These characteristics should be used for the analysis of the both most extensive spread methods for increasing the ion charge states in ECR ion sources.

5.3.1   Ion Cooling

In the description of ECR sources the notation gas mixing effect traditionally is used in the literature for ion cooling. In the middle of the Eighties in different laboratories an increase of the output of highly charged ions from the injection of two different gases in the source was observed [209, 211]. The nature of this phenomenon was not clear at this time and the outer side of this effect was used to designate the phenomenon. Meanwhile it is fully clear, that the reason of the gas mixing effect is based on the principle of ion cooling common for all ion sources [41]. Independent of it the first designation has lasted up to the present day.

Antaya [194] was probably the first who proposed to insert in ECR sources mixtures of light and heavy ions, whereby the heavier ions transfer energy to the lighter ions and cool so them due to elastic collisions. However in this work the role of the temperature of different ion species and of the negative plasma potential for the ion cooling was not considered.

In principle the ion cooling in ECR sources does not differ from the process, which takes place EBIS at the cooling of heavy ions by light low charged ions (chapter 4.3.) and can be explained on the basis of the above formulated principles.

In ECR sources the multiply charged heavy ions are located in a potential trap, which is formed by the negative plasma potential. Thereby the electrons heat the ions (5.10) and their life times (equations (5.14) and (5.15)) are determined by the ion temperature (5.12). With the addition of a light gas, this gas is ionized by electron collisions and remains in the source too. Light ions display more lower charge states and thus have a low rate of electron heating (5.10). Simultaneously the light ions receive transferred energy by intense elastic collisions with the multiply charged ions (5.11). Because of their low potential barrier and small life time (equations (5.14) and (5.15)) the light ions can escape from the source and carry away energy. This process reduces the ion temperature in the source and increases the ion life time τ_i and also the ion charge states of heavy ions.

In contrast to the EBIS and EBIT sources, where a certain amount of ions of the working gas is found in the course of the entire working cycle in the source and where the avoidance of ion losses the ion temperature must be held about ten times more lower than the amount of the potential barrier, in ECR sources the ions are formed permanently and there do not exist such strong demands on the adjusting temperature and on the type of ions used for the cooling process. The lightest ions (Hydrogen or Helium) are quickly emitted from the source and quickly carry away energy, however they have a small effectiveness for the energy exchange with ions of the working gas. Heavier ions have an potential well comparable with the amount of the potential barrier of the working gas and thus they can not decrease the ion temperature in the source essentially. Optimally for the cooling process in ECR sources are ions, which have charges and an amount of the potential barrier well considerably lower than that of the ions of the working gas but charge states of coolant ions are preferable higher than 1.

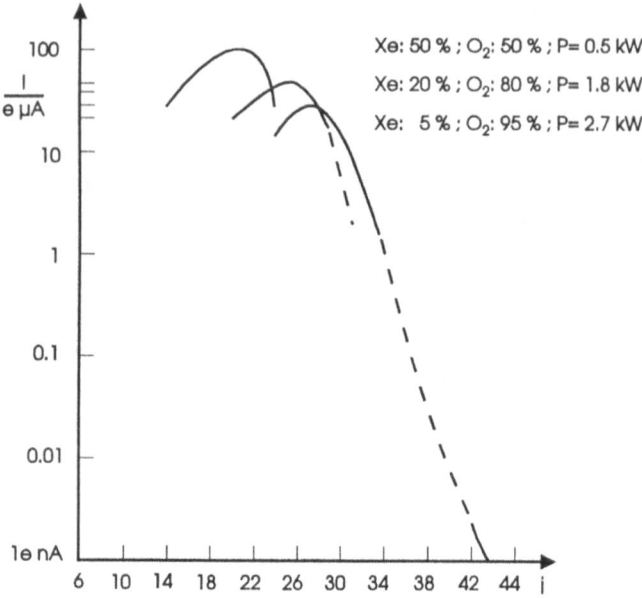

Figure 5.15   Charge distribution of Xenon ions at the extraction from the 16.6 GHz MINI-MAFIOS source for cooling by Oxygen ions for different concentrations of Xenon and Oxygen [198].

For the production of light ions (Nitrogen, Oxygen, Neon) Helium can be used as cooling medium. For the cooling of heavier elements (Argon, Krypton) Carbon, Nitrogen or Oxygen are applied. For the heaviest elements (Xenon, Uranium) elements from Nitrogen to Neon are best suited as cooling the medium. Thereby it is important, that the relationship of the ion charges of the working gas to the charges of the cooling medium is as large as possible.

In Ref.[212] different combinations of gases from Helium to Argon in reference to their effectiveness for the cooling process were explored. Thereby it was shown, that ions up to Neon are cooled unusually well by Helium, but for cooling of Argon it is already more favorable to use Nitrogen ions.

In [194] a reduction of the energy spread of extracted Argon ions was described for the case of cooling by Oxygen ions. This indirectly indicates a reduction of the ion energy in the source and can be examined as a confirmation for the mechanism of ion cooling.

In Fig.5.15 the measured current of Xenon ions from the 16.6 GHz MINIMAFIOS source for different concentrations of Oxygen ions acting as coolant is presented.

Numerical calculations for charge state distributions for ion mixtures of different elements were accomplished in Refs.[42, 194, 204, 205, 213]. In the Figs.**5.16** and **5.17** calculated charge state distributions of Krypton in the source and at the ion extraction for different ratios of the Krypton/Nitrogen mixture [205] are shown. The calculations were accomplished for an ECR ion source of the MINIMAFIOS type. Thereby it

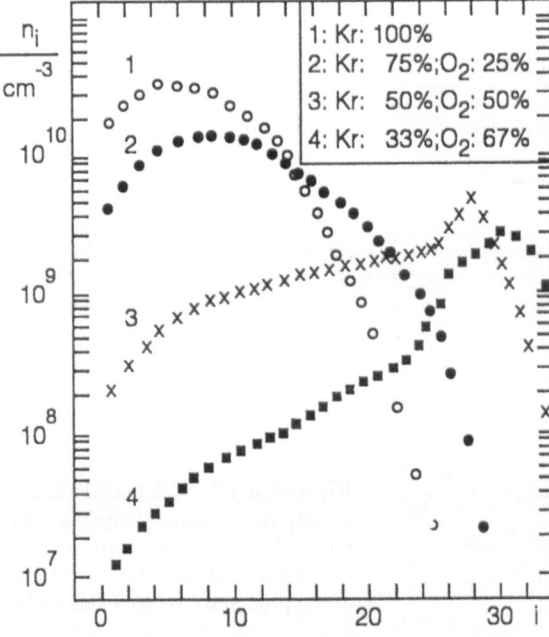

Figure 5.16   Calculated charge density distribution n_i for Krypton ions for the ion extraction at different concentrations of Krypton and Nitrogen ions [205].

was assumed, that the electrons have a temperature $T_e = 5000$ eV and a density of $n_e = 2 \cdot 10^{12}$ cm^{-3}. It is recognized, that Fig.**5.17** almost completely repeats the measured dependence of the charge distribution of the ion current at the ion extraction from the 16.6 GHz MINIMAFIOS source for different ratios of Krypton and Oxygen gases as they are presented in Fig.**5.15**.

One of the unusual effects of gas mixing was an anomalous charge state distribution for Oxygen isotopes observed by Drentje [216]. It was found that the ratios of the extracted beam currents $^{18}O^{i+}/^{17}O^{i+}$ and $^{18}O^{i+}/^{16}O^{i+}$ increase with the charge state i as it is shown in Fig.**5.18**. Let us introduce the explanation of this effect according to the above discussed model of ion confinement and ion losses.

The electron heating rate of the ions depends on the atomic mass of the ion or on the mass number A (5.10). Thereby the energy mixing effect via elastic collisions among the ions decreases the temperature difference between various ion components or atomic isotopes. For very small temperature differences (as we it have in the case of Oxygen isotopes) the energy redistribution is in accordance with (1.31) a very slow process. Hence, we can suppose that various isotopes have a little difference in the temperatures. For example, the temperature of $^{18}O^{i+}$ is lower by approximately 5% in comparison to $^{17}O^{i+}$ and 10% in comparison to the temperature of $^{16}O^{i+}$. The ion confinement time (equations (5.14) and (5.15)) depends strongly on the ion temperature and on the mass number A. So the different isotopes have another charge state distribution due

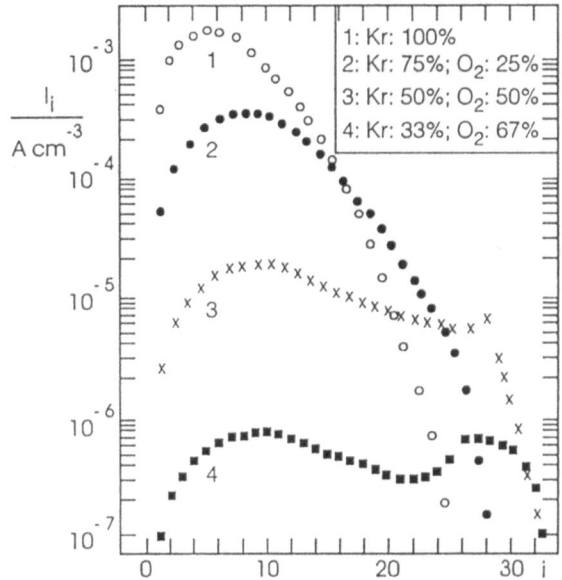

Figure 5.17 Calculated charge density distribution of the ion current of Krypton ions for the ion extraction at different concentrations of Krypton and Nitrogen ions [205].

to different confinement times. If Oxygen ions are cooled with Helium ions then the difference in confinement conditions for various isotopes is smaller and the isotopic effect practically disappears. A numerical simulation of the isotopic effect shows similar results (Fig.**5.19**).

Calculations of the ion charge states for mixtures of Krypton and Argon ions showed, that an addition of Nitrogen ions in the plasma reduce the ion temperature from 16 eV to 3 eV and increase the mean ion charge of Krypton from 7 to 28 [205]. Simultaneously the output of multiply charged ions increases from the source, however not in the same manner as in the source itself. In Figs.**5.16** and **5.17** the difference in the ion charge distribution in the trap and at the ion extraction is clearly to recognize. This difference is especially high for cooled ions. This important result is explained thereby that the multiply charged ions are found at the bottom of the potential well and have according to equations (5.16) and (5.17) a higher life time in the source and can leave the source with small probability. An increase of the output of multiply charged ions from ECR sources can be expected if the potential in the plasma allows it to extract multiply charged ions. One of these methods is the *pulsed ECR regime*.

5.3.2   Pulsed Working Regime

A short-term increase of the current of multiply charged ions can be achieved if the ion storing plasma potential is switched off. Then the stored ions are pulled out from

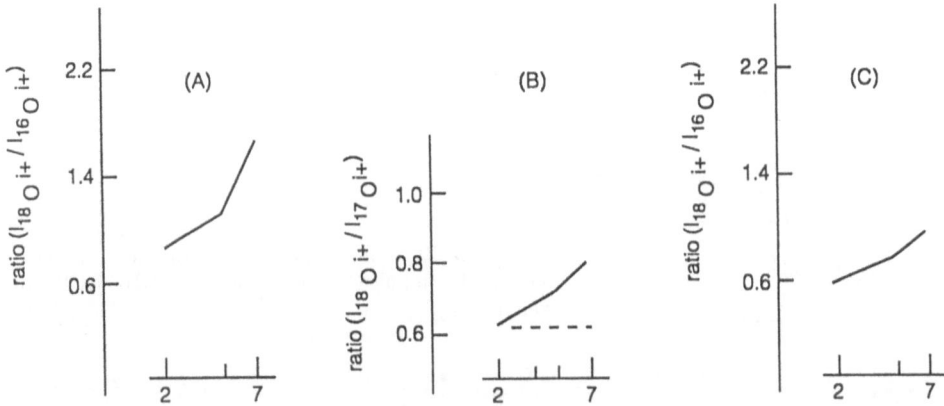

Figure 5.18   Ratios of output currents of Oxygen isotopes (A and B) and for O_2/He mixtures (C) [216].

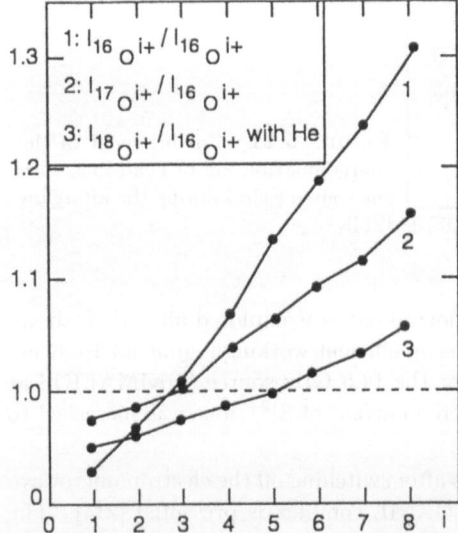

Figure 5.19 Ratios of calculated output currents for Oxygen isotope and Oxygen/Helium mixtures [217]

the source and the charge distribution of the ion current impulse will correspond to the distribution as it realized in the source. The same effect appears at the cut-off of the electrons heated by microwaves. During the cooling down of the electrons the intensity of the collisions between the electrons and ions increases. Thus many electrons arrive in the loss cone and are lost from the source. A reduction of the electron number hurts the charge balance in the plasma and causes a disappearance of the negative potential. Thus the ions are not held longer in the source and leave it. For this process the characteristic time is determined by the rate of the electron losses and is close to the electron life time τ_e in the source (100 μs up to 1 ms).

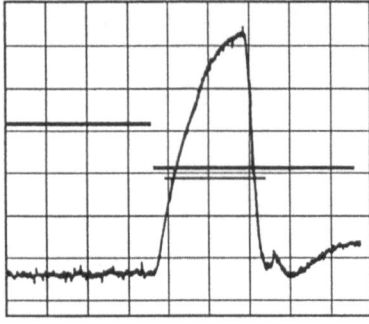

Figure 5.20   Time dependence of the Ar^{14+} ion current in the impulse regime at an 14.5 GHz ECR ion source [214].

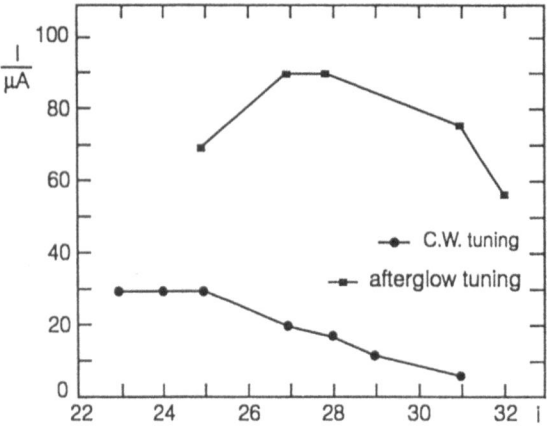

Figure 5.21   Enhancement of the charge distribution of Lead ions with the source ECR4 during the afterglow [214].

The casual observed pulse regime (or afterglow mode) was explored after their discovery inclusively and found further application as an efficient working regime for ECR ion sources at ion accelerators. So, for instance, for the 14.5 GHz source MINIMAFIOS at the accelerator complex SPS (CERN) the injected current of S^{12+} ions was increased to 1 mA [215].

In Fig.**5.20** the current impulse of Ar^{13+} ions after switching off the electron microwave heating in the 14.5 GHz source ECR4 of the GANIL complex is presented [214]. The impulse length totals about 1 ms. In Fig.**5.21** the ion charge state distribution of Lead ions in the 14.5 GHz source in the impulse regime and at continuous work is presented.

Measured results [214] show, that in accordance with the theory the effect of raising ion current is distinctive in the pulsed regime for the highest charged ions. In Fig.**5.22** for Argon ions the charge dependence of the ratioof the ion current in the pulsed regime to the current in the continuous regime is presented [214]. The measurements were accomplished at the 18 GHz source MINIMAFIOS-18 GHz and at the 14.5 GHz source ECR4.

For a better understanding of the physical processes in ECR ion sources a mathematical modelling of the pulsed working regime [204, 205] was undertaken. The solution of the balance equations (5.18) made it possible to determine the time dependence of processes in ECR ion sources.

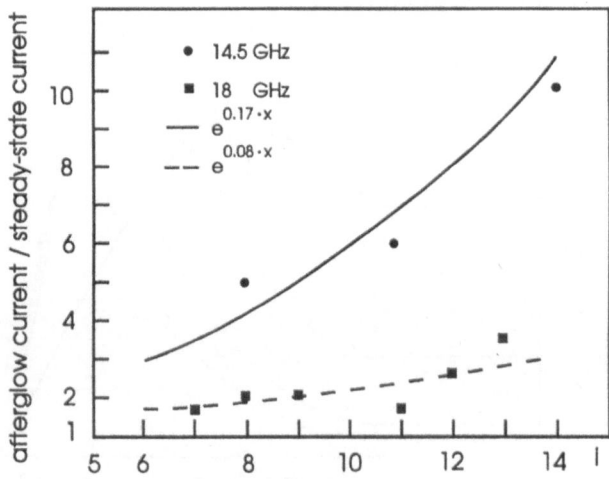

Figure 5.22   Ratio of the afterglow current to the steady-state current for Argon ions in ECR4 (14.5 GHz) and in MINIMAFIOS (16.6 GHz) [214].

In Figs.**5.23** and **5.24** the calculated changes of Argon and Krypton ion currents after switching off of the microwave heating of the electrons are presented. The calculations were accomplished for a source of the MINIMAFIOS type. For the determination of the ion life time equations (5.13), (5.14) and (5.15) were used. It was assumed that in the stationary regime up to the moment of the switching off of the microwave heating the electrons have a temperature $T_e = 5000$ eV and a density $n_e = 2 \cdot 10^{12}$ cm^{-3}.

It shows that the calculated dependences are found in good agreement with measured values, which can be examined as a further reference on the applicability of the model used for the storage and the ion loss from the description of processes occuring in ECRIS.

For the operation of an ECR ion source in the pulsed regime the repetition time is determined for individual impulses by the time in which all source parameters pass over into a stationary state after switching off the microwave for the electron heating. This time depends on the time for the production and storage of ions of the highest charge states in the potential well. Usually this time totals some ten milliseconds.

Beside the switching off the electron microwave heating evidently other processes for the production of pulses of highly charged ions must also exist. For it it is already sufficient, to destroy the negative potential well, which stores the ions. For instance, this can result by additional coils which are localized beside the coils S1 in Fig.**5.1**. If a short current pulse is given on these additional coils with opposed polarity the holding of the electrons magnetic barrier can be eliminated, which leads to the similar effect like the elimination of microwave heating. This idea was realized at the PuMa-ECR [210].

An analogous result can be achieved by the injection of ions in the working area of the source which eliminates also the positive potential barrier necessary to the storage of the ions. In this case the duration of the ion pulse extracted from the source is not determined by the time requiring by the electrons for the exit from the source, but by

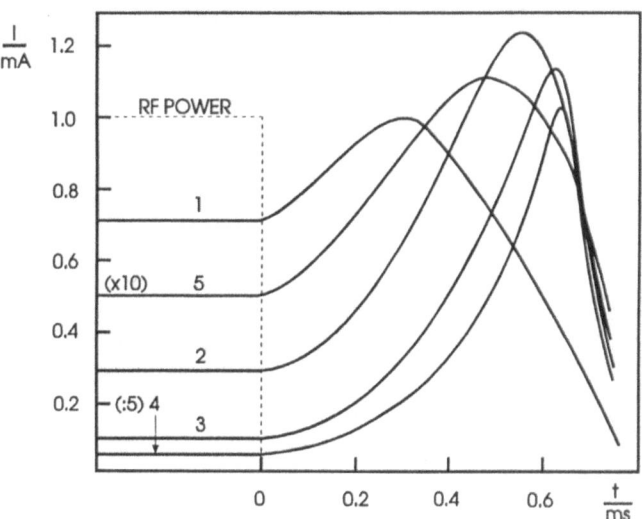

Figure 5.23   Calculated time dependence of the current density of Ar^{5+} (1), Ar^{9+} (2), Ar^{11+} (3) and Ar^{13+} (4) ions and of the total current density (5) of all charges in the pulsed regime of an ECR source [204].

the time of the exit of the ion component, which can total for $U = 0$ V in (5.13) some ten microseconds leading to an increase of the ion current amplitude.

5.3.3   Pulsed Regime with Ion Cooling

Calculations of the charge distributions show, that for ion cooling the differences between the charge distributions within the source and at the ion extraction increase (Figs.**5.16** and **5.17**). It can proceed from the assumption that the largest effect at a pulsed source regime appears by simultaneous ion cooling. For an estimation of the amplitude of this effect simple calculations can be provided.

The electric charge Q of the stored ions is in an ECR source equal to the electron charge

$$Q = \bar{i} N_i = N_e = \bar{i} n_i V = n_e V \tag{5.25}$$

with $\bar{i}$ as mean ion charge state, N_e and N_i as electron and ion numbers and V as plasma volume.

The assumption of equidistributed electron and ion densities does not correspond in full degree to the reality but it is however permissible for qualitative estimations.

In accordance with (5.17) the total density of the ion current from the source totals

$$j_{\text{tot}} = e\,\bar{i}\,n_i\,\frac{V}{2S\tau_i} = \frac{n_e l}{2\tau_e} \tag{5.26}$$

with l as length and S as area of the plasma occupied cylinder.

In the pulsed regime the entire ion charge Q is extracted in a time τ_p from the source, which is equal to the length of the ion current pulse. Thus a mean current density of the

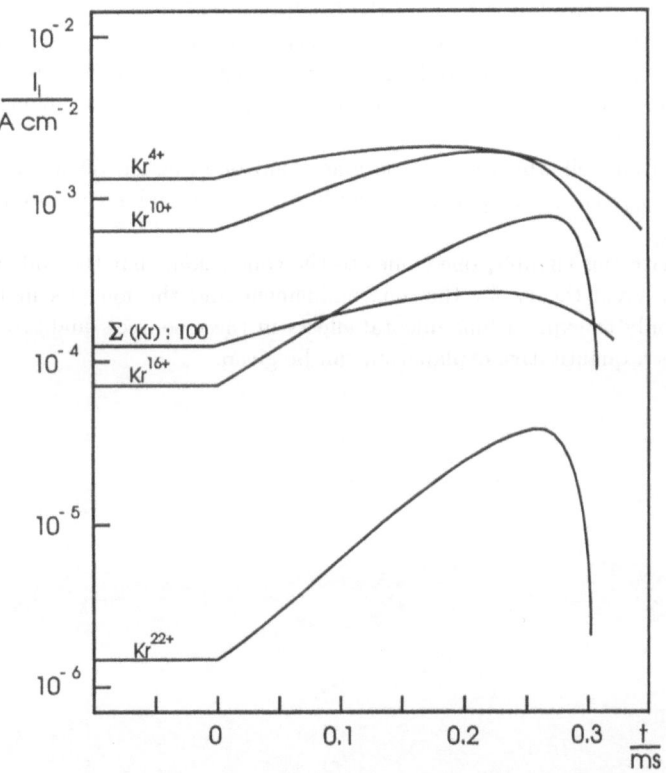

Figure 5.24   Calculated time dependence of the current density of Kr^{4+}, Kr^{10+}, Kr^{16+} and Kr^{22+} ions and of the total current density in the pulsed regime of an ECR source [205].

ion impulse can be indicated

$$j_{tot} = \frac{e\, n_e\, l}{2\tau_p} \tag{5.27}$$

The charge state distribution usually displays opposite to the mean charge state a dispersion of some charge states. For heavy ions the width of the charge distribution totals ten to twenty electron charges. Thus it can approximately be written

$$j_i = \frac{e\, n_e\, l}{40\tau_e}$$

Assuming as an example the MINIMAFIOS source parameters

$$\nu_{RF} = 15 \text{ GHz}; \quad n_e = 2 \cdot 10^{12} \text{ cm}^{-3}; \quad V = 10^3 \text{ cm}^3; \quad S = 40 \text{ cm}^2; \quad l = 25 \text{ cm}$$

and in accordance with calculations $\tau_e = (0.3 \ldots 1.0) \cdot 10^{-3}$ s and $\tau_i = (10^{-4} \ldots 10^{-2})$ s for different charge states counts

$$Q = 2 \cdot 10^{15}; \quad j_{tot} = (5 \ldots 10) \text{ mA cm}^{-2} \quad \text{and} \quad j_i = (0.3 \ldots 1.0) \text{ mA cm}^{-2}$$

These values are found in accordance with numerical values and correspond to optimal results which are received at sources of the MINIMAFIOS type.

Ion cooling shifts the charge state distribution in direction of highly charged ions (Figs.**5.16** and **5.18**). At an pulse length of $\tau_p = (0.3 \dots 0.5)$ ms can be proceeded on the assumption that Kr^{30+} or Pb^{40+} ions can be achieved with a current up to $j_i = (0.1 \dots 0.3)$ mA cm^{-2}.

If the potential well which stores the ions could be switched off for a shorter time, then still shorter pulses of the ion current with an accordingly higher amplitude could be achieved.

To summarize this chapter, one comes to the conclusion, that the sufficiently simple and physically vivid theory for the ion confinement and the ion loss in ECR sources allows it not only to explain fundamental effects in these sources qualitatively but also for many cases a quantitative explanation can be given.

Chapter 6

Plasma Diagnostics Related to Ion Sources

6.1 Introduction

Ion sources are usually plasma sources where each component (ionic as well as electronic) plays an important role. The present chapter deals with the experimental techniques which can be used to get information on the plasma state in electron impact ion sources. Thereby, the most important plasma parameters and correlated diagnostics are summed up in Table **6.1**.

Overviews about relevant plasma diagnostics on ECR ion sources are given by Girard [218] and for EBIT by Schmieder [220]. Plasma effects are secondary in ion traps like EBIS and EBIT because quasineutrality is violated. On the other hand, the description of an ECR plasma requires the use and concepts of statistical physics. The complete

Table 6.1 Plasma parameters and correlated diagnostics [218].

plasma parameters	diagnostics
electron density	interferometry, Langmuir probe, linewidth, line ratios
electron temperature or energy	Langmuir probe, bremsstrahlung, electrostatic analyzer, diamagnetism, spectroscopy
ion density	passive and active spectroscopy, Langmuir probe
ion temperature	Doppler width, electrostatic analyzer
neutral density	visible spectroscopy, laser absorption
neutral temperature	laser absorption
instabilities	microwave emission, end-loss analyzers
ion charge state	X-ray spectroscopy, visible spectroscopy, magnetic analyzers
plasma geometry	pin-hole camera
spatial particle distribution	pin-hole camera

statistical description of a plasma would then be achieved by the knowledge of each ion distribution function $f_i(x, \vec{v}, t)$ and the electron distribution function $f_e(x, \vec{v}, t)$ at any point x and any time t. In practice, only a few moments of these distributions are necessary. The most important are

- the densities (zero-order moments),
- the currents or velocities (first-order moments),
- the pressure or temperature (second-order moments) and
- heat fluxes (third-order moments).

The description of the plasma by the quantities listed above can give valuable information on the plasma state as long as the plasma is not too far from the thermal equilibrium, e.g. the distribution function has to remain close to a Maxwellian.

The *electron plasma density* is directly related to the electron current that can be extracted; a high flux ion source requires a high electron density. The *ion density* n_i can be deduced from the electron density in a single charge state ion source but it requires a detailed knowledge of the relative abundances of each charge state in an ion source of highly charged ions. The *electron temperature* T_e is of crucial importance in high charge state ion sources because of the high ionization energy of high ion charge states. On the other hand, the *ion temperature* T_i generally affects the emittance of the ion source. Moreover, in magnetically confined ion sources higher ion temperatures lead to a worser ion confinement. The plasma pressure is related to the plasma temperature

$$p = \frac{3}{2} \left(n_e k T_e + n_i k T_i \right) \tag{6.1}$$

and it is sometimes easier to measure the temperature, especially in magnetically confined ion source plasmas. Energy fluxes are not so commonly measured; however the walls of the plasma chamber can be directly affected by large energy fluxes which influence the plasma itself.

As we have mentioned above, in ion traps like EBIS and EBIT plasma effects are secondary. Thus in the following we will at first discuss plasma diagnostics in ECR ion sources and will than add some special features for other electron impact ion sources.

6.2   Diagnostics of ECR Ion Sources

6.2.1   Plasma Models

6.2.1.1   Introduction

To describe the fundamental connections in ECR plasmas for diagnostic purposes, models must be chosen, which contain special characteristics of the plasma source considered.

The plasma theory distinguishes between equilibrium and non-equilibrium plasmas, whereby the validity of plasma models is determined by the electron density and the temperature. Equilibrium plasmas are distinguished in *global* and *local thermodynamic equilibrium*, in *collision-radiation balance* and in the *Corona model*. Since different diagnostics come from a suitable plasma model, in the following at first we give a short overview about often used plasma models. A detailed description of the different models however should be explained in the specialized literature.

6.2.1.2   Complete Thermodynamic Balance

A complete thermodynamic balance exists, if each basic process is found in balance with its inverse one, i.e., the plasma is found in a balance between collisions and radiation processes. The model describes completely self-contained systems, which must be optically thick for all wavelengths. Thereby the plasma is completely determined by the thermodynamic quantities temperature and density without any temperature and density gradients. If the radiation field is isotropic, it can be described by the *Planck radiation law* as a spectral energy density $u(\nu, T)$ per volume and solid angle in the frequency interval from ν to $\nu + d\nu$ of a black body emitter

$$u(\nu, T)\, d\nu = \frac{8\pi h \nu^3}{c^3} \frac{1}{\exp\left(h\nu/kT\right) - 1}\, d\nu \tag{6.2}$$

whereby k is the Boltzmann constant, ν the frequency and T the absolute temperature.

The velocity distribution $f(v)$ of all particles is described by a Maxwellian distribution

$$f(v)\, dv = \left(\frac{m}{2\pi kT}\right)^{3/2} 4\pi v^2 \exp\left(-mv^2/2kT\right) dv\ . \tag{6.3}$$

Thereby $f(v)\, dv$ gives the probability, that a particle has a velocity distribution in the interval between v and $v + dv$. The quantity m describes the mass and v the velocity of the plasma particles.

The application of the law of mass action on the balance between ionization and recombination results in the *Saha-Eggert equation* for the occupation distribution of the ground states of adjacent ionization stages

$$\frac{n_{i+1}}{n_i} = \frac{1}{n_e} \frac{2u_{i+1}(T)}{u_i(T)} \frac{2m_e \pi k T^{3/2}}{h^3} \exp\left(-(E_i - \Delta E_i)/kT\right) \tag{6.4}$$

with n_i and n_{i+1} as the densities of i- and (i+1)-fold charged ions. n_e describes the electron density and m_e the electron mass. u_i and u_{i+1} are the sums over all states of the respective ionization stage, which can be calculated by the relation

$$u_i = \sum_{j=1}^{jmax} g_{ij} \exp(E_{ij}/kT)\ . \tag{6.5}$$

E_i is the ionization energy of an i-fold charged ion for the ground state and ΔE_i the decrease of the ionization energy of an i-fold charged ion and describes the interaction of the particles in the plasma. A quantitative estimation for ΔE_i yields [221]

$$\Delta E_i = \frac{(i+1)^{3/2}}{4\pi \varepsilon_0} \frac{3e^2}{r_M} \tag{6.6}$$

with

$$r_M = \sqrt[3]{\frac{3}{4\pi n_e}} \tag{6.7}$$

and ε_0 as electric field constant and r_M as mean distance of the electrons.

The particle densities in each excited level within an ionization stage i are then calculated by the *Boltzmann relation*

$$n_{ij} = n_{i0} \frac{g_{ij}}{g_{i0}} \exp(-E_{ij}/kT) \ . \tag{6.8}$$

Thereby the n_{ij} and n_{i0} are the particle densities in the excited and in the ground state of ions with the ionization stage i, the g_{ij} and g_{i0} are the statistical weights of the excited state j and of the ground state of the ionization stage i and the E_{ij} the excitation energies of the state j of the ionization stage i from the ground state.

In a laboratory plasma a complete thermodynamic equilibrium never exists, since radiation escapes from the plasma and temperature and density gradients exists. Therefore, the Planck radiation law is not longer valid with it. Thus a thermodynamic equilibrium in a plasma can exist only *locally*.

6.2.1.3   Local Thermodynamic Equilibrium

Local thermodynamic equilibrium refers to a plasma model for the description of optically thin plasmas. In a collision dominated plasma the electron impact ionization is in balance with the threefold collision recombination and the electron impact excitation is in balance with the electron impact deexcitation. The plasma is optically thin, since absorption processes can not appear in the plasma interior and radiation can escape. Thus it leads to a change of the spectral distribution and the radiation is no longer described by the Planck radiation law. Temperature and density gradients appear. As a condition for the existance of a collision dominated plasma, which can be described through a local thermodynamic equilibrium, a sufficiently high electron density is necessary, since electron collisions dominate in comparison to radiation processes. Using the electron density a necessary condition for the existence of a local thermodynamic equilibrium was approximated by McWhirter [222]

$$n_e \geq 1.7 \cdot 10^{14} \sqrt{T_e} \, (\Delta E_{ij})^3 \quad \mathrm{cm}^{-3} \ . \tag{6.9}$$

In this estimation the electron temperature T_e is inserted in eV. ΔE_{ij} is the excitation energy for the level j starting from level i. The inequality (6.9) must be satisfied for the largest energy difference of single states in the term scheme of the ion. For hydrogen-like ions this is the 1s $\rightarrow$ 2p transition, whose excitation potential ΔE_{ij} with $i = 1$ and $j = 2$ can be estimated by

$$\Delta E_{ij} = E_H \left(\frac{1}{i^2} - \frac{1}{j^2} \right) \ . \tag{6.10}$$

E_H describes here the ionization energy of the Hydrogen atom. For $i{=}1$ and $j{=}2$ results

$$\Delta E_{12} = \frac{3}{4} \, 13.6 \, Z^2 \, \mathrm{eV} = 10.2 \, Z^2 \, \mathrm{eV} \ . \tag{6.11}$$

Thus for hydrogen-like ions it follows

$$n_e \geq 10^{17} \sqrt{T_e} \, Z^6 \quad \mathrm{cm}^{-3} \tag{6.12}$$

with the electron energy T_e in eV.

Furthermore, the velocity distribution $f(v)$ of the plasma particles is described by a Maxwellian distribution. The particle densities of the individual ion components are described by the Saha-Eggert equation and the occupation of excited ionic states by the Boltzmann relation.

6.2.1.4 Impact-Radiation Equilibrium

The basic assumptions for the impact-radiation equilibrium described by McWhirter [222] are:

- free electrons have a Maxwellian velocity distribution;
- electron impact ionization is balanced only partially with the threefold impact recombination;
- electron collisions lead to transitions between two arbitrary bound states;
- there are only two radiation processes, whereby radiation is emitted by spontaneous emission between bound states or by impact recombination as collision-free transitions of free electrons in bound states;
- the plasma is optically thin for emitted resonance lines.

In the impact-radiation equilibrium each occupied level of an ion component is described by a differential equation, which contains an algebraic sum of the rates of those processes, which populate the level. There are likewise as many differential equations as energy levels. With increasing principal quantum number the distances between two atomic states decreases, so that the probability of electron collision processes increases while simultaneously the probability for radiative processes decreases. Thus a critical energy level exists, for which the radiative transitions are neglected between bound states [222]. Since the solution of the differential equation system describing the level occupation of different ion components is complicated, the number of equations can be reduced by the consideration of the impact boundary, which is described by a critical energy level q. For larger quantum numbers than q the Saha-Eggert equation can be used to calculate the occupation of ion components of adjacent charge states, because impact processes dominate. The ion ground state changes, and a redistribution in the occupation of the excited levels follows. With it the number of equations of the differential equation system can be reduced for quantum numbers lying under the impact boundary to one differential equation, which describes the ground state occupation and for quantum numbers between the ground state and the impact boundary as a system of ordinary equations due to quasistationary conditions.

6.2.1.5 Corona Model

Laboratory plasmas of small density can be described with the Corona model, which was first used to interpret spectra of the solar corona. The corona model is based on the assumption, that the rate for spontaneous emission is for all levels larger than the collision rate, that the plasma is optically thin and that for the free electrons a Maxwellian velocity distribution exists [223].

The corona model is derived from a balance of electron impact ionization and radiative recombination as well as from a balance of electron impact excitation and spontaneous emission. In relation to the ion ground state only a small number of ions is found in excited states. Then the Saha-Eggert equation (6.4) can not be used for the calculation of the population distribution of the ground states of adjacent ion charge states. The population of each ionization stage is described by a *rate equation*, where the calculation of the population and depopulation of adjacent levels results accordingly to the situation

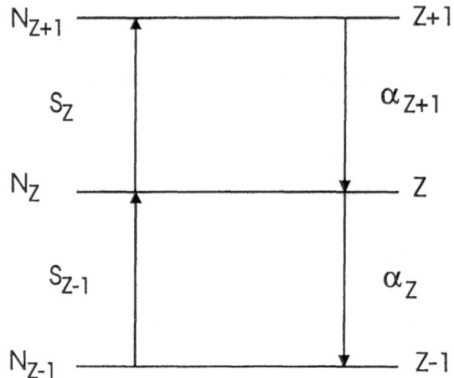

Figure 6.1   Level population of adjacent ionization stages.

presented in Fig.**6.1**

$$\frac{dN_i}{dt} = n_e \left(N_{i-1}S_{i-1} + N_{i+1}\alpha_{i+1} - N_iS_i - N_i\alpha_i\right) \tag{6.13}$$

where the S_i and S_{i-1} represent rate coefficients for the ionization from stage i to $i+1$ and from $i-1$ to i respectively. The quantities α_i and α_{i+1} describe the rate coefficients for recombination processes from i to $i-1$ and from $i+1$ to i. The population numbers for the corresponding ionization stages are indicated by N_{i-1}, N_i and N_{i+1} and n_e is the electron density. The rate coefficients depend on the electron temperature. Since the threefold impact recombination at small electron densities is negligible, the rate coefficient for recombination processes is a sum of contributions from radiative recombination and dielectronic recombination with following radiative stabilization.

For two adjacent ionization stages for a stable state the *corona ionization equilibrium* results in

$$\frac{N_{i+1}}{N_i} = \frac{N_i}{\alpha_{i+1}} \ . \tag{6.14}$$

The population densities of individual energy levels of an ionization stage can not be determined by the Boltzmann relation. Level population can result like indicated in Fig.**6.2**.

A temporal change in the population density of the energy level i results from the balance of electron impact excitation and cascades from higher levels as well as by the level depopulation by spontaneous emission. Thus, for the level i the following differential equation results

$$\frac{dn_i}{dt} = n_0 n_e < \sigma_{ex}v_e > + \sum_{p>i} A_{ip}n_p - n_i \sum_{q<i} A_{qi} \tag{6.15}$$

The quantities n_0 and n_i describe the occupation densities of the ground state and of the state i. The quantities A_{ip} and A_{qi} are transition probabilities for the spontaneous emission from p to i (p > i) and from i to q (q < i). With $< \sigma_{ex}v_e >$ the electron impact excitation coefficient for the state i is described. Neglecting cascades for the stationary case the population of the state i follows from the solution of (6.15)

$$n_i = n_e n_0 \frac{< \sigma_{ex}v_e >_i}{\sum\limits_{q<i} A_{qi}} \ . \tag{6.16}$$

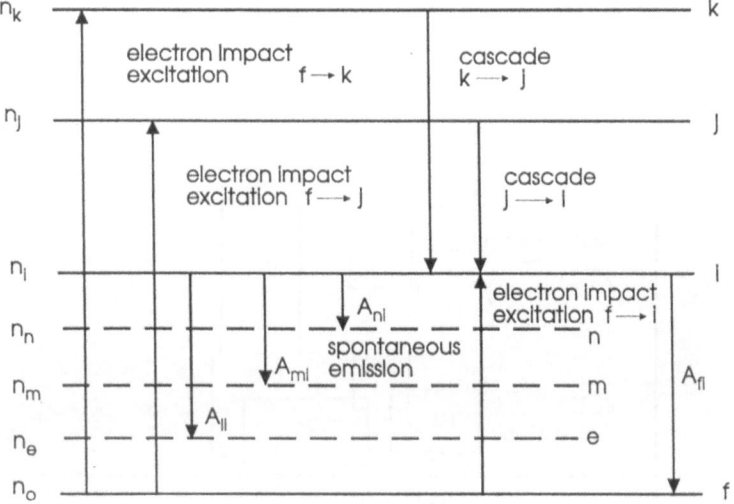

Figure 6.2   Population diagram for the calculation of the population densities of a selected ionization stage in the corona model.

The validity of the corona model is determined by the condition, that for the depopulation of the excited states the electron impact deexcitation can be neglected in comparison to the spontaneous emission. With this a critical value for the electron density of hydrogen-like ions with the nuclear charge Z can be determined [222]

$$n_e \leq 6 \cdot 10^{10} Z^6 \sqrt{T_e} \exp(0.1 \, Z^2/T_e) \tag{6.17}$$

where T_e describes the electron temperature in eV and Z the nuclear charge of the hydrogen-like ion. From the estimation of McWhirter [222] a principal quantum number n = 6 as a collision boundary was derived. The corona model works for plasmas with small electron densities. Thereby, ECR ion sources are in the $N_e T_e$ plasma map located as shown in Fig.**6.3**.

A critical value for the electron density results in

$$n_e \leq 1.7 \cdot 10^{14} \sqrt{T_e} \, \Delta E_{ij}^3 \quad cm^{-3} \, . \tag{6.18}$$

If for the excitation energy ΔE_{ij} the impact boundary from j to i is determined to j = 6, then it follows

$$\Delta E_{56} = 13.6 \text{ eV } Z^2 \left(\frac{1}{i^2} - \frac{1}{j^2} \right) = 13.6\text{eV } Z^2 \left(\frac{1}{5^2} - \frac{1}{6^2} \right) \tag{6.19}$$

and

$$\Delta E_{56} = 0.166 \text{ eV } Z^2 \, . \tag{6.20}$$

For hydrogen-like ions the critical value of the electron density for the applicability of the corona model is

$$n_e \leq 7.8 \cdot 10^{11} \sqrt{T_e} \, Z^6 \quad cm^{-3} \, . \tag{6.21}$$

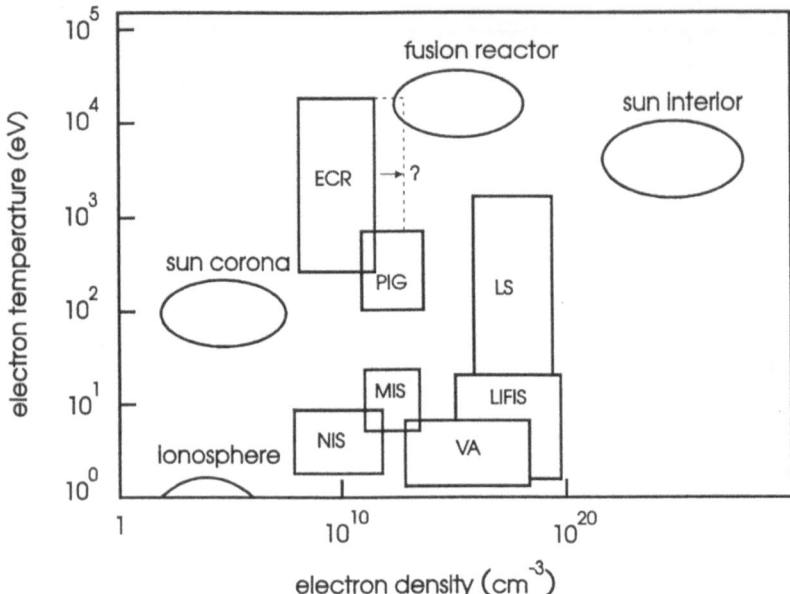

Figure 6.3 $N_e T_e$ plasma map related to ion sources. NIS – negative ion sources, MIS – microwave ion sources, LIFIS – light ion fusion sources, LS – laser sources, VA – vacuum arcs [218].

The value of the critical electron density is already determined from hydrogen-like Helium for electron temperatures $T_e > 10$ eV larger than the estimated electron density in the ECR plasma, so that this model would be applicable, if the assumption of a Maxwellian electron distribution would be not presupposed . Thus, this plasma model is only conditionally applicable to the description of ECR plasmas.

6.2.2 Magnetic and Electrical Diagnostics

Among other diagnostics, magnetic and electrical diagnostics are probably the most popular ones. They are relatively easy to operate, although their results require a careful interpretation. In the following we discuss different diagnostic methods, following Ref.[218].

6.2.2.1 Diamagnetic Loop

Diamagnetic loops can be used in any magnetically confined axisymmetrical plasma. The measurement principle, as shown in Fig.**6.4**, is based upon MHD equilibrium of the plasma: the pressure gradient is balanced by a Lorentz force due to the diamagnetic current $\vec{j}$ which induced a magnetic field opposite of the main confinement magnetic field

$$\nabla \vec{F} = \vec{j} \times \vec{B} \ . \tag{6.22}$$

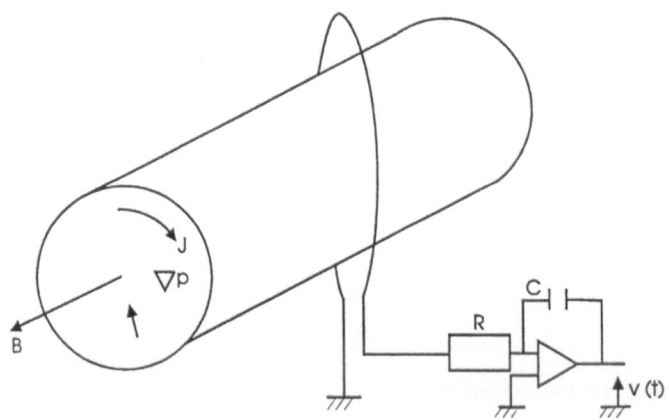

Figure 6.4 Operation principle of the measurement of the diamagnetic effect in a axisymmetrical plasma.

With a loop of N turns surrounding the plasma any variation of the magnetic flux can be measured. If the plasma is created the diamagnetic current increases and the flux decreases. This variation induces a voltage $V(t)$ which can be time integrated

$$\int V(t)\, dt = -N\Delta\Phi \approx \frac{1}{2} N \pi r_0^2 \frac{\sum nkT}{B/2\mu_0} \tag{6.23}$$

where the sum is performed over the ions and electrons, r_0 is the radius of the plasma and B the magnetic field. Therefore, the voltage is proportional to the plasma pressure. Such measurements which are commonly be used in magnetic fusion experiments have also been successfully employed in ECR ion sources [224].

For instance, in Ref.[224] the electron cyclotron emission (ECE) and the diamagnetism signals are measured in two different ways in a transient regime, which consists in a pulsed RF power at a given constant level, but variable from pulse to pulse:

1. When running in a steady state regime, the RF power is turned off for about 200 ms and then turned on again.
2. The RF power is turned alternatively on and off every 200 ms.

In these transient regimes, the diamagnetism and the total ECE intensity signals are simultaneously recorded.

A scheme of the location of these diagnostics is given in Fig.**6.5**. Both signals shown in Fig.**6.6** behave approximately the same time behaviour. Thereby, the major difference is the high instability level superimposed on the ECE signal. The diamagnetism signal is obtained after integration, so that these rapid instability spikes do not appear.

In Fig.**6.7** the rise time at RF turn-on (τ_{on}) and the decay time at the RF turn-off (τ_{off}) of both the signals for varying the RF power is shown. Because of the instability noise of the ECE signal, only the decay time of the diamagnetism is deduced. The rise time is always smaller than the decay time.

In the transient experiments described there are only small differences between the ECE total intensity and the diamagnetism signals. The diamagnetism signal is proportional to $n_e E_e$, while this is not fully true for the ECE signal, due to the better response at higher frequencies of the InSb detector used and the filtering of the first harmonic.

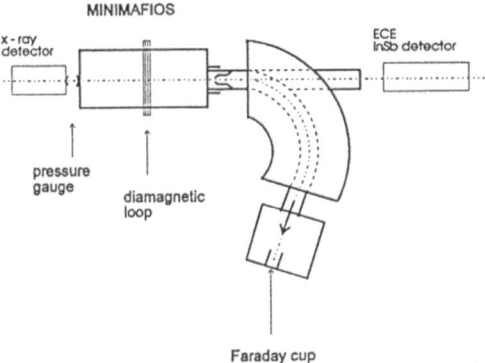

Figure 6.5   Scheme of a typical location of diagnostics components.

Since the heating of electrons is much shorter than the characteristic time of the signal $n_e E_e$ [225], the signal may be described by

$$\frac{d(n_e E_e)}{dt} \cong E_e \frac{dn_e}{dt} = E_e \left(-\frac{n_e}{\tau_e} + S \right) \qquad (6.24)$$

where τ_e is the electron life-time and S a particle source term. According to this crude model, the electron life-time during the RF τ_{on} is obtained by

$$\tau_{on} = \frac{(n_e E_e)_{t=\infty}}{\left(\dfrac{d(n_e E_e)}{dt} \right)_{t=0}} \qquad (6.25)$$

After the RF turn-off, assuming the energy has not yet changed, the model may again be used with $S = 0$ and τ_{off} is approximately calculated as close as possible to the RF turn-off. The experimental results show that τ_{off} is greater than τ_{on}. Several reasons may explain this behavior:

1. The source term could changed during the buildup of the discharge, therefore the model is no longer valid.
2. The confinement time could be dependent on whether RF is on or not [225].

In [224] from the calibration of the diamagnetic loop the electron density was calculated since the electron energy was measured: $n_e \approx 10^{12}$ cm^{-3} for 1 kW injected RF power. However, the diamagnetic signal may be enhanced by important electron drift currents caused by the complicated magnetic structure. Thus a measurement of the electron density by using only the diamagnetism may not be quite correct.

Furthermore, it should be noted, that when the pressure is anisotropic (which is the case for ECR sources) only the perpendicular pressure with respect to the magnetic field must be considered. Information about the plasma energy confinement can be also obtained [208].

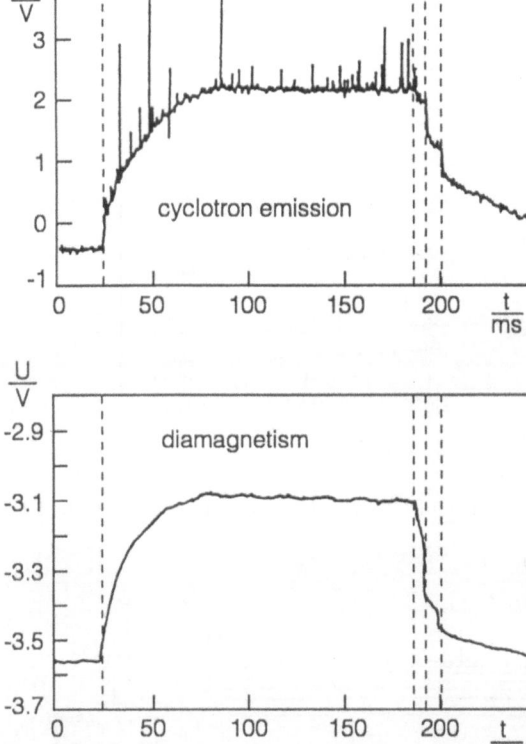

Figure 6.6 ECE and diamagnetism signals versus time [224].

6.2.2.2 Electrical Flux Analysis

6.2.2.2.1 Langmuir Probe Among the diagnostics which measure charged particle fluxes (i.e., first order moments) *Langmuir probes* are the most commonly used [226, 227]. The construction of a cylindrical Langmuir probe is shown in Fig.**6.8** and the operation scheme in Fig.**6.9**.

The working principle is, that a conducting wire of area A, whose potential V (measured with respect to the plasma potential) can be controlled is immersed in a plasma. When sweeping its potential and measuring the current I through the probe a *probe characteristic* as shown in Fig.**6.10** is obtained, which can be interpreted as follows:

Region 1: When the voltage is made very negative all electrons are repelled and the ion current is [228]

$$I_i = 0.6\, n_i A e \sqrt{\frac{kT_e}{m_i}} \tag{6.26}$$

with m_i as the ion mass (the formula holds for a singly charged ion). Is V made less negative a few higher energetic electrons are collected and this electron current partially cancels the ion current. The voltage for equal electron and ion currents is calling the *floating potential* V_f. On the probe electrode the positive ions are collected, so that

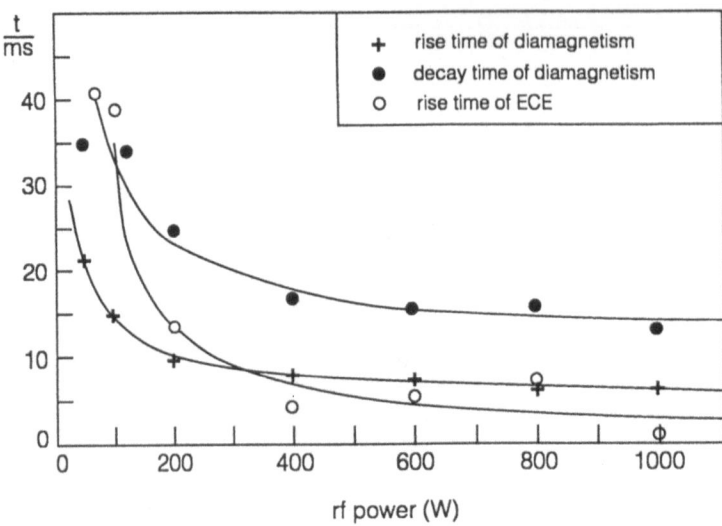

Figure 6.7   Rise time τ_{on} and decay time τ_{off} from ECE and diamagnetism signals versus RF power [224].

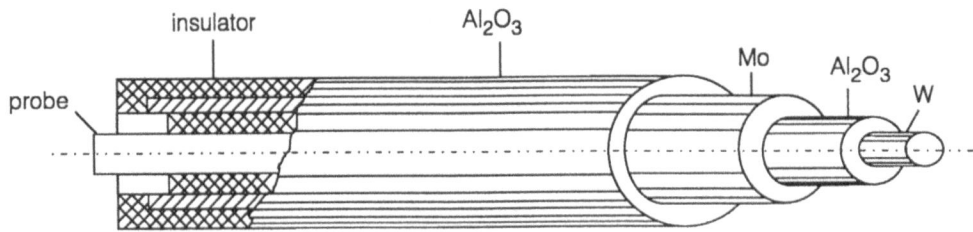

Figure 6.8   Construction of a cylindrical Langmuir probe.

around the probe a positive spatial charge is created. This charge screens the probe, which is on negative potential, from all other particles in the plasma. Just through this spatial charge layer, which surrounds the probe, the application of a Langmuir probe for plasma diagnostics becomes possible because the disturbance created is small and localized. The spatial charge boundary layer joins an undisturbed plasma.

Since the probe does not influence the plasma because of the spatial charge layer on the probe surface only an ion diffusion current (6.26) flows. This situation is presented in Fig.**6.11**.

In the region 1 the ion current is constant over a wide voltage range, since the diffusion current is determined completely by the plasma parameters (density n_i and ion velocity v_i) and it does not act up to the probe potential. A change of the potential causes an influence on the spatial charge layer. If the negative potential increases the expansion of the spatial charge layer also increases, but the current density is further determined by diffusion processes. For the existence of a spatial charge in the layer and under the

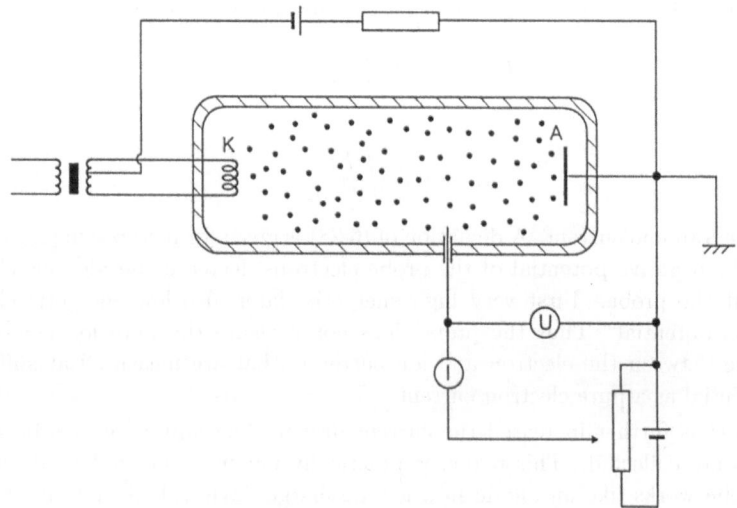

Figure 6.9   Operation scheme of a Langmuir probe.

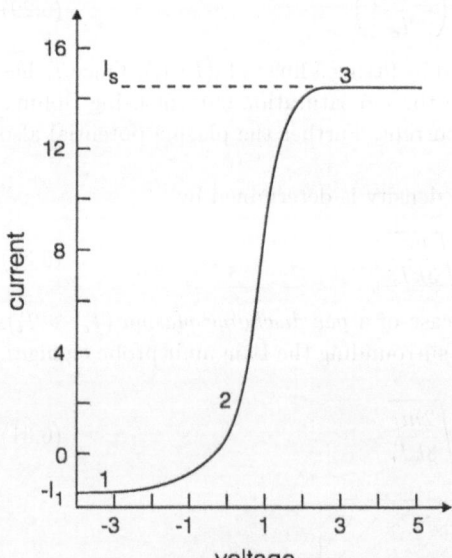

Figure 6.10   Langmuir probe U-I characteristic.

action of electrical forces the ion current density j_i totals

$$j_i = \frac{1}{9\pi} \sqrt{\frac{2e}{m_i}} \frac{\Delta U^{3/2}}{d^2} \tag{6.27}$$

with m_i as ion mass, ΔU as potential difference in the layer and d as layer thickness.

Region 2: When the voltage keeps increasing the current varies according to

$$I = I_e \exp\left(\frac{eU}{kT_e}\right) - I_i \tag{6.28}$$

with

$$I_e = neA \sqrt{\frac{kT_e}{2\pi m_e}}$$

as the electron random current. A deviation of (6.28) is given for instance in [227]. In this manner as the negative potential of the probe electrons decreases beside ions electrons also arrive at the probe. First very high energetic, later also low energetic electrons penetrate the potential. Thus the probe does not measure the pure ion current, but the difference between the electron and ion currents, that are measured at sufficiently positive potential as a pure electron current.

Region3: As U is further increased the current drawn eventually saturates because all the electrons are collected. This region is practically not used for analytical purposes. Here the probe works like an anode in a gas discharge device: high, purely electronic currents arrive at the probe, causing intense secondary processes on the probe surface.

Region 2 gives the electron temperature

$$kT_e = -eU \ln\left(\frac{I + I_i}{I_e}\right) \tag{6.29}$$

so that the electron temperature can be obtained by fitting a line to $\ln(I+I_i)$. Once T_e has been calculated n_e can be deduced either from the ion saturation current using Bohm's formula [228] or from the electron saturation current. Further the plasma potential also can be derived.

In the case of an *isothermic plasma* the ion density is determined by

$$n_i = \frac{4I_i A}{e} \sqrt{\frac{m_i}{3kT_i}} . \tag{6.30}$$

The situation becomes complicated in the case of a *gas discharge plasma* $(T_e \gg T_i)$, because in this case the structure of the layers surrounding the Langmuir probe changes. In this case yields

$$n_i = \frac{4I_i A}{e} \sqrt{\frac{2m_i}{3kT_i}} . \tag{6.31}$$

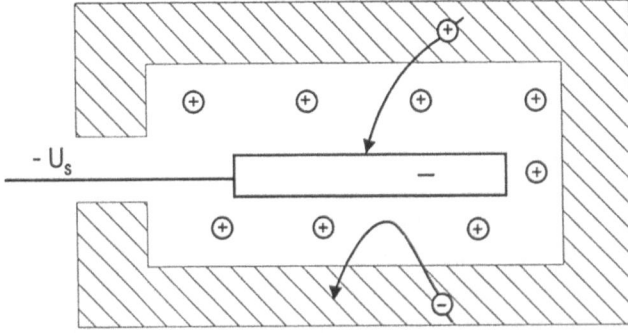

Figure 6.11 The charge distribution in the plasma at negative probe potential.

The determination of the quantity A in (6.29) is met with difficulties, since quantitative criteria for the determination of the effective area are connected with considerable uncertainties.

Accurate measurements are possible only under the following conditions:

1. no magnetic field;
2. the probe dimensions are small compared to the electron and ion mean free paths;
3. the electron temperature is not too high;
4. the Debye length is much smaller than the dimensions of the probe.

When a magnetic field is present so that the electron cyclotron radius becomes comparable to probe dimensions drastic modifications appear. However, the development above is still valid when the surface of the probe is perpendicular to the magnetic field lines. A lot of studies have been carried out using Langmuir probes in ion sources [229, 230, 231, 156]. The approximations mentioned above are usually valid so that measurements of density and temperature are possible.

However, some questions arise when using Langmuir probes in ECR sources, where energetic electrons are present, as well as an anisotropic electron distribution function.

Langmuir probe measurements can also be carried out to study the electron distribution function [233]. The technique can be improved by using double probes [234, 235, 71]. However, electrical probes immersed in plasmas always perturb the discharge and the probe can be destroyed due to heavy heat fluxes.

6.2.2.2.2   Double Retarding Field analyzers

If the plasma is confined in a magnetic bottle by superimposing axial and radial magnetic fields and heated by microwaves then electrons have a two-dimensional anisotropic distribution function. Measurements with double retarding field probes can be evaluated without special assumptions about the electron distribution function. The basic ideas of this method were worked out by Mausbach and Wiesemann [237].

Double retarding field analyzers are inserted in ECR plasmas as probes for the determination of two-dimensional anisotropic electron distribution functions analyzing electrons escaping from the loss cone of the magnetic mirror. Thereby the probe is positioned outside the plasma in the loss cone of the magnetic mirror field of the ECR plasma. The scheme of a double retarding field analyzer including the variations of the potential and of the magnetic field is shown in Fig.**6.12**.

The probe has cylindrical symmetric geometry in reference to the z-axis and is located in a weak, axial symmetric magnetic field. The value of the magnetic induction $B(z)$ decreases with growing distance from the plasma.

The plasma electrons from the loss cone of the magnetic mirror field with an anisotropic velocity distribution function $f(v_{\|0}, v_{\perp 0})$ arrive at z_0 the entrance aperture of the retarding field analyzer. In the interiors of the cylinder segments of the probe the trajectories of the magnetic field are assumed approximately as parallel to the z-axis. The plasma electrons penetrating at z_0 in the retarding field analyzer with the velocity distribution function $(v_{\|0}, v_{\perp 0})$ are collected by a Faraday cup at z_3. The parallel and vertical velocity components of the electrons change for energy conservation reasons because of the movement of the electrons in the interior of the probe and the invariance of the magnetic momentum of the electrons.

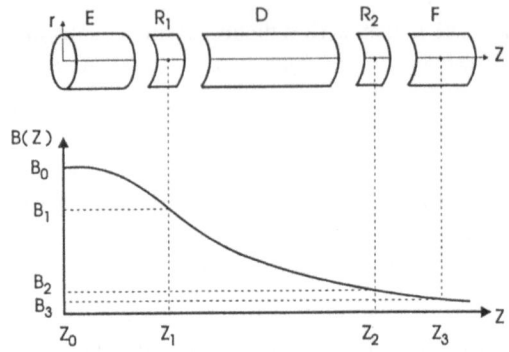

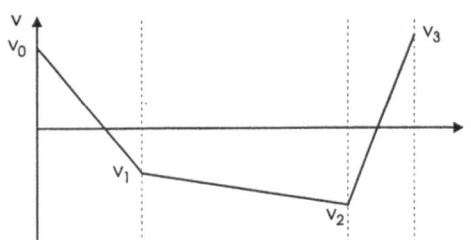

Figure 6.12 Scheme of a double re-tarding field analyzer including the course of the magnetic field and of the potential [238]. E – entrance aperture; R$_1$, R$_2$ – first and second grid analyzer; D – drift tube; F – Faraday cup.

From the energy conservation law

$$E_{tot} = E_\perp(z) + E_\parallel(z) = E_{\perp 0} + E_{\parallel 0} \tag{6.32}$$

and from the invariance of the magnetic moment

$$\mu = \frac{E_\perp(z)}{B(z)} = \frac{E_{\perp 0}}{B_0} \tag{6.33}$$

with $E_{\perp 0} = E_\perp(z=0)$, $E_{\parallel 0} = E_\parallel(z=0)$ and $B_0 = B(z=0)$ follows

$$E_\parallel(z) = E_{\parallel 0} + E_{\perp 0}\left(1 - \frac{B(z)}{B_0}\right) . \tag{6.34}$$

Since the magnetic induction $B(z)$ decreases with growing distance z from the plasma, the parallel component of the electron velocity simultaneously increases, so that the electrons can reach the Faraday cup at z_3, if the following conditions are fulfilled

$$\infty > E_\parallel(z_1) \geq eV_1 \tag{6.35}$$

$$\infty > E_\parallel(z_2) \geq eV_2 \tag{6.36}$$

with V_1 and V_2 as retarding voltages of the retarding field analyzers 1 and 2.

The electron current in a double retarding field probe is given by

$$I = -2\pi eA \int\int f(v_{\parallel 0}, v_{\perp 0})\, v_{\parallel 0}\, v_{\perp 0}\, dv_{\parallel 0}\, dv_{\perp 0} . \tag{6.37}$$

Equation (6.37) describes the current in the entrance aperture into the analyzer at z_0. For the calculation of the electron distribution function $f(v_{\|0}, v_{\perp 0})$ a transformation instruction for the current registered at the Faraday cup dependant on the retarding voltages of the electron distribution function $f(v_{\|0}, v_{\perp 0})$ is necessary. Using the result (6.34) relationships for the parallel and the vertical velocity components at z_0 are

$$v_{\|0} = \sqrt{\frac{B_0 - B_2}{B_1 - B_2} v_{\|0}^2 - \frac{B_0 - B_1}{B_1 - B_2} v_{\perp 0}^2} \tag{6.38}$$

$$v_{\perp 0} = \sqrt{\frac{B_0}{B_1 - B_2} \left(v_{\|0}^2 - v_{\perp 0}^2 \right)} . \tag{6.39}$$

With (6.34) the parallel and the vertical velocity components are obtained at z_0 through the parallel velocities and through the values of the magnetic induction at the geometric places of the retarding field analyzers. By a coordinate transformation the electron current can be transformed in a function of the parallel velocity components at z_1 and z_2

$$I = -2\pi e A \frac{B_0}{B_1 - B_2} \int_{u_{\|1}}^{\infty} \int_{u_{\|2}}^{\infty} f(v_{\|1}, v_{\|2}) \, v_{\|1} \, v_{\|2} \, dv_{\|1} \, dv_{\|2} \tag{6.40}$$

where the lower integration boundaries $u_{\|1}$ and $u_{\|2}$ result from (6.35) and (6.36)

$$\infty > v_{\|1} \geq u_{\|1} = \sqrt{\frac{2}{m_e} eV_1} \tag{6.41}$$

$$\infty > v_{\|2} \geq u_{\|2} = \sqrt{\frac{2}{m_e} eV_2} \tag{6.42}$$

If the electron current from (6.40) is differentiated after V_1 and V_2, this results in

$$\frac{d^2 I}{dV_1 \, dV_2} = \frac{e^3}{m_e^2} 2\pi A \frac{B_0}{B_1 - B_2} f(u_{\|1}, u_{\|2}) . \tag{6.43}$$

Using (6.38) and (6.39) the electron distribution function $f(U_{\|1}, u_{\|2})$ can be written as a function of the parallel and the vertical electron energy $E_{\|0}$ and $E_{\perp 0}$

$$E_{\|0}(V_1, V_2) = e \frac{B_0 - B_2}{B_1 - B_2} V_1 - e \frac{B_0 - B_1}{B_1 - B_2} V_2 \tag{6.44}$$

$$E_{\perp 0}(V_1, V_2) = e \frac{B_0}{B_1 - B_2} (V_1 - V_2) . \tag{6.45}$$

For the electron distribution function yields

$$f(E_{\|0}, E_{\perp 0}) = \frac{1}{2\pi A} \frac{m_e^2}{e^3} \frac{B_1 - B_2}{B_0} \frac{d^2 I}{dV_1 \, dV_2} . \tag{6.46}$$

For a double retarding field analyzator at fixed retarding voltages V_1 and V_2 all electrons contribute to the current which can overcome with their kinetic energy the potential thresholds. Since the energies $E_{\|0}$ and $E_{\perp 0}$ are zero at the minimum voltage, for which the electrons reach the Faraday cup (see (6.44) and (6.45)), this can be used for the calculation of the electron distribution. Threfore, the limit of the analyzed voltage region is given by

$$V_1 \leq V_2 \leq \frac{B_0 - B_2}{B_0 - B_1} V_1 \ . \tag{6.47}$$

Equation (6.47) defines the area of validity of (6.46) in the V_1 – V_2 plane. The interpretable area is shown in Fig.**6.13**. The area is shaded between the limiting lines 2 and 3 corresponding to $E_{\parallel 0} = 0$ and $E_{\perp 0} = 0$. A current-voltage characteristic for a constant potential V_1 is shown in Fig.**6.14**.

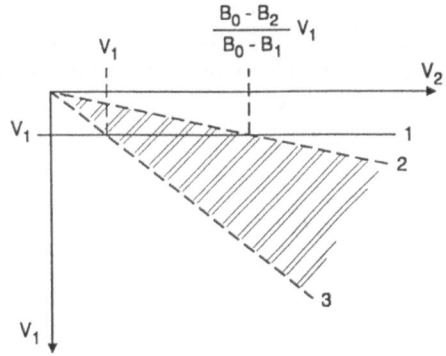

Figure 6.13 The interpretable area of the current-voltage array [231]. Curve 1: course for a current-voltage characteristic; curve 2: boundary for $V_2 \leq (B_0 - B_2) V_1 / (B_0 - B_1)$; curve 3: boundary for $V_2 \geq V_1$.

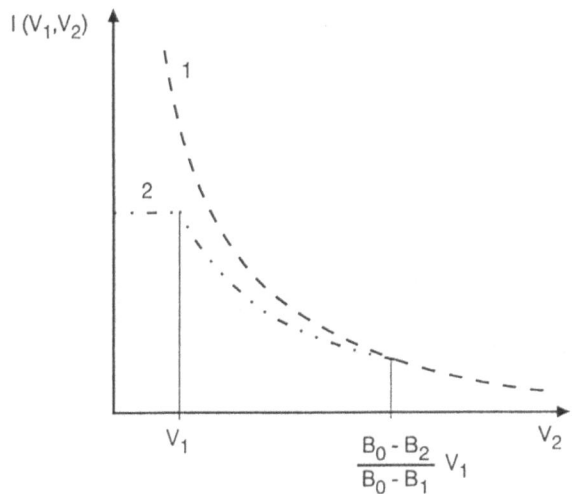

Figure 6.14 The course of the current-voltage characteristics for a constant potential V_1 [231]. Curve 1: $I(0, V_2)$; curve 2: $I(V_1, V_2)$.

For the construction of a retarding field analyzer grids as well as the electrode structure shown in Fig.**6.12** can be used.

Fig.**6.15** shows a typical experimental set-up. In the experiment the electron current is measured at the Faraday cup as a function of a variable and of a constant retarding voltage. A continuous change of the voltage makes it possible to measure a two-dimensional

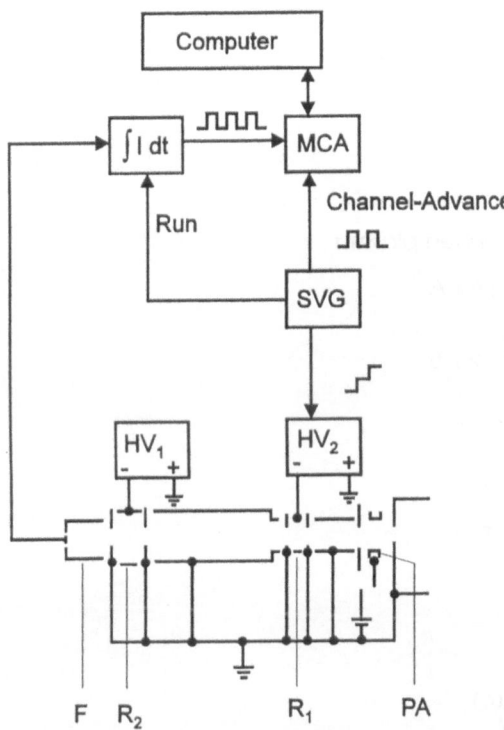

Figure 6.15 Experimental measuring set-up [231]. F: Faraday cup; R_i: electrode for the ith retarding field; PA: pre-aperture; HV_i: ith voltage supply; SVG: step voltage generator; $\int I\,dt$: current integrator; MCA: multichannel analyzer.

family of curves $I = I(V_1, V_2)$. The two-dimensional electron distribution function then results from the differentiation of the current after both retarding voltages. The influence of the construction of the electrostatic lenses on the potential distribution of a cylindrical analyzer is indicated by Mausbach and Wiesemann [231] and can be given by the following equation

$$V_i(r = 0, z_i) = c_i U_i \tag{6.48}$$

for $i = 1, 2$ with the instrument constant c_i.

Because of (6.48) with $dV_i = c_i\, dU_i$ for the velocity distribution function from (6.46) follows

$$f(E_{\parallel 0}, E_{\perp 0}) = -\frac{1}{2\pi A}\,\frac{m_e^2}{e^3}\,\frac{1}{c_1 c_2}\,\frac{B_1 - B_2}{B_0}\,\frac{d^2 I}{dU_1\, dU_2}\,. \tag{6.49}$$

6.2.2.3   Ion Temperatures from Ion Profiles

The ion temperature can be studied by the measurement of the width of ion beam profiles [240, 208]. As an example Fig.**6.16** shows the evolution of ion profiles for different source conditions.

Generally, the energy distribution of extracted ions and its dependence on different plasma and beam parameters can be measured by using a double hemispherical energy analyzer [242]. Energy spectra were measured by scanning the acceleration voltage in front of the analyzer. In Fig.**6.17** the counting rate as a function of the acceleration

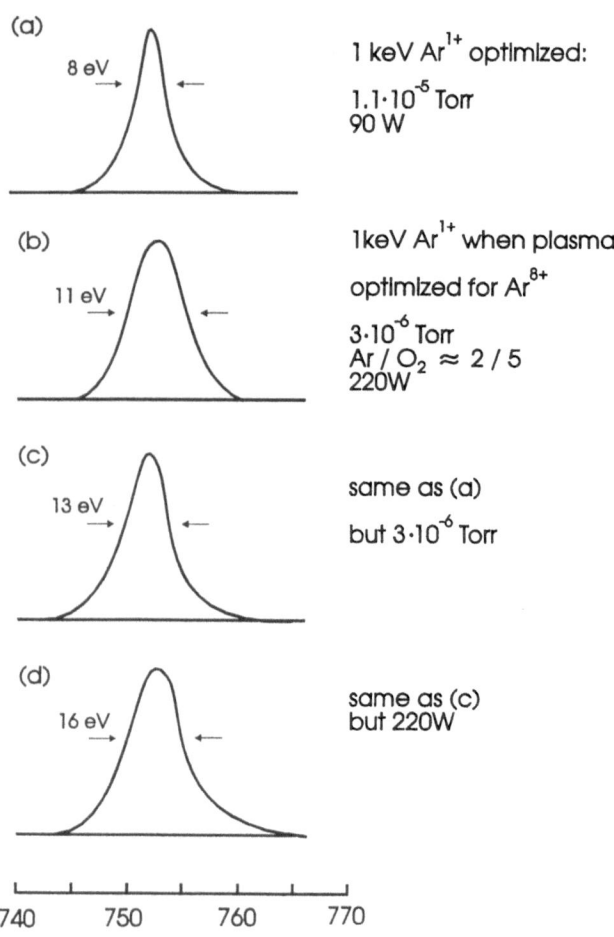

(a)

8 eV

1 keV Ar^{1+} optimized:

$1.1 \cdot 10^{-5}$ Torr
90 W

(b)

11 eV

1keV Ar^{1+} when plasma

optimized for Ar^{8+}

$3 \cdot 10^{-6}$ Torr
Ar / O$_2$ $\approx$ 2 / 5
220W

(c)

13 eV

same as (a)

but $3 \cdot 10^{-6}$ Torr

(d)

16 eV

same as (c)
but 220W

740   750   760   770

analyzing magnetic field (G)

Figure 6.16 Ar^{1+} ion profiles measured for different source conditions and different ion temperatures.

voltage is shown for doubly charged Nitrogen ions [242] on the left side. On the right side of Fig.**6.17** the same curve is shown on a logarithmic scale. The energy distribution is rather asymmetric, showing the shape of an accelerated Maxwellian. Thus from the logarithmic slope of the high energy tail the ion temperature inside the plasma can be evaluated.

To distinguish between different mechanisms, which may influence the energetic half-width of the extracted ion energy distributions, in [242] the ion temperature was determined as a function of the charge state from the exponential decay of the distribution function. The result is shown in Fig.**6.18**.

Assuming an ideal plasma boundary and a Maxwellian velocity distribution of the ions inside the plasma, the energetic halfwidth of an accelerated half Maxwellian distribution should be given by $2.5kT_i$ [243]. In Fig.**6.19** the reduced energy ΔE per ion charge i is shown. Here one can expect that the corrected halfwidth ΔE should increase linearly

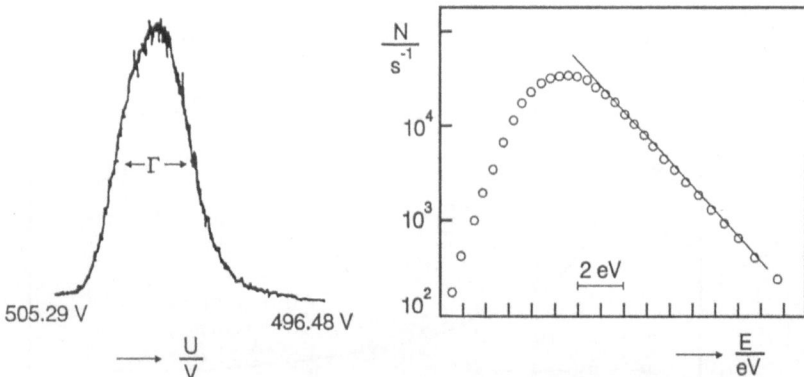

Figure 6.17 Energy distribution of doubly charged Nitrogen ions at P = $2 \cdot 10^{-4}$ Pa and injected microwave power of 230 W at a frequency of 5 GHz [242]. Left side: linear plot of the counting rate versus the deceleration voltage; right side: logarithmic plot of the energy distribution.

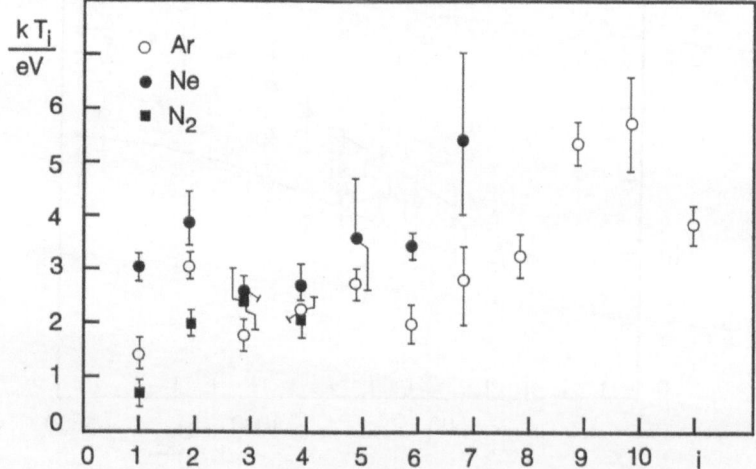

Figure 6.18 Ion temperature kT_i as a function of the ion charge state i.

with i, if variations of the space potential in the extraction region are considered. This behavior was found for N_2 molecules, but for rare gases $\Delta E / i$ shows an increase towards lower charge states because of nonlinear energy transfer processes in the extraction region.

Further, an increase in the aperture diameter of the extraction hole and of the gap results in a slight increase of the halfwidths when equal voltages are considered (Fig.**6.20**).

Another example of the measurement of ion and electron energy spectra from a duoplasmatron is given by Volk et al. [244], where the spectra are measured with an electrostatic 127° cylinder spectrometer, like early measurements by Gautherin et al. [245].

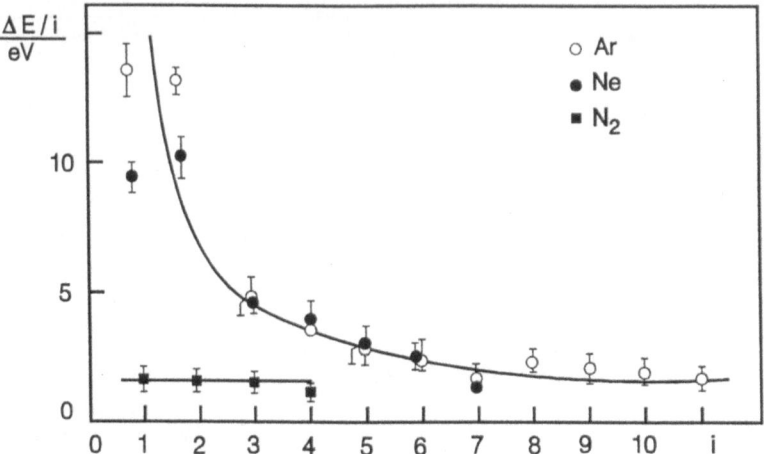

Figure 6.19   Reduced energy width ΔE versus ion charge i.

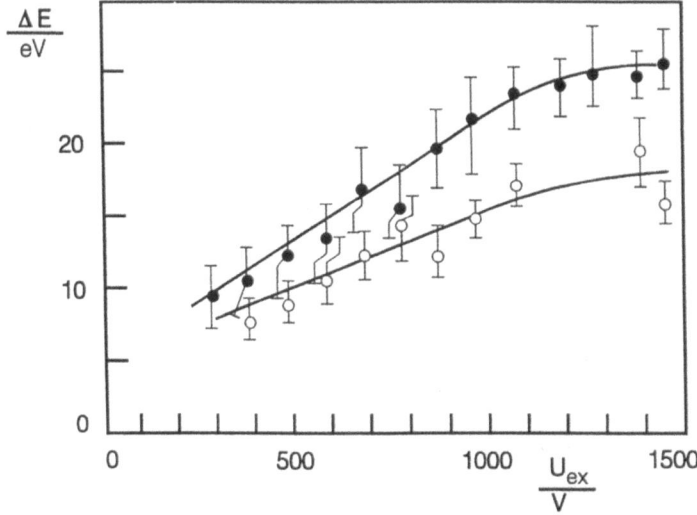

Figure 6.20   Variation of the energetic beam width ΔE for Ar^{4+} ions with the extraction voltage. ○: extraction hole diameter 2 mm, gap 3 mm, acceleration aperture diameter 4 mm; ●: extraction hole diameter 4 mm, gap 6 mm, acceleration aperture diameter 8 mm.

6.2.2.4   Other Electrical Flux Analyzers

Precise measurements of the ion beam current may be necessary for

- the optimization of ion source parameters;
- the determination of ion charge state distribution;
- the optimization of time structures in pulsed ion sources;
- the determination of the beam transmission through a beam transport system;
- the precise measurement of particle flux to an experiment.

Thereby, the technique for the beam current measurement depends on

- the beam energy,
- the beam intensity range,
- the beam power,
- the time structure of the beam.

For concrete experiments it is possible to use *destructive* and *nondestructive* measuring techniques. A review about this techniques is given by Strehl [246]. In the following paragraphs we follow explanations given in Ref.[246].

6.2.2.4.1   Faraday Cup   An Faraday cup collects the ions in an electrical insulated cup. The charge is measured by a calibrated resistor in an appropriate electronic circuit. The construction and operation of such a *Faraday cup* is complicated by the following effects [246]:

- generation of electrons and ions from the ionization of the residual gas;
- emission of secondary electrons or of other charged particles (from secondary electron emission see for instance [247, 248]);
- leakage currents due to the deterioration of insulating material by sputtering or high temperatures;
- leakage currents arising from the conductance of the cooling medium;
- formation of galvanic elements due to the use of different materials in the case of water-cooled cups;
- heating up by high beam power.

For getting realistic results, an important problem is the suppression of secondary electrons. If ϑ is the angle of a secondary electron trajectory to the beam axis, then the flux of secondary electrons varies with $\cos\vartheta$. The portion of electrons passing out of the cup depends on the ratio of cup aperture R to cup length L_F [249]

$$f \sim \frac{\sin^2\varphi}{2} = \frac{R^2}{2(R^2 + L_F^2)}$$

with

$$\varphi \sim \arctan\frac{R}{L_F} \ .$$

This implies $L_F > R$, which will not always be possible.

Fig.**6.21** shows the schematic view of an uncooled Faraday cup. The distortion of the measured result can be reduced by a repelling electrical field using a suppressor electrode in front of the cup or by biasing the cup together with the measuring electronics. The efficiency of the repelling electric field can be increased by combining it with a magnetic field which forces the electrons to have circular orbits inside the aperture of the cup. By a non-relativistic approximation the bending radius r_b is given as

$$r_b = \sqrt{\frac{2m_e E_e}{eB}} \approx 3.37 \frac{\sqrt{E_e[\text{eV}]}}{b[\text{mT}]} \quad [\text{mm}] \tag{6.50}$$

with E_e as the kinetic energy of the secondary electrons and B as the magnetic field strength.

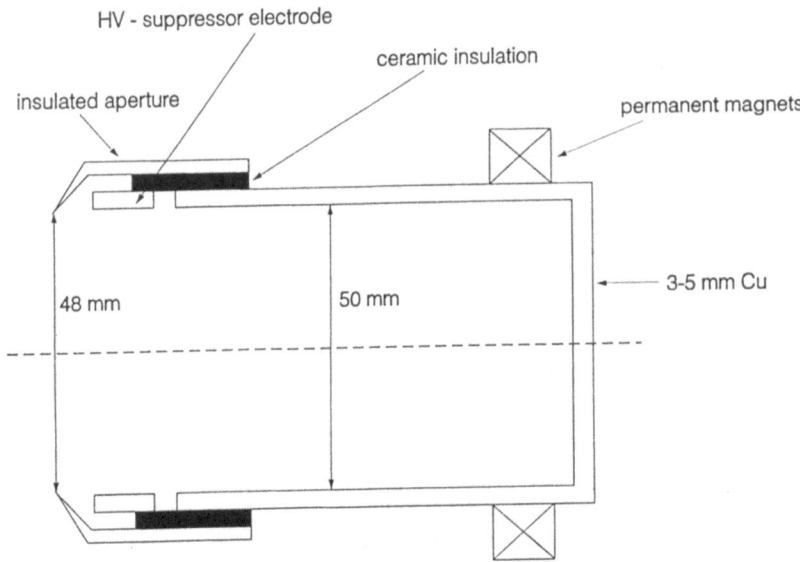

Figure 6.21   Scheme of a uncooled Faraday cup [246].

For the design of a Faraday cup the effect of sputtering should also be taken into account. To avoid leakage currents, insulators have to be shielded against the deposition of conducting materials. Since the sputtering rate shows a maxima at low energies it may be necessary to estimate the desposition of material by sputtering. The amount of material which will be removed by sputtering yields

$$R_s = 0.36 \, \frac{N_s A}{i \varrho} \, \frac{I}{\Delta F} \quad [\frac{\mu m}{h}] \tag{6.51}$$

with N_s as sputtering rate [atoms/incident ion], A as the atomic weight, ϱ as the density of the bombarded material [g/cm^3], i as the ion charge and $I/\Delta F$ as the electrical current density [mA/cm^2]. The sputtering rate for various materials is shown in Fig.**6.22** for a 45 keV Krypton beam [250].

For sufficiently high beam power an effective water cooling system must be used. More details to this problem are given by Strehl [246].

6.2.2.4.2   Calorimetric Measurements   Indirectly, the beam current may be determined by a calorimetric measurement of the beam power (deposited energy). The measurement method is illustrated in Fig.**6.23**. The deposited energy which melts the material within the penetration range volume is determined by ΔE_d to heat up to this range from T_0 to the melting point T_m

$$\Delta E_d = m \, C_p(T) \, (T_m - T_0) = \varrho \, C_p V \, (T_m - T_0) \tag{6.52}$$

with m as the mass in g, $C_p(T)$ as the heat capacity in Ws/g·K, ϱ as the specific weight in g·cm^{-3} and V as the volume in mm^{-3}. $\Delta T = T_m - T_0$ is measured as the resulting rise of temperature to determine the exact change in ΔQ.

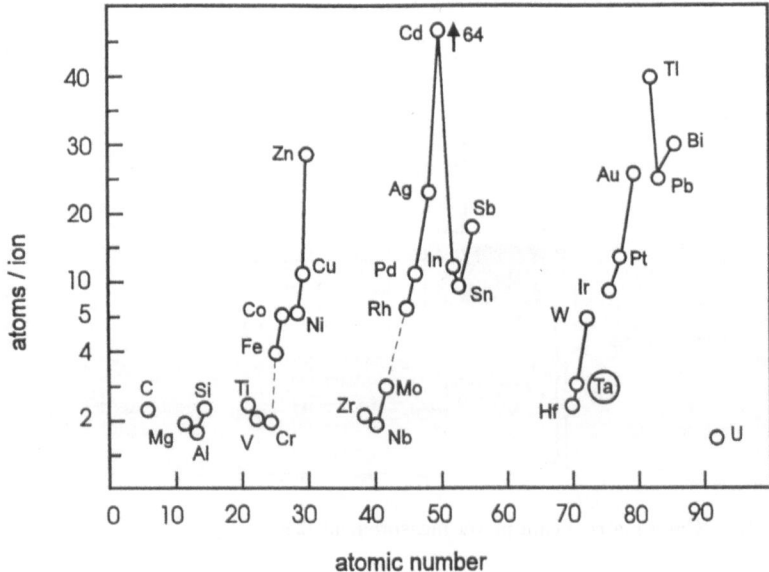

Figure 6.22 Sputtering rate for a 45 keV Krypton beam [250].

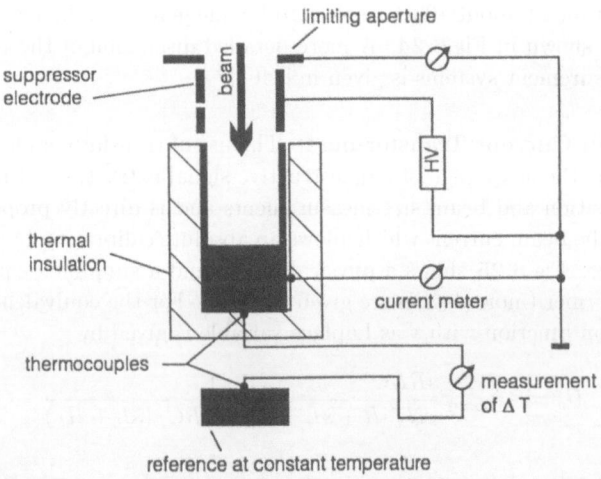

Figure 6.23 Scheme for a calorimetric measurement [246].

The relation between energy deposition and the resulting change of temperature is given approximately by

$$\Delta T[\text{K}] \approx \frac{1}{3} \frac{\Delta Q \, [\text{Ws}]}{V \, [\text{cm}^3]} \tag{6.53}$$

In this approximation the time of the temperature change is in the order of seconds.

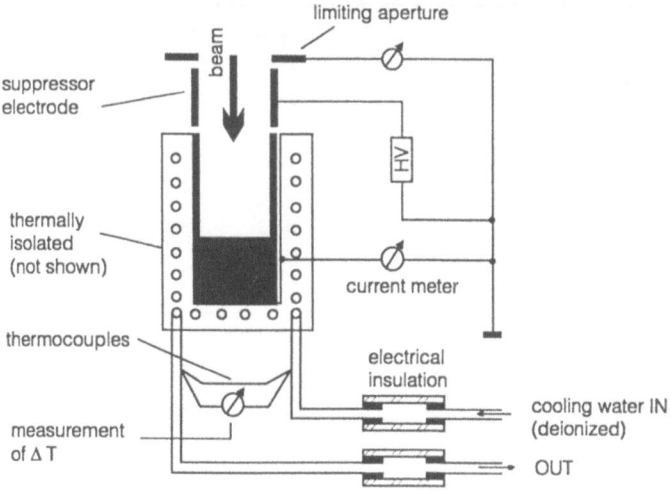

Figure 6.24   Calorimetric beam power measurement [246].

A caloric device can also be used for monitoring and optimizing the continuous particle
flow from an ion source. For this application the calorimeter has to be cooled and the
alterations of input power can be recorded by changing of the cooling water temperature.
Here a response time of about 100 ms seems to be adequate. A scheme of the measurement principle is shown in Fig.**6.24**. A more detailed discussion of the requirements of
calorimetric measurement systems is given in [246].

6.2.2.4.3   Beam Current Transformers   The use of transformers for beam current
measurements has the advantage of nondestructive signal extraction, of nearly independent of beam position and beam size measurements and is directly proportional to the
output signal of the beam current which allows an absolute calibration by using an external current source. Fig.**6.25** shows a physical model and a simplified equivalent circuit
of a beam transformer (more details are given in [251]). For the equivalent circuit shown
the transformation function with s as Laplace variable is given by

$$U_a = -i_{\text{beam}} \frac{sRL}{N} \frac{1}{R + sL + R_L + sRC_L(sL + R_L)} \tag{6.54}$$

A more detailed discussion of beam transformers with corresponding citations is given
by Strehl [246].

6.2.2.5   Mass Analyzers

Mass spectrometers for ion beam diagnostics have been extensively used, so that we
give here only a short description of the operation principle. Details of this kind of
spectrometer one can find for instance in [252, 253, 254, 255].

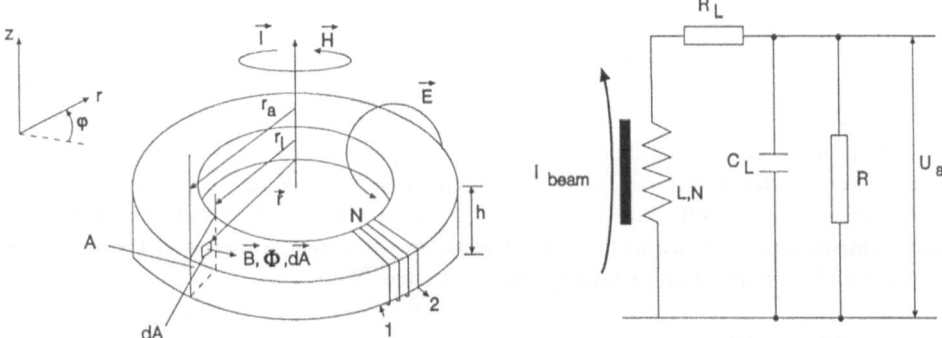

Figure 6.25 Physical model of a current transformer and simplified equivalent circuit of a beam current transformer [246].

The deflection of ions in a perpendicular magnetic field is proportional to the particle momentum per unit charge. If all particles entering a magnetic field have the same energy per unit charge, then the field will separate according to their masses. On the other hand, if the particles have the same mass, the field will separate according to the charge to mass ratio of the particles and we get spectra characterizing the particles according to their masses or charge states, respectively.

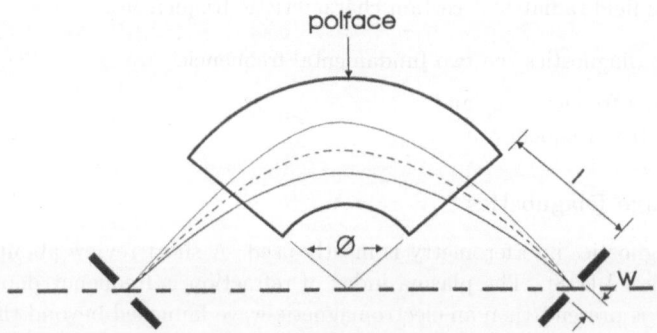

Figure 6.26 Focussing properties of a magnetic sector analyzer.

In the usual configuration a magnetic field is produced between two sector-shaped polefaces of an electromagnet (Fig.**6.26**). The focussing properties of this field are the same as those of spherical electrical analyzers and the locations of the entrance and exit slits lie on a common line of the object, at the centre of curvature of the spheres and at the image. The radius of curvature of ions of energy E in a magnetic field B and of charge i is

$$R[\text{mm}] = \frac{143.9}{iB[\text{T}]}\sqrt{m_i[\text{amu}]E[\text{eV}]} \tag{6.55}$$

In terms of parameters specified in Fig.**6.26** the resolution is

$$\frac{\Delta m_i}{m_i} = \frac{w}{R(1 - \cos \Phi) + l \sin \Phi} \tag{6.56}$$

To first order, there is no focussing in the plane perpendicular to the plane of deflection. However, when the curvature of the fringing field is taken into consideration, some focussing in the perpendicular plane results. Incident ion trajectories should be normal to the entrance plane to avoid focussing effects. For instance, a review of the focussing properties of corresponding magnets is given by Enge [256].

6.2.2.6   Time-of-Flight spectroscopy

The time-of-flight (TOF) method, which is also used for the measurement of neutron energies and for the determination of the mass of fission fragments and other heavy ions is also successfully applied to the charge state analysis of ions extracted from a source. But because this kind of diagnostic is basically applied on EBIS [6] we will discuss this method in a later subsection.

6.2.3   Microwave diagnostics

There are two kinds of diagnostics for ECR ion sources

- *active diagnostics*, when a microwave beam is launched into the plasma;
- *passive diagnostics* using the phenomena that the plasma when immersed in a magnetic field radiates at certain characteristic frequencies.

For microwave diagnostics the two fundamental frequencies are

- the plasma frequency ω_p and
- the cyclotron frequency ω_c.

6.2.3.1   Active Diagnostics

For active diagnostics interferometry is mostly used. A short review about this method is given by Girard [218]. The plasma index of refraction is frequency dependent: if no magnetic field is present, then an electromagnetic wave launched beyond the plasma frequency (to penetrate into the plasma) experiences a phase shift related to the plasma refractive index. This is still true with the plasma in a magnetic field when the electromagnetic wave couples to the ordinary mode (i.e., the wave vector is perpendicular to the (static) magnetic field and the electric vector is parallel to the (static) magnetic field.

In that case the refractive index depends upon the plasma density

$$n^2 = 1 - \frac{\omega_p^2}{\omega^2} \tag{6.57}$$

with $\omega_p^2 = n_e e^2 / m_e \varepsilon_0$. If we define $n_c = \omega^2 m_e \varepsilon_o / e^2$ follows

$$n_c = \frac{4\pi^2 c^2 m_e \varepsilon_0}{e^2} \frac{1}{\lambda^2} \, .$$

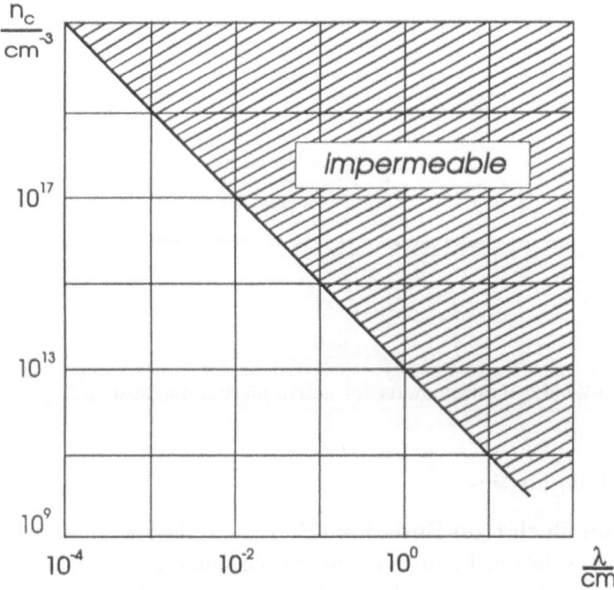

Figure 6.27   Transparency of microwave radiation of the wavelength λ at a critical plasma density n_c.

Thus, at increasing electron density for plasma diagnostic purposes radiation of quite short wavelengths must be chosen. In Fig.**6.27** the dependence between the wavelength of radiation used for diagnostic purposes and the critical density n_c is shown. For instance the plasma becomes non-transparent for the radiation of a Neodymium laser ($\lambda = 1.06$ μm) only at about $1 \cdot 10^{21}$ cm^{-3}. The plasma induced phase shift along a path L in the plasma is

$$\Delta\Phi = \frac{\omega}{c} \int_{0}^{L} \sqrt{1 - \frac{n_e}{n_c}}\, dr \tag{6.58}$$

The operation principle of a Mach-Zender interferometer is presented in Fig.**6.28**. The wave generated from a microwave generator is splitt into two beams: the first beam passes the plasma and the second one serves as a reference. These two beams are then allowed to interfere. A square-law detector measures the resulting intensity. When no plasma is present the beams have no phase shift. If a plasma is present, then the intensity is multiplied by $\cos\Phi$. Unfortunately, the cosine does not only define the phase shift, e.g. a second output is in quadrature with the previous one to provide $\sin\Phi$.

Another possibility consists of modulating the phase of the interferometer [258] which can be compared to heterodyne detection.

For high density plasmas interferometric techniques in the visible range are also used.

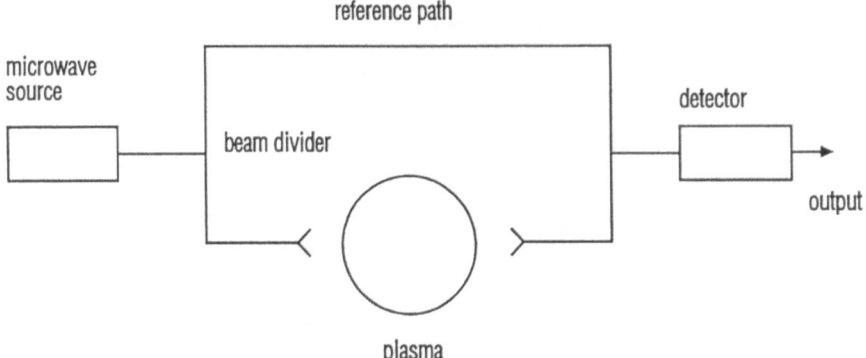

Figure 6.28   Mach-Zender interferometer for active plasma diagnostics.

6.2.3.2   Passive Diagnostics

6.2.3.2.1   Electron Cyclotron Emission (ECE)   A plasma immersed in a magnetic field radiates at the cyclotron frequency and its harmonics. The emission of a single radiating electron (velocity $v_\perp$ perpendicular to the magnetic field) at a given harmonic a is

$$I_a = \frac{e^2\omega_c^2}{6\pi\varepsilon_0 c^3}\frac{1-\beta_o^2}{\beta_0}\left[a\beta_0^2 J_{2a}'(2a\beta_0) - a^2\left(1-\beta_0^2\right)\int_0^{\beta_0} J_{2a}(2at)\,dt\right] \qquad (6.59)$$

with

$$\beta_0 = \frac{\beta_\perp}{\sqrt{1-\beta_\parallel}}\;;\quad \beta_\perp = \frac{v_\perp}{c}\;;\quad \beta_\parallel = \frac{v_\parallel}{c}$$

where $v_\parallel$ is the velocity of the electron parallel to the magnetic field, c the speed of light, J_a a Bessel function of order a. Using (6.59) it is possible to get information on the electron energy.

When the plasma density increases or the plasma length is not negligible it can become optically thick. Then the measured intensity reaches the level of blackbody radiation at the electron temperature (Rayleigh-Jeans law) for $h\nu \ll kT_e$.

Electrons with an energy spectrum typically for an ECR ion source generate intense emission at high harmonics. Electron cyclotron emission has been used to study microwave generated energetic electrons with a Michelson type interferometer [224]. An example for a frequency spectrum measured with an InSb crystal is shown in Fig.**6.29** (see also Fig.**6.5**).

6.2.3.2.2   Instabilities   In ECR ion sources instabilities can occur, for instance caused by the anisotropy of the electron distribution function. For this case it is well known that the instability induced enhanced diffusion is related to the growth of waves either within ω_p or ω_c. Measurements of microwave emission in an ECR source show a strong correlation between microwave emission and enhanced particle losses [259].

In [259] the *whistler instability* of an ECR heated plasma is observed. The whistler instability is an electron microinstability which may occur in a plasma with an anisotropic

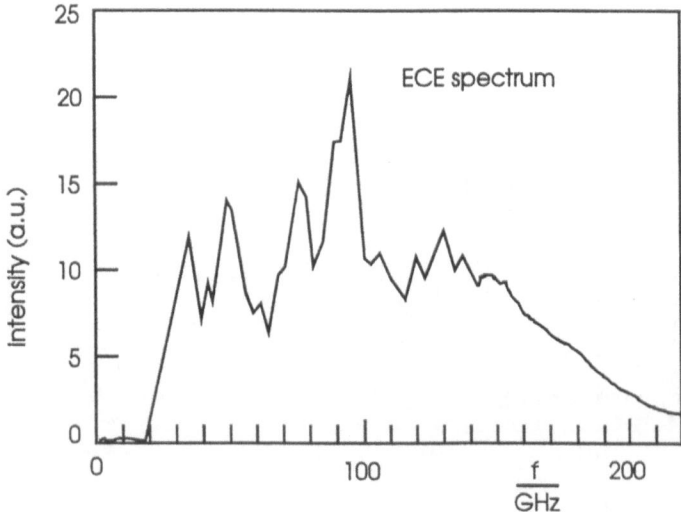

Figure 6.29   ECE spectrum measured in the 18 GHz ECR ion source MINIMAFIOS [224] at an input power of 700 W. Shown is the harmonic emission up to 11 ω_c.

velocity-space distribution. The existence of this instability is demonstrated by different authors [259, 260, 261, 262, 263]. In these experiments radiofrequency emission in the ECR frequency range is observed and typically accompanied by enhanced particle end loss induced by the unstable wave. In the experiment by Garner et al. [259] (Hydrogen plasma heated by microwaves of 10.5 GHz and 1 kW) it was shown, that the whistler instability is driven by the warm electron component (2 keV), while the hot electron component is stable. The identification of the instability is based upon good agreement between the observed emission frequency and the prediction of the most unstable frequency obtained from dispersion-relation calculations.

For the mentioned above conditions the radiofrequency emission associated with the microinstability occurs in regular bursts with frequencies in the range from 6.7 GHz to 8.7 GHz (Fig.**6.30**). The radiofrequency bursts correlate with electron end loss bursts as well as bursts of ion end loss, diamagnetism and potential fluctuations.

Investigations by Garner et al. [259] have shown that the hot electron component is stable and has little effect on microinstability. In general, microinstabilities are driven by the warm electron component.

A plasma containing a hot electron component can be produced by turning off the gas source during the shot while leaving the ECR heating on as shown in Fig.**6.31** or by tuning of the ECR heating. In the latter case the unstable radiofrequency emission bursts sporadically for several milliseconds and then completely stops. The hot electron temperature increasing after ECR heating is turned off and the diamagnetic signal together with results from X-ray spectroscopy indicate that the hot electron density decays exponentially (Fig.**6.32**).

Unstable emission is constant in time during ECR heating, whereas the hot electron temperature varies in time. Although the hot electron component does not drive the

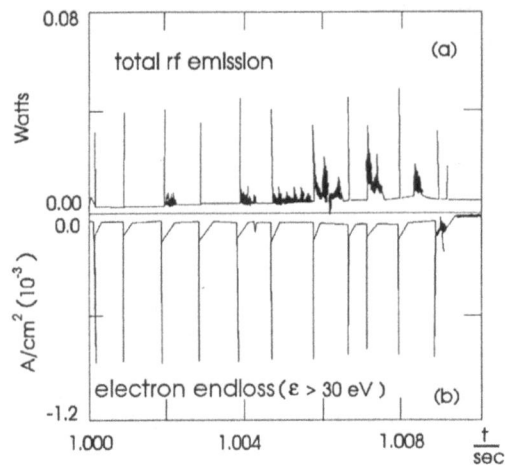

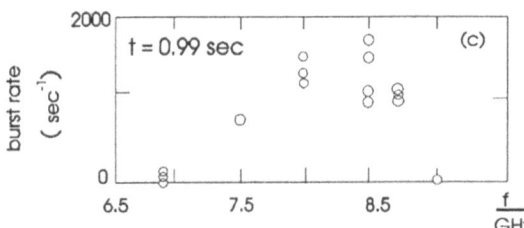

Figure 6.30 Burst of radiofrequency emission with frequencies in the range from 5.25 GHz to 12 GHz (a), bursts of electron end loss (b) and frequency spectrum of the radiofrequency emission (each point corresponds to $\Delta f = 1$ MHz [259].

whistler instability there is still hot electron end loss induced by the instability when a warm component is present. However, for hot electrons a net energy loss exists because the hot electrons diffuse downward in energy due to the unstable waves and so have a chance to enter the loss cone and carry their total energy deposited of the plasma.

Dispersion relation calculations [259] using an ECR heating distribution function

$$f_0(e,\mu) = \exp\left[-\frac{\chi}{T_\chi} - \Theta(\eta)\,\frac{\eta}{T_{\eta+}} + \Theta(-\eta)\,\frac{\eta}{T_{\eta-}}\right] \tag{6.60}$$

with
$\chi = \frac{1}{2}\left(E + \mu B_h\right)$ and $\eta = \frac{1}{2}\left(E - \mu B_h\right)$;
$E = mc^2(\gamma - 1)$ – particle kinetic energy;
$\mu = \frac{1}{2}\,mu^2/B$ – magnetic moment;
$u/c = \sqrt{\gamma^2 - 1}$ – relativistic velocity;
$\Theta(\eta)$ – unit step function;
$B_h = \omega_h m_0 c/e$;
ω_h – applied ECR heating frequency.

This function describes the response of warm electrons to ECR heating and predicts unstable whistler wave frequencies which match the experimentally measured unstable frequencies. Furthermore, calculations with (6.60) suggest, that relativistic effects are stabilizing.

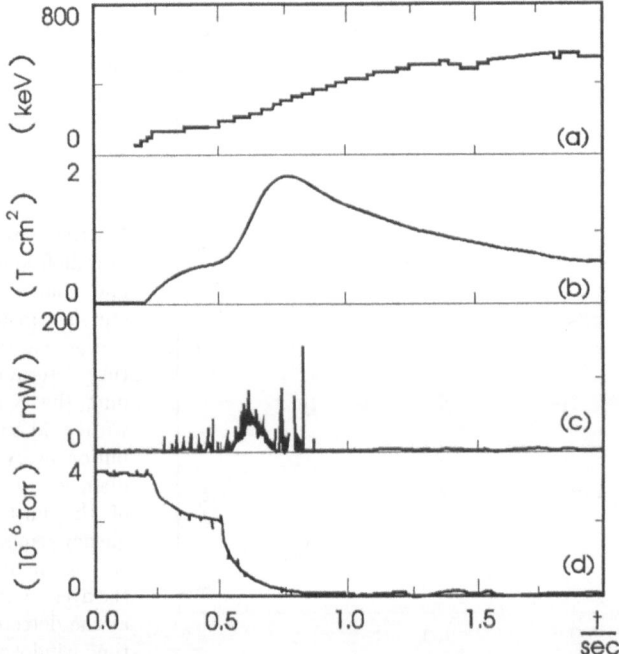

Figure 6.31 Data showing the stability of the hot electron component [259]. The gas is tuned off after 0.6 s while the ECR heating is left until 2 s. The hot component has the longest confinement and remains for several seconds after the cold and warm components have decayed. During this time is not unstable radiofrequency emission. a – hot electron temperature; b – diamagnetic flux; c – power of unstable radiofrequency emission; d – gauge pressure.

6.2.4   Spectroscopic Diagnostics

6.2.4.1   Electron Bremsstrahlung

6.2.4.1.1   Deconvolution of Electron Bremsstrahlung Spectra   For a definite evaluation of the *electron energy distribution function* $F(E)$ it is necessary to measure electron volume bremsstrahlung [274]. In the case of measurements on an ECR plasma wall bremsstrahlung is thick target bremsstrahlung, i.e. scattering and energy loss of the electrons inside the wall are involved in shaping the X-ray spectrum. The electron energy distribution function can be obtained from volume bremsstrahlung spectra by a two step deconvolution taking into account the energy dependence of the bremsstrahlung cross-sections and the detector response function. To obtain distribution functions from measurements, containing secondary radiation or wall bremsstrahlung, a further deconvolution step must be done to take into account the additional scattering of electrons and of radiation. This step is ambiguous since the fate of the scattered electrons and photons is not very well known.

An example for a suitable X-ray collimator is shown in Fig.**6.33** for suppressing the influences of secondary scattered radiation and wall bremsstrahlung. The ray path and the detector are shielded by Lead against large-angle scattered photons. Thereby, the

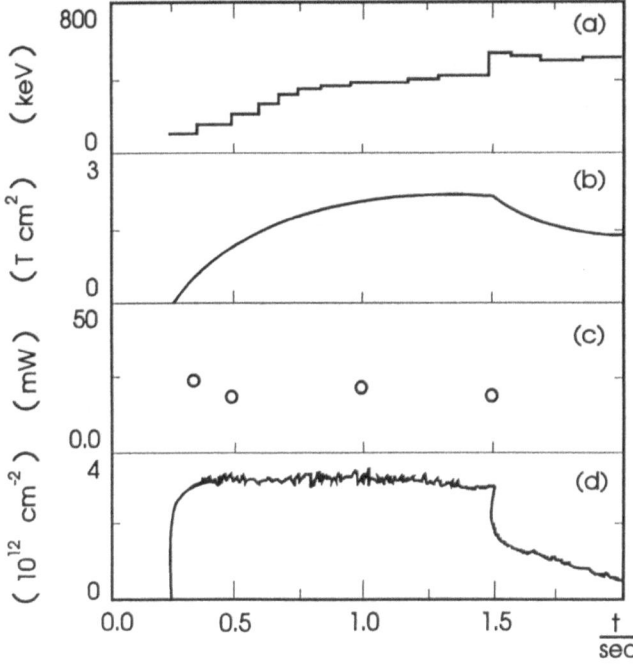

Figure 6.32 Unstable radiofrequency emission and course of characteristic parameters [259]. a – hot electron temperature determined by unfolding the X-ray spectrum with a Maxwellian distribution every 100 ms; b – diamagnetic flux; c – power of the unstable radiofrequency emission whereby each point corresponds to the average emission power at the detector in a 40 ms time window; d – line density.

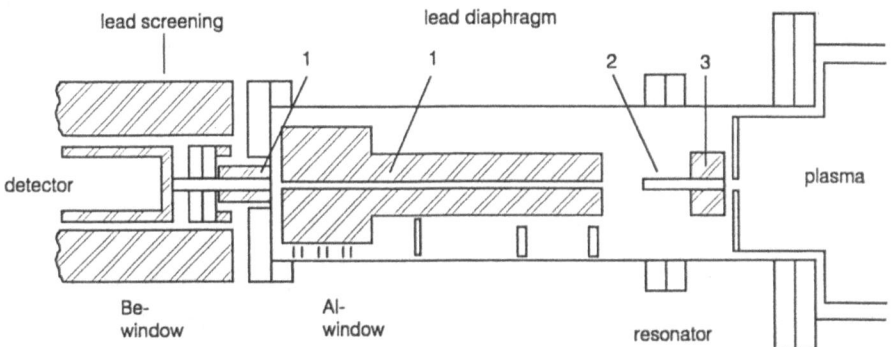

Figure 6.33 X-ray collimator system for measuring X-rays from an ECR plasma volume [274]. 1: Lead shield for the detector on the ray path; 2: acceptance defining diaphragm; 3: pre-diaphragm.

main shield is the long Lead tube 1. The angular acceptance of the collimator system is determined by the small Lead tube 2. Therefore, its inner diameter is smaller than the inner diameter of the other components of the collimator system.

Secondary Compton scattered radiation or wall bremsstrahlung emitted from the entrance of tube 2 can also distort the measurements. This tube is therefore protected against plasma electrons and against radiation which arises from outside the geometric acceptance of the collimator system by a pre-diaphragm consisting of the Lead tube 3.

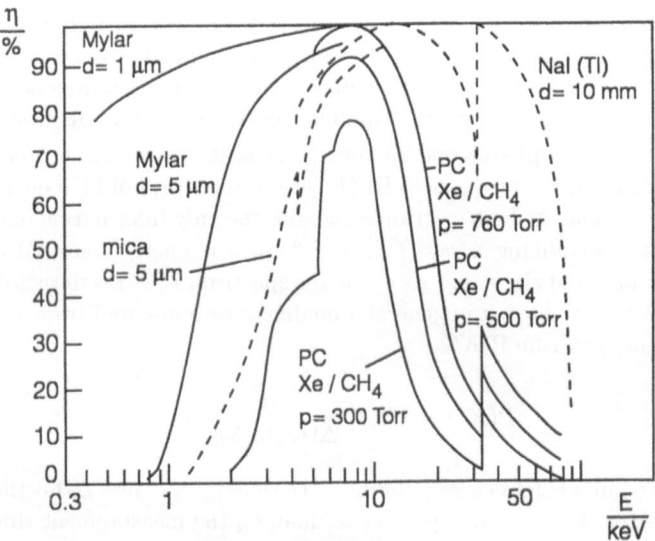

Figure 6.34 Efficiency η of different gas ionization and scintillation detectors as a function of the photon energy E.

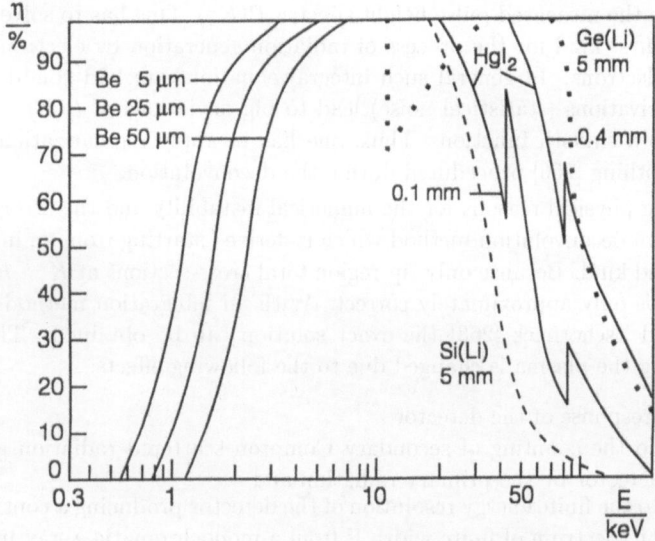

Figure 6.35 Efficiency η of different semiconductor detectors as a function of the photon energy E.

As radiation detectors NaI(Tl) scintillation detectors and semiconductor ones are applicable. Typical curves for detector efficiency in the interesting energy range from several keV up to some hundreds keV are shown in Figs.**6.34** and **6.35**. The high energy resolution of semiconductor detectors allows an accurate measurement of bremsstrahlung spectra and thus a precise determination of electron energy distribution functions. The low energy boundary for this kind of measurement is determined by radiation attenuation in the window materials and by air and the upper limit by the course of the detector efficiency. High Z materials such as Germanium allows one to measure energetic quanta up to the high energy end of the bremsstrahlung spectrum in an ECR plasma.

In plasma physics, astrophysics and ion source diagnostics arises the problem of measuring the energy distribution function $F(E)$ (E: electron energy) of high energy electrons in an isotropical plasma. At high electronic energies the only information one can obtain on $F(E)$ is by bremsstrahlung X-rays $I(K^o)$ (K^o: photon energy) emitted in electronic interactions with ions and electrons. As a rule the spectrum $I(K^o)$ is measured by energy sensitive X-ray detectors having in general a nonlinear response and transforming $I(K^o)$ into a pulse height spectrum $P(K^o)$

$$P(K^\circ) = I(K^o) \frac{4\pi\eta}{\Delta\Omega V_a t_m \Delta K^\circ \epsilon} \qquad (6.61)$$

where K° is is the pulse height corresponding to the energy K. Here $\Delta\Omega$ is the solid angle seen by the detector, V_a the active plasma volume, t_m the measurement time, ΔK° the channel width, ϵ the absolute efficiency of the detector and η the radiation attenuation coefficient of the vacuum window. Thus, $P(K^\circ)$ is the number of photons per active plasma volume, time and energy interval [$cm^{-3} s^{-1} keV^{-1}$]. Hence the problem is to evaluate $F(E)$ from the measured pulse height spectra $P(K^\circ)$. One has to solve an integral equation of the first kind for the process of radiation generation by electronic collisions with ions and electrons. In general such integral equations are bad conditioned. This means small derivations (statistical noise) lead to big deviations of $F(E)$. A resonable $F(E)$ should be a smooth function. Thus, one has to apply regularization [189, 265], filtering or smoothing [266] procedured during the deconvolution.

Considerating physical reasons for the numerical instability and the errors Bernhardi [266] has given an deconvolution method which is derived starting from an integral equation of the second kind. Because only tip-region total cross-sections at $K = E$ is used the obtained $F(E)$ is only approximately correct. With an integration method developped by Friedlein and Zschornack [268] the exact solution can be obtained. The radiation spectrum leaving the plasma is changed due to the following effects:

1. nonlinear response of the detector
    - due to the counting of secondary Compton scattered radiation generated in the detector by the primary radiation and
    - due to the finite energy resolution of the detector producing a continuous pulse height spectrum of finite width Γ from a monochromatic X-ray line;
2. wall bremsstrahlung and secondary radiation due to scattering at the collimator system.

The influence of these effects can be described by a response function $L(K^o, K)$ obtained as a pulse height spectrum when measuring a monochromatic X-ray line with photon energy K. The pulse height spectrum obtained from a continuous X-ray spectrum $I(K^o)$ can be considered as a superposition of the contributions of an ensemble of many monochromatic X-ray lines $I(K^o)\delta(K^o - K)$. $L(K^o, K)$ can be understood as the probability of a registered photon of the energy K leaving the plasma with the energy K^o.

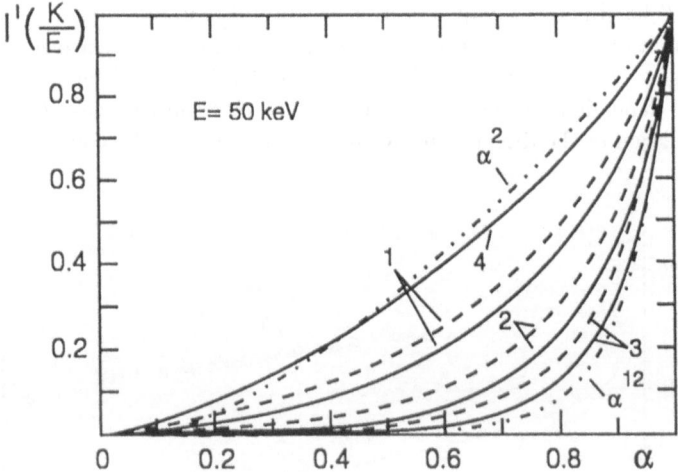

Figure 6.36 Dependence of the ionization factor I' from the ion charge state i for $E = 50$ keV [272].

For a given K this function is normalized as

$$\int_0^\infty L(K^o, K) \, dK^o = 1. \tag{6.62}$$

Thus, the kernel $M(K^o, E)$ in the integral equation

$$P(K^o) = c \int_{E_{min}}^\infty M(K^o, E) \, dE \tag{6.63}$$

is an integral over the product of total bremsstrahlung cross-sections

$$n_n \frac{d\sigma_{ei}}{dK} + n_e \frac{d\sigma_{ee}}{dK}$$

weighted by the total neutral and ion densities n_n and the total electron density n_e (for data see e.g. [296, 269] and $L(K^o, K)$:

$$M(K^o, E) = \int_{K=0}^\infty \beta(E) \left(n_n \frac{d\sigma_{ei}}{dK} + n_e \frac{d\sigma_{ee}}{dK} \right) L(K^o, K) \, dK \tag{6.64}$$

Here β is the relativistic factor, c the velocity of light, $d\sigma_{ei}/dK$ the total electron-ion interaction cross-section and $d\sigma_{ee}/dK$ the total electron-electron interaction cross-section.

The screening of the nucleus by atomic electrons is not very strong for high electron energies and for high ratios K/E, e.g. $d\sigma_{ei}/dK$ is not much different between the different ion charge states. But for energies $E < 100$ keV the cross-section can be scaled between $d\sigma_{ei}/dK$ for neutrals and for bare nuclei using ionization factors I' [272]

$$\frac{d\sigma_i}{dK} = (1 - I') \frac{d\sigma_n}{dK} + I' \frac{d\sigma_e}{dK} \tag{6.65}$$

A scaling for I' for different charge states is shown in Fig.**6.36**.

The contribution of the electron-electron bremsstrahlung is relatively small at low electron energies ($E < 400$ keV) and is negligible for high Z elements in the plasma due to the Z^2 dependence of $d\sigma_{ei}/dK$. At very low electron energies ($E < 10$ keV) the continuous radiation of the direct radiative recombination must be considered too [273].

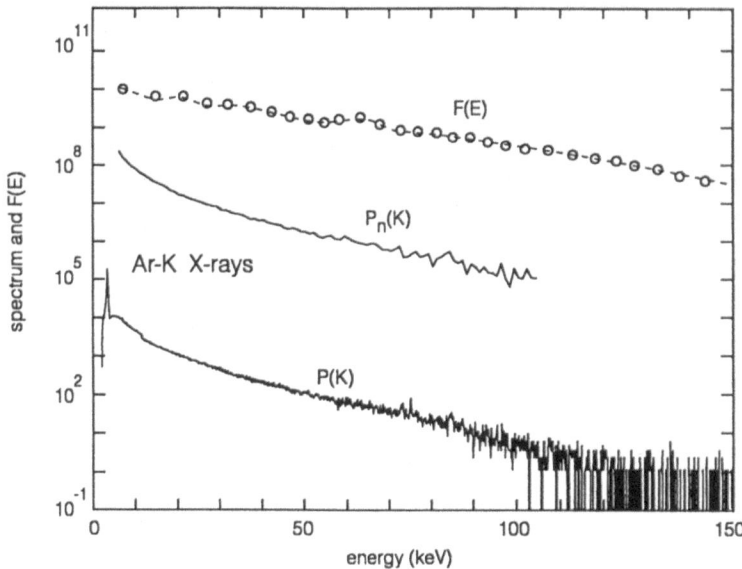

Figure 6.37   Example for the de-convolution of a bremsstrahlung spectrum $P(K)$. Considering a different fraction of scattered radiation the shown electron energy distributions $F(E)$ were obtained from the corrected spectrum $P_n(K)$. $F(E)$ is a two component Boltzmann distribution with the temperatures T_{e1} and T_{e2}, respectively.

With the integration method [268] the integral equation (6.63) can be solved. The method is based on the comparison of the measured spectrum $P(K_i^o)$ integrated in the energy interval i of the width κ with the integrated calculated spectrum $S_i(K_i)$ in i, that one obtains with an arbitrary function $F_0(E_i)$

$$S_i(K_i^o) = n_n c \int_{K^o=K_i^o-\frac{\kappa}{2}}^{K_i^o+\frac{\kappa}{2}} \int_{E=K_i^o-\frac{\kappa}{2}}^{K_i^o+\frac{\kappa}{2}} M(K^o, E) F_o(E)\, dE\, dK^o \qquad (6.66)$$

$$\tilde{S}(K_i^o) = n_n c \int_{K^o=K_i^o-\frac{\kappa}{2}}^{K_i^o+\frac{\kappa}{2}} \int_{E=K_i^o+\frac{\kappa}{2}}^{K_{max}^o} M(K^o, E) F(E)\, dE\, dK^o$$

$S_i(K_i^o)$ contains all photons being produced in i by electrons in the same interval, and $\tilde{S}(K_i^o)$ those photons being produced by the electrons with higher energy. The procedure has to start at the high energy end of the spectrum, because one has to know $F(E)$ for all higher energies before calculating $S_i(K_i^\circ)$

$$F(E_i) = F_\circ(E_i) \frac{\int\limits_{K_i^\circ - \frac{\kappa}{2}}^{K_i^\circ + \frac{\kappa}{2}} P(K^\circ)\, dK^\circ - \tilde{S}_i(K_i^\circ)}{S_i(K_i^\circ = E_i)} \tag{6.67}$$

Simulations show, that the error is mainly due to the finite upper integration limit and converges to zero when starting at energies being high compared to the mean energy of $F(E)$. Other errors are due to the statistical error of the spectrum, to the quality of the collimator system and the detector, to the error of the interpolation of the cross-section data, to the knowledge of the sum of the neutral and ion densities. Nonphysical oscillations of $F(E)$ due to the bad condition of the integral equation can be supressed by choosing an κ big enough.

An example for the processing of electron bremsstrahlung spectra is shown in Fig.**6.37**.

6.2.4.1.2 Approximative Calculation of the Anisotropical Electron Energy Distribution Function f(E;Θ)   The method described above can be extended for the deconvolution of bremsstrahlung spectra produced from hot electron components $f(p, \Theta)$ (p: relativistic electron momentum; Θ: angle between z-axis and momentum vector; see also Fig.**6.38**) being localized in momentum space in cylindrical symmetry as it is the case in ECR ion sources and approximately in Tokamak fusion devices.

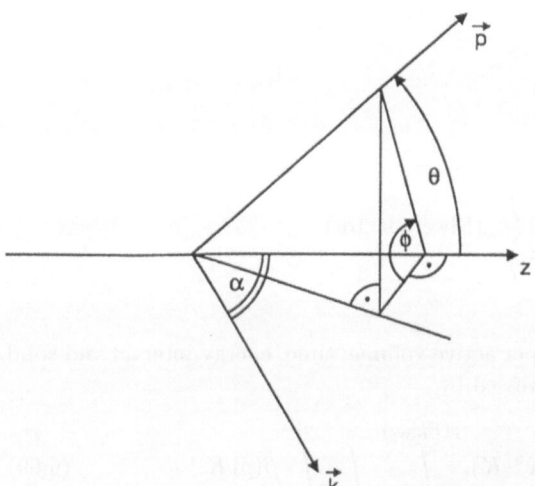

Figure 6.38 Geometry of motion of hot electrons localized in momentum space in cylindrical symmetry.

The calculation of $f(p, \Theta, \dots)$ is impossible because the cross-section $d^2\sigma/dK\, d\Omega$ is doubly differential. For theoretical data of these cross-sections, resp. of the shape function of bremsstrahlung emission, see for instance Refs.[269, 296].

Analytical expressions like the Bethe-Heitler formula (nonrelativistic with the Elwert factor as an relativistical correction) are given by Koch and Motz [270]

$$\frac{d^2\sigma}{dK\, d\Omega} = \frac{Z^2 r_0^2 \alpha}{8\pi} \frac{1}{K} \frac{p_2}{p_1} F_E \left[\frac{8\sin^2 \Theta (2E_1^2 + 1)}{p_1^2 \Delta_1^4} - \frac{2(5E_1^2 + 2E_1 E_2 + 3)}{p_1^2 \Delta_1^2} \right.$$

$$-\frac{2(p_1^2 - k^2)}{Q^2\Delta_1^2} + \frac{4E_2}{p_1^2\Delta_1} + \frac{L}{p_1 p_2}\left(\frac{4E_1 \sin^2\Theta(3k - p_1^2 E_2)}{p_1^2\Delta_1^4} + \frac{4E_1^2(E_1^2 + E_2^2)}{p_1^2\Delta_1^2}\right.$$

$$+\frac{2 - 2(7E_1^2 - 3E_1 E_2 + E_2^2)}{p_1^2\Delta_1^2} + \frac{2k(E_1^2 + E_1 E_2 - 1)}{p_1^2\Delta_1^2}\right)$$ (6.68)

$$\left.-\frac{4\varepsilon}{p_2\Delta_1} + \frac{\varepsilon_r}{p_2 Q}\left(\frac{4}{\Delta_1^2} - \frac{6k}{\Delta_1} - \frac{2k(p_1^2 - k^2)}{Q^2\Delta_1}\right)\right]$$

with

$$
\begin{array}{lll}
E_0 & : & \text{electron rest energy} \\
Z & : & \text{nuclear charge} \\
r_0 & : & \text{classical electron radius} \\
\alpha & : & \text{fine structure constant} \\
k & = & K/E_0 \\
E_1 & = & (E_0 + E)/E_0 \text{ (total kinetic energy of the incident electron)} \\
E_2 & = & (E_0 + E - K)/E_0 \\
p_1 & = & \sqrt{E_1^2 - 1} \\
p_2 & = & \sqrt{E_2^2 - 1} \\
Q & = & \sqrt{p_1^2 - k^2 - 2p_1 k \cos\Theta} \\
L & = & \ln\dfrac{E - 1E_2 - 1 + p_1 p_2}{E_1 E_2 - 1 - p_1 p_2} \\
\varepsilon & = & \ln\dfrac{E_2 + p_2}{E_2 - p_2} \\
\varepsilon_r & = & \ln\dfrac{Q + p_2}{Q - p_2} \\
F_E & = & \dfrac{E_2 p_1}{E_1 p_2}\dfrac{1 - \exp\left(-2\pi\dfrac{\alpha Z E_1}{p_1}\right)}{1 - \exp\left(-2\pi\dfrac{\alpha Z E_2}{p_2}\right)} \quad \text{(Elwert factor)} \\
\Delta_1 & = & E_1 - p_1 \cos\Theta
\end{array}
$$

Following [274] the number of photons per active volume, time, energy interval and solid angle $[\text{cm}^{-3}\,\text{s}^{-1}\,\text{keV}^{-1}\,\text{sr}^{-1}]$ can be expressed by

$$P(K^\circ, \alpha) = n_n c \int_{K=0}^{K_{\max}} L(K^\circ, K) \int_{p(E=K^\circ)}^{p(E_{\max})} \int_{\Theta=0}^{\pi} \int_{\Phi=0}^{2\pi} \beta(p)\, K \qquad (6.69)$$

$$\times \frac{d^2\sigma_{eff}(p(E), K, \Theta, \Phi, \alpha)}{dK\, d\Omega}\, p^2 \sin(\Theta) f(p, \Theta)\, d\Phi\, d\Theta\, dp\, dK$$

The angle α gives the direction of observation with respect to the symmetry axis. It is easily to see, that in special case of isotropy one obtains the energy distribution function

$$F(E)\, dE = 2\pi \int_{\Theta=0}^{\pi} f(p, \Theta)\, p^2 \sin\Theta\, dp\, d\Theta \qquad (6.70)$$

The symmetry can be considered when transforming the cross-section into an effective cross-section $d^2\sigma_{eff}(E,K,\Theta,\Phi,\alpha)/dK\,d\Omega$. Equation (6.69) then becomes

$$P(K^\circ,\alpha) \;=\; n_n c \int\limits_{K=0}^{K_{max}} L(K^\circ,K) \int\limits_{E=K^\circ}^{E_{max}} \int\limits_{\Theta=0}^{\pi} \int\limits_{\Phi=0}^{2\pi} \beta(E)$$

$$\times\; \frac{d^2\sigma_{eff}(E,K,\Theta,\Phi,\alpha)}{dk\,d\Omega}\, f(E,\Theta)\, d\Phi\, d\Omega\, dE\, dK \tag{6.71}$$

When observing at $\alpha = 0$ (6.71) can be simplified due to the integration over Φ and Θ from $\pi/2$ to π

$$P(K^\circ,\alpha) \;=\; 2\pi n_n c \int\limits_{K=0}^{K_{max}} L(K^\circ,K) \int\limits_{E=K^\circ}^{E_{max}} \int\limits_{\Theta=0}^{\frac{\pi}{2}} \beta(E)K \tag{6.72}$$

$$\times\; \left(\frac{d^2\sigma(E,K,\Theta)}{dK\,d\Omega} + \frac{d^2\sigma(E,K,\pi-\Theta)}{dK\,d\Omega} \right) f(E,\Theta)\, d\Theta\, dE\, dK$$

The quantities $d^2\sigma(E,K,\Phi)/dK\,d\Omega$ are tabled cross-sections, whereby Φ is the emission angle of the photon and must be substituted [269, 296]. With at least two spectra measured at different angles one can try to find a distribution function that produces the given spectra. Measurements of the degree of polarization of the emitted radiation allows a selection of the distribution function. To achieve a fast convergence of the numerical procedure both angles should be clearly different.

Like in the isotropical case one starts at high photon energies and finds a solution for $f(E_i,\Theta)$ in the first interval of the width κ. As an ansatz an superposition of Legendre-Polynomials in $\cos\Theta$ can be used. These functions satisfy the boundary conditions at $\Theta = 0$ and $\Theta = \pi/2$ where the first derivation in Θ must be zero due to the symmetry.

The functional

$$\int\limits_{K_i^\circ - \frac{\kappa}{2}}^{K_i^\circ + \frac{\kappa}{2}} [P_j(K^\circ,\alpha_j) + P_k(K^\circ,\alpha_k)]\, dK^\circ - |s_i(K_i^\circ,\alpha_j)| - |s_i(K_i^\circ,\alpha_k)| \;\to\; 0 \tag{6.73}$$

is a hyper surface with the parameters a_i. Thus, it follows

$$s_i(K_i,\alpha_l) \;=\; n_n c \int\limits_{K_i^\circ - \frac{\kappa}{2}}^{K_i^\circ + \frac{\kappa}{2}} \int\limits_{K=0}^{K_{max}} L(K^\circ,K) \int\limits_{E=K_i^\circ - \frac{\kappa}{2}}^{K_i^\circ + \frac{\kappa}{2}} \int\limits_{\Theta=0}^{\pi} \int\limits_{\Phi=0}^{2\pi} \beta(E)K\, \frac{d^2\sigma(E,K,\Theta,\Phi,\alpha_l)}{dK\,d\Omega}$$

$$\times\; f(E_i,\Theta)\, d\Phi\, d\Theta\, dE\, dK\, dK^\circ \tag{6.74}$$

$$+\; n_n c \int\limits_{K_i^\circ - \frac{\kappa}{2}}^{K_i^\circ + \frac{\kappa}{2}} \int\limits_{K=0}^{K_{max}} L(K^\circ,K) \int\limits_{E=K_i^\circ + \frac{\kappa}{2}}^{K_{max}^\circ} \int\limits_{\Theta=0}^{\frac{\pi}{2}} \int\limits_{\Phi=0}^{2\pi}$$

$$\times\; \beta(E)\, \frac{d^2\sigma(E,K,\Theta,\Phi,\alpha_l)}{dK\,d\Omega}\, f(e,\Theta)\, d\Phi\, d\Theta\, dE\, dK\, dK^\circ; \quad l = j,k$$

The coordinates of the minimum can be calculated with a gradient method and a step width regulation [275]. For every new step one has to find the direction of the steepest descent and to carry out a line search to minimize the functional (6.73) in this direction. Supposing the interval i is not to wide and $f(E, \Theta)$ does not change too much in the energy interval one can take the final parameter values of the interval i as the initial values of the interval $i + 1$.

6.2.4.1.3   Polarization of Bremsstrahlung from an ECR Plasma   ECR plasmas have an anisotropic velocity distribution [276, 277] and the polarization of bremsstrahlung from an ECR plasma was reported for low temperature plasmas [278] and high temperature plasmas [279]. Thereby, the degree of polarization was related to the anisotropy of the plasma.

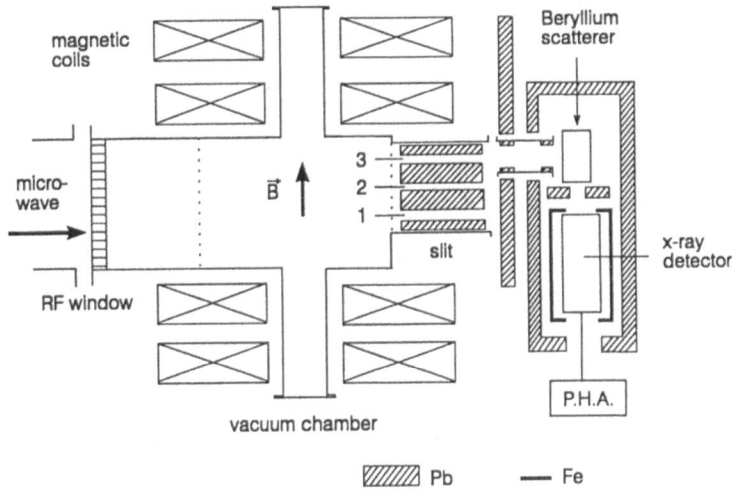

Figure 6.39   Principal scheme of the experimental apparatus for measuring the polarization of electron bremsstrahlung.

The degree of polarization P is defined as

$$P = \frac{J_\perp - J_\parallel}{J_\perp + J_\parallel} \tag{6.75}$$

where $J_\perp$ and $J_\parallel$ are the relative flux of bremsstrahlung quanta having a polarization vector perpendicular and parallel to the magnetic mirror axis, respectively. The measurements of $J_\perp$ and $J_\parallel$ can be performed as follows [278, 279]: As shown in Fig.**6.39** the bremsstrahlung beam emitted perpendicularly to the magnetic axis interacts with a Beryllium scatterer. Beryllium is often chosen, because single Compton scattering may occur more frequently than multiple ones in low Z materials [280]. If photons are scattered by 90° single Compton scattering their polarization vectors are conserved. Therefore, $J_\perp$ is scattered in a direction parallel to the magnetic axis and $J_\parallel$ is scattered in the perpendicular direction. Thus, by placing the face of the detector in the direction normal to the magnetic axis as shown in Fig.**6.39**, $J_\perp$ can be measured and placing it parallel

$J_{\parallel}$ can also be measured. It has been shown from Compton scattering theory [278, 281] that the true degree of polarization P_T is related to the measured degree of polarization P_{exp} by

$$P_T = \frac{R+1}{R-1} P_{exp} \tag{6.76}$$

where R is the asymmetry ratio of the detection apparatus. R is defined as the ratio of the intensity of detected photons having, before being scattered, a polarization vector parallel to the surface of the detector, to those photons having a polarization vector perpendicular to this surface.

In Ref.[279] the polarization of electron bremsstrahlung from a 2.45 GHz microwave heated plasma according to the geometry shown in Fig.**6.39**, is analyzed. Generally, $J_{\perp}$ is larger than $J_{\parallel}$. At the resonance zone, their difference becomes distinctive for high energy photons, while in the case at the off-resonance zone it becomes small.

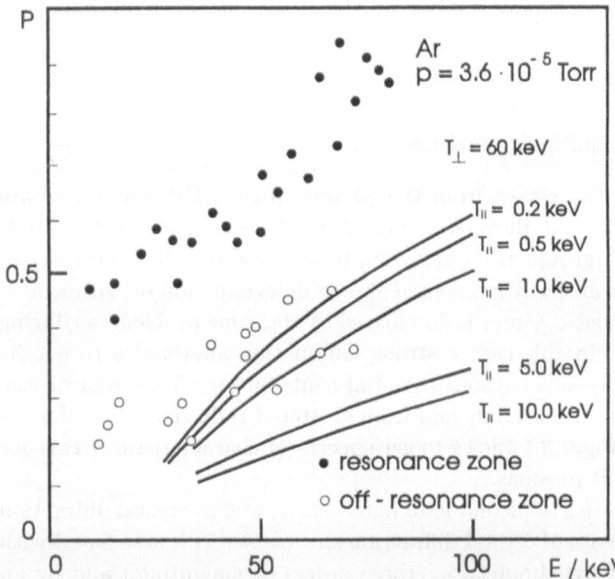

Figure 6.40 Comparison of the degree of polarization P between experimental and numerically calculated results [279].

According to (6.75) the degree of polarization is shown in Fig.**6.40**. The dots and circles represent the degree of polarization at the center and the edge of the cavity, respectively. At the center of the cavity (resonance zone), the degree of polarization is very high and becomes nearly unity for high energy photons, while at the edge of the cavity the degree of polarization is low compared with the above.

For a given electron energy distribution function $F(E, \Theta)$ the quantities $J_{\perp}$ and $J_{\parallel}$ can be calculated using (6.71) but splitting into $J_{\perp}$ and $J_{\parallel}$ just taking $d\sigma_{\parallel eff}/dK\, d\Omega$ and $d\sigma_{\perp eff}/dk\, d\Omega$ instead of $d\sigma_{eff}/dK\, d\Omega$. These relativistic cross-sections were derived by Glückstern [283, 284].

It was shown, that the degree of polarization decrease with increasing gas pressure, because the collisions of electrons increase with the gas pressure and the plasma becomes isotropic. Further, the provided studies show, that the plasma is extremely anisotropic in the resonance zone and the electron temperatures perpendicular to the magnetic axis are an order of magnitude higher than the parallel ones.

6.2.4.2   Spectroscopy of Characteristic X-Rays

6.2.4.2.1   Measurement of Characteristic X-Rays   The spectroscopy of characteristic X-rays from an ECR plasma allows one to get information on

- the ion charge distribution inside the source plasma which obviously differs from the charge state distribution of ion beams emitted by the plasma;
- the working conditions of the ion source;
- the relation between low and high energetic electrons in the source (ions as a threshold detector);
- an anisotropy estimation of $f(E, \Theta)$.

Furthermore, in the case of highly charged ions and wavelength dispersive measurements an analysis of X-ray spectra gives also information on the atomic structure of ions, such as

- X-ray transition energies;
- X-ray line intensities;
- X-Ray line asymmetries (multiplet coupling).

Measurements of characteristic X-rays from the plasma of an ECR ion source are known from Briand et al. [286] and Herpich et al. [287]. Spectroscopy in the VUV region was provided by different groups [288, 289, 290] to measure the characteristics of cold electron populations inside an ECR ion source and to determine ion densities.

The spectroscopy of characteristic X-rays is dominated by the same problems as during bremsstrahlung measurements. In this case a strong collimation allows also to get X-ray spectra from the working gas ions without essential contributions from fluorescence excitation of wall and construction materials and from scattered radiation. Therefore, a collimator system as shown in Fig.**6.33** allows to get spectra of characteristic X-rays for diagnostic investigations on ECR plasmas.

For successful plasma diagnostics semiconductor detectors as well as crystal diffraction spectrometers can be used. The use of crystal diffraction spectrometers is associated with a decreasing detector efficiency (to about two... three orders of magnitude) and by an increasing resolving power. A comparison of the energy resolution of semiconductor detectors and crystal diffraction spectrometers is given in Fig.**6.41**.

Detailed information on detector systems, which can be used for X-ray diagnostics, are given for instance in [291, 292, 293, 294, 295].

6.2.4.2.2   Production of Characteristic X-Rays   To observe characteristic X-ray lines a certain number of electrons must have energies higher than the ionization potential of the shell in which the primary vacancy leading to the observed X-ray transition is created. Taking photoionization into account the intensity of K transitions is given as

$$I_K = n\,\omega_X\,(\ \int\limits_{E_{B,K}}^{\infty}\ F^\star(E)\,\sigma_i^K(E)\,v(E)\,dE\ +\ c \int\limits_{E_{B,K}}^{\infty}\ P^\star(K)\,\sigma_{i,\text{photo}}^K(K)\,dK\ ) \quad (6.77)$$

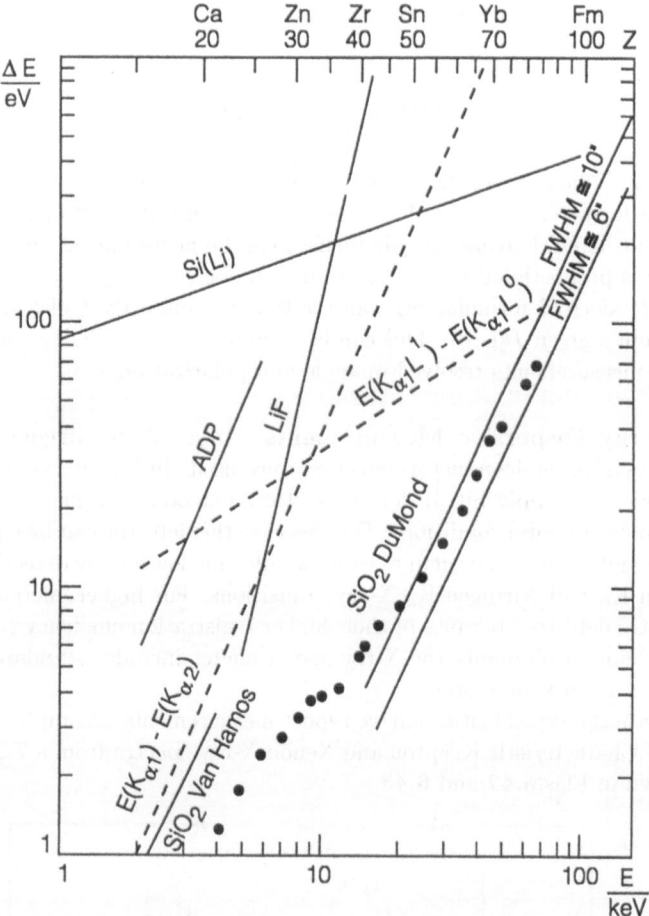

Figure 6.41   Energy resolution of semiconductor detectors and crystal diffraction spectrometers in the energy range of characteristic X-rays.

with n – total density of the ions and atoms, K – photon energy, I_K – summed intensity of all K transitions, ω_X – fluorescence yield, c – velocity of light, $E_{B,K}$ – binding energy of the K shell, $\sigma_i^K(E)$ – K-shell electron impact ionization cross-section, $\sigma_{i,photo}^K(K)$ – K-shell photoionization cross-section, $c\,P^\star(K)$ – bremsstrahlung photon flux inside the plasma and $P^\star(K)$ – bremsstrahlung spectrum integrated over all emission angles. For other shells the formulae are similar.

The bremsstrahlung photon flux inside of an anisotropical plasma depends on the angle of emission relative to the axis of symmetry and is itself a function of the electron energy distribution function. In the isotropical case the spectrum is given as

$$P(K) = n \int_K^\infty F(E) \frac{d\sigma_{Br}}{dK}(K,E)\, v(E)\, dE \qquad (6.78)$$

with $\dfrac{d\sigma_{Br}}{dK}(K,E)$ being the total bremsstrahlung cross-section [296]. For an anisotropical plasma yields

$$P^\star(K) \;=\; \int\limits_{0}^{\frac{\pi}{2}} p(K,\theta)\,d\theta, \qquad (6.79)$$

as the real total spectrum integrated over the θ-dependent emission $p(K,\theta)$. Due to the geometry computing $p(K,\theta)$ using differential cross-section is difficult and requires much time. However, for light elements the photoeffect can be neglected because bremsstrahlung production is proportional to Z^2 (Z: atomic number).

Looking at the derived formulae one can see that the intensity I of a transition can be calculated from a given $f(E,\theta)$. This can be used as a test for $f(E,\theta)$ obtained from bremsstrahlung measurements resolved in angle and polarization.

6.2.4.2.3 Energy Dispersive Measurements As a rule for diagnostic purposes often energy dispersive semiconductor detectors were used. In the energy region beyond 30 keV it is obvious to apply Si(Li) detectors. Here detectors are known, which allow measurements under vacuum conditions. For this case the detector can be equipped with an conflat flange and a soft X-ray vacuum window, allowing low energy detection of X-rays down to Oxygen K$_\alpha$ and Nitrogen K$_\alpha$ X-ray transitions. For higher energies it is more common to use Ge detectors because of their higher registration efficiency ($\sigma_{\text{photo}} \sim Z^5$) in this energy region. Commonly the X-ray spectrometer includes standard electronics and a PC as data acquisition system.

To demonstrate the typical situation we report measurements accomplished by Friedlein et al. [268]. Characteristic Krypton and Xenon X-ray spectra from a 7.25 GHz ECR plasma are shown in Figs.**6.42** and **6.43**.

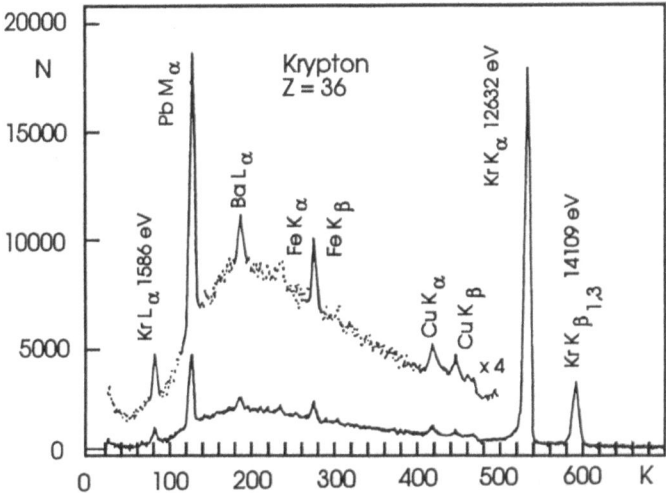

Figure 6.42 Krypton X-ray spectrum emitted by a 7.25 GHz ECR plasma.

In the spectra besides the lines from the working gas characteristic X-rays from wall and collimator materials appear.

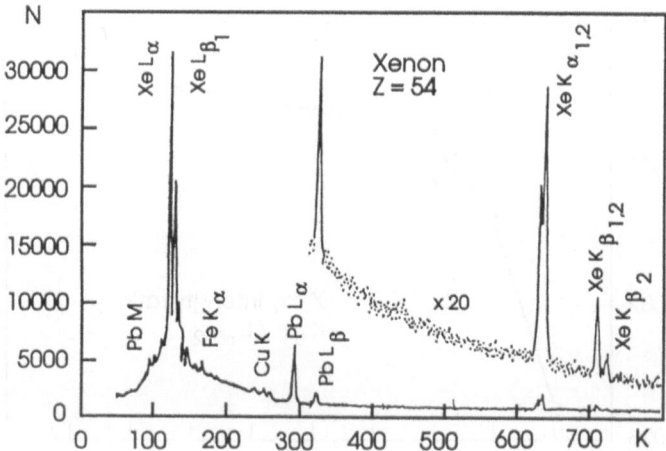

Figure 6.43   Xenon X-ray spectrum emitted by a 7.25 GHz ECR plasma.

In the Xenon spectrum a very low intensity of the K transitions and strong L transition lines were found. This occurs despite the values of the fluorescence yields: $\omega_K(\mathrm{Xe}) = 0.89$ and $\omega_{L_{III}}(\mathrm{Xe}) = 0.09$. On the other hand, in the case of the Krypton spectrum strong K lines and much weaker L transition lines were observed. The reason is the behavior of the electron distribution function, e.g. there are far less energetic electrons to ionize electrons from the Xenon K shell. Thus, the spectroscopy of different characteristic lines serves as a *threshold detector* for the excistence of energetic electrons. As threshold acts here the electron binding energy corresponding to the initial electronic state of the observed X-ray transition. Electron binding energies for the K and L_{III} levels of the neutral atom are shown in Figs.**6.44** and **6.45**.

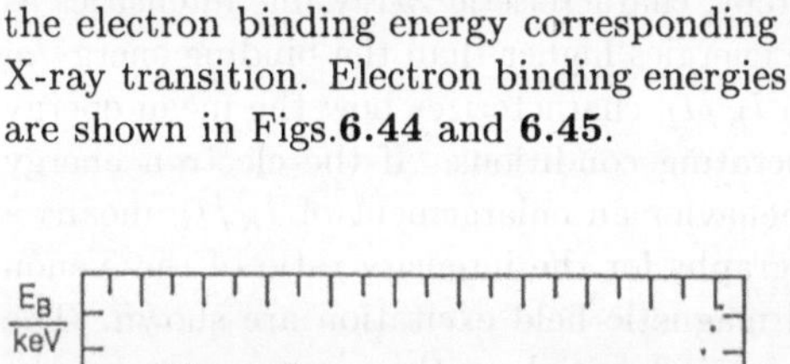

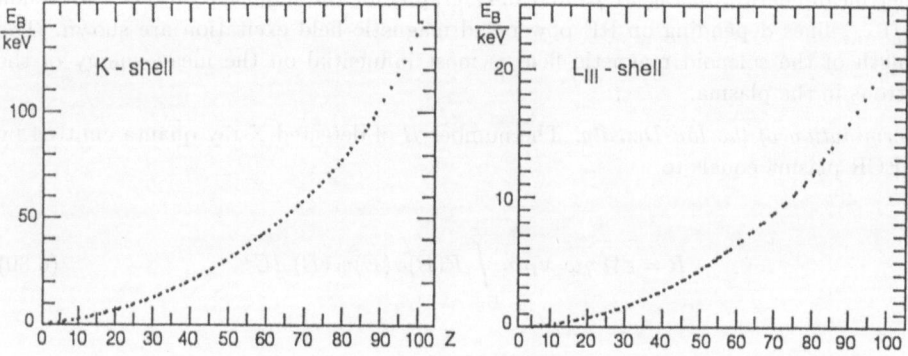

Figure 6.44   K shell electron binding energy as a function of the atomic number Z for neutral atoms.

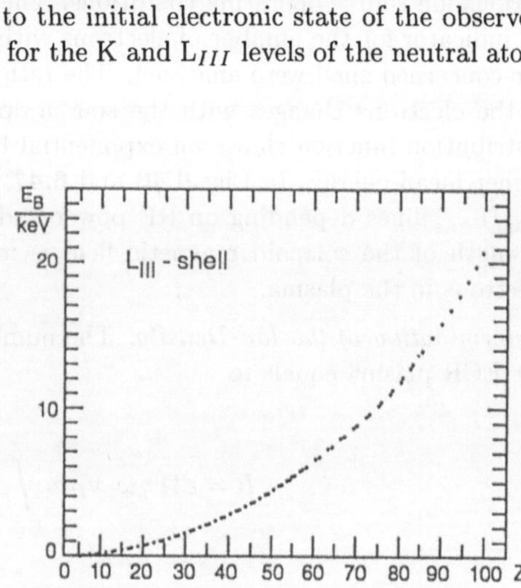

Figure 6.45 L_{III} shell electron binding energy as a function of the atomic number Z for neutral atoms.

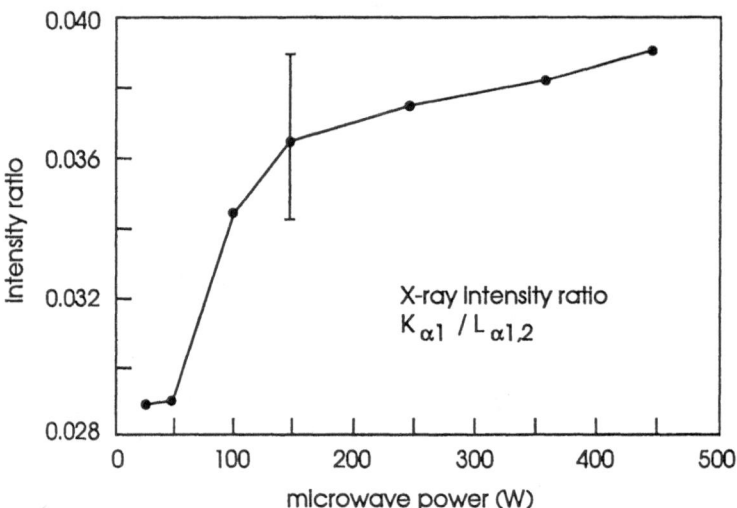

Figure 6.46 $K_{\alpha_1}/L_{\alpha_{1,2}}$ intensity ratio of Xenon as a function of the RF power measured at the ECR ion source of the TU Dresden.

Due to line energies, line widths and relative transition intensities information about the occupation of different atomic subshells of the ions inside the plasma can be deduced.

Line Intensity Ratios In RefZi96 X-ray spectra of an Xenon plasma were measured under various ion source operating conditions. Thereby, characteristic X-ray line intensities as an indicator for the number of electrons with energies higher than the binding energy of the concerned shell were analyzed. The ratio I_K/I_L characterizes how the mean energy of the electrons changes with the source operating conditions. If the electron energy distribution function shows an exponential behavior an enlargement of I_K/I_L means a higher mean energy. In Figs.**6.46** and **6.47** graphs for the intensity ratio of the Xenon $K_{\alpha_1}/L_{\alpha_{1,2}}$-lines depending on RF power and magnetic field excitation are shown. The strength of the solenoid magnetic field is most influential on the mean energy of the electrons in the plasma.

Determination of the Ion Density. The number R of detected X-ray quanta emitted by the ECR plasma equals to

$$R = \varepsilon \, \Omega \, \eta \, \omega_i \, V_P n_i \int_{E_B}^{\infty} F(E)\sigma(E)v_e(E) \, dE \qquad (6.80)$$

with n_i – ion number; n_e – electron distribution function; E_B – electron binding energy of the K,L shells; σ – ionization cross-section of the corresponding subshell; v_e – electron velocity; η – radiation attenuation; ε – detector efficiency; Ω – solid angle; ω_i – fluorescence yield; V_P – plasma volume. Therefore, we get information about the ion density

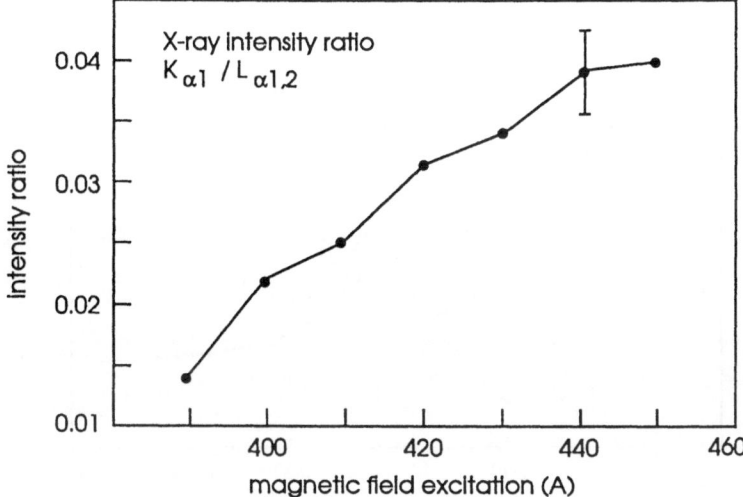

Figure 6.47 $K_{\alpha_1}/L_{\alpha_{1,2}}$ intensity ratio of Xenon as a function of the exciting current in the magnetic coils, measured at the ECR ion source of the TU Dresden.

by

$$n_i = R \left[\varepsilon\, \Omega\, \eta\, \omega_i\, V_P \int\limits_{E_B}^{\infty} F(E)\sigma(E)v_e(E)\, dE \right]^{-1} \qquad (6.81)$$

if the electron energy distribution function is determined by an independent alternative way.

Energy dispersive measurements in the low energy region. For different purposes (e.g. for gas mixing) it is important to analyze X-rays emitted from light elements acting as coolants such as Oxygen and Nitrogen [297] and also from light noble gases. For these elements spectroscopic measurements are more difficult due to radiation attenuation in window materials and in some cases also in air.

Using a Si(Li) detector Argon X-ray spectra were measured under different source operating conditions. An example for an Argon spectrum is given in Fig.6.48. Besides of the characteristic ArK_α line (E = 2957 eV) lines from Oxygen and Nitrogen as plasma components and fluorescence lines from wall and collimator materials are observed.

Using X-ray lines excited by fluorescence excitation (Al, Cr, Fe) the spectra were energy calibrated. Spectra processing was done with a code written by Küchler [299]. Thus, an energy shift of the ArK_α transition line of 70.5 eV was calculated, corresponding to X-ray emission from Argon ions with a mean ionization stage of 10+...11+ comparing the measured energy shift with calculated ones [10].

The K_β/K_α intensity ratio for Argon was determined to 0.38±0.03. This deviates strongly from the value of 0.1088 for the neutral atom [300]. The physical reason for this behavior is connected to the different populations of atomic subshells by multiple ionization processes and vacancy cascades.

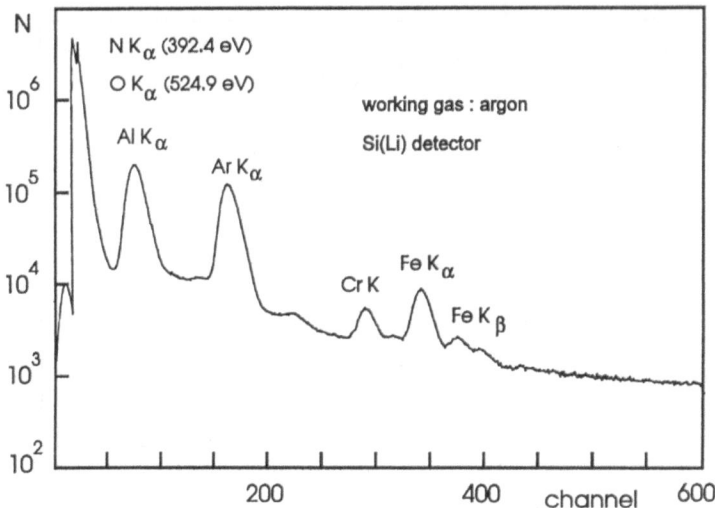

Figure 6.48   Argon X-ray spectrum of an 7.25 GHz ECR plasma measured by a Si(Li) detector.

X-rays and electron anisotropy. In plasmas the motion of hot electrons is strongly correlated to the general plasma behavior. How the momentum vector is directed effects during the coupling of electromagnetic waves e.g. due to ECR heating and losses if these electrons are confined in magnetic or electrostatic traps. In other words knowing their location in phase space provides information on the stability of the plasma. In the case of cylindrical symmetry this location is described by the anisotropical electron energy distribution function $f(E, \theta)$ (θ: angle between the momentum vector and the axis of symmetry).

The method of obtaining $f(E, \theta)$ at high energies is to measure continous radiation (electron-ion bremsstrahlung and direct radiative recombination radiation) resolved in angle and polarization followed by an simulation with model functions or by an angle dispersive de-convolution of the spectra [268, 301].

Often the access to the plasma chamber is restricted to one port or it could be desired to observe only from one direction to avoid the problem of different active volumes. As a rule in this case the information about the anisotropy and so about the absolute height of the distribution function is limited. One had to do simulations or to assume isotropy in momentum space and to perform an isotropical de-convolution for the electron energy distribution function $F(E)$ using total bremsstrahlung cross-section data.

A complementary way to get information about the distribution function inside the plasma is the spectroscopy of characteristic X-rays. These photons provide values for

$$F^\star(E) = \int_0^{\pi/2} f(E, \Theta)\, d\Theta, \tag{6.82}$$

the real electron density, and furthermore in combination with bremsstrahlung measurements information on the anisotropy of $f(E, \Theta)$.

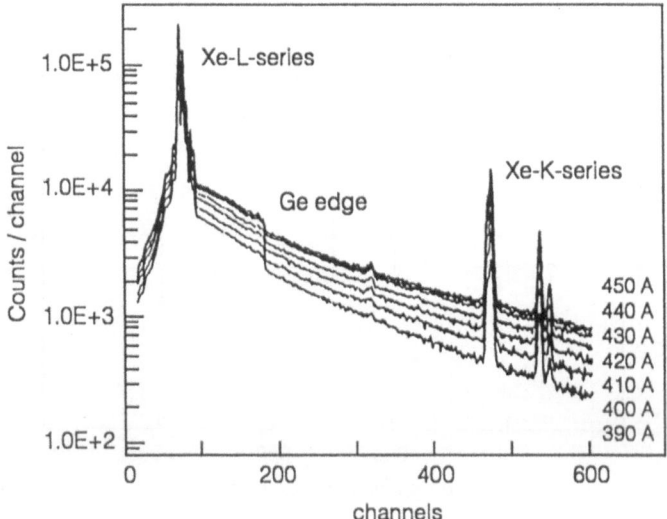

Figure 6.49   Xenon X-ray line spectra for different magnetic excitation conditions of an ECR plasma.

Series of Xenon X-ray line spectra emitted from the 14.5 GHz ECR ion source of the TU Dresden [302] at different working conditions (variation of RF power input, magnetic field excitation and gas pressure) have been recorded with a HP Ge-detector. Details on the experimental set-up are given in Ref.[303]. At the 7.25 GHz ECR ion source of the Forschungszentrum Rossendorf [303] measurements on an Argon plasma were performed with the same experimental apparatus [305]. The Argon K_α line at 2.96 keV could be detected but the efficiency of the detector at this energy and therefore the absolute intensity is poorly known. From the corrected spectrum $P(K)$ (K: photon energy) $F(E)$ was calculated and fitted with a two component distribution [268]

$$F(E) = a_{\text{hot}}\, E\left(1 + \frac{E}{2mc^2}\right)\left(1 + \frac{E}{mc^2}\right)\exp\left(-E/T_{\text{hot}}\right)$$
$$+ \; a_{\text{warm}}\, E\left(1 + \frac{E}{2mc^2}\right)\left(1 + \frac{E}{mc^2}\right)\exp\left(-E/T_{\text{warm}}\right)$$

with a_{warm}, a_{hot}, T_{warm} and T_{hot} as fitting parameters.

At a given angle of observation (e.g. at the z-axis) one gets for a given bremsstrahlung spectrum one gets different distribution functions for assuming isotropy in phase space (using total bremsstrahlung cross-section data) or assuming anisotropy (using differential cross-section data). There exists a set of anisotropical functions and the correct one has to be chosen by line intensity observation and/or measurements of spectra at different angles and of the degree of polarization (not at the z-axis where the degree of polarization is exact 0). The degree of anisotropy is correlated to

$$A = \frac{F^\star(E)}{F(E)} \tag{6.83}$$

the so-called anisotropy ratio. A must be multiplied with $F(E)$ to get the real electron density per energy interval. In other words A corresponds to the scaling of the spectrum

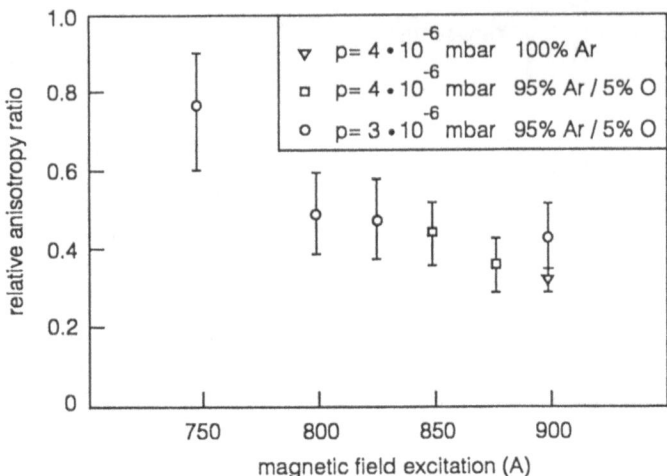

Figure 6.50 Measured anisotropy ratios for changing magnetic field excitation for Argon at the 7.25 GHz ECRIS of the FZ Rossendorf (microwave power 40 W).

due to the distribution of the electrons in θ. The energy dependence of the function used is:

$$F(E) = E^{\frac{\chi}{2}} (1 + \frac{E}{2\,mc^2})^{\frac{\chi}{2}} (1 + \frac{E}{mc^2}) \exp\left(-\frac{E}{T_\perp}\right) \tag{6.84}$$

with χ as a parameter. The temperatures T are perpendicular temperatures respectively. The slope of A depends not only on $T_\perp$ but of course also strongly on the direction of observation. If the anisotropy ratio would be approximately independent from the electron energy one could calculate A from the bremsstrahlung spectrum and the line intensity. This can mostly be the case especially for distribution functions with an exponential behavior, because electrons with energies just above the considered binding energy are strongly weighted.

For the determination of A the quantity $F(E)$ from the isotropical de-convolution of the spectrum is used. If the photoeffect is neglected follows

$$F^\star(E) = \int\limits_0^{\frac{\pi}{2}} f(E, \theta)\, d\theta \tag{6.85}$$

from the inversion of the line intensity I_K:

$$A = \frac{I_K}{n\,\omega_X \int\limits_{E_{B,K}}^{\infty} F(E)\, \sigma_i^K(E)\, v(E)\ dE}. \tag{6.86}$$

If there would exist more than one characteristic line it is possible to split the electron energy interval and to obtain A for each interval. For an Argon plasma results for the determination of A are shown in Fig.**6.50**. For the observation at the axis a higher anisotropy ratio A indicates electrons at lower θ meaning $f(E,\theta)$ to be more isotropic. For Fig.**6.50** it can therefore be conducted that the anisotropy of $f(E,\theta)$ changes with the magnetic field excitation.

X-ray emission profiles. For a more complete description of the ECR plasma the use of a pin-hole camera with an energy and position sensitive CCD-array allows it to analyze the geometrical size and spatial X-ray emission profiles in different energetic regions of the plasma. Similar methods were used in the past to measure the electron beam diameter of the LLNL EBIT [306], the geometric size of an ECR plasma by using a film registration technique [307] and for diagnostics of heavy ion induced dense plasmas [308].

Spectroscopic imaging means that position and energy of an incident X-ray photon are detected simultaneously. In Ref.[309] a CCD camera with a large depletion depth (about 50 μm) was used to detect X-ray photons directly, without any X-ray phosphor. The device comprises 576*385 front-side illuminated pixels, each 22 μm square. The sensor chip and the preamplifier were cooled to 140 K to lower the readout noise and reduce the dark signal. Temperatures below 140 K were not used, because that would deteriorate the charge transfer efficiency within the device. For profound details concerning the X-ray performance of CCD devices see [310, 311].

The usage of the charge image at a CCD sensor for spectroscopic purposes requires that no more than one photon is incident on each pixel in any image frame. The charge generated by one X-ray photon is commonly not localized in only one pixel, but splitted over a few neighbouring ones. Therefore, a computer code that collects the charge of a whole continuous pixel cloud and interprets its center of gravity as the registration point of one photon must be applied. The energy resolution of the CCD-device allows a FWHM of the full-energy peak of 180 eV for Mn K-radiation.

CCD detection devices are profitably applicable as pinhole cameras. Such a technique has been used for several imaging applications without the need of a spectroscopic resolution [307, 308]. In Ref.[309] a CCD-detector was used in a spectroscopical pinhole camera for X-ray imaging at the 14.5 GHz ECR source of the Technical University of Dresden. A crossed-slit diaphragm with a 50 μm square aperture was mounted in 2 m distance from the ECR discharge on the axis of the source. About 40 cm behind it was mounted the CCD-camera, seeing the whole plasma volume out to the wall of the vacuum chamber (see also Fig.**6.51**). The ECR source was operated with Krypton under a pressure of $5*10^{-4}$ Pa and 100 W microwave power.

As the construction of the ECR source allows a detection of the axial radiation only, the described investigations were restricted to images that integrate over the plasma length.

Fig.**6.52** shows the full X-ray emission intensity in the energy range from 5 to 50 keV. Photons below 5 keV could not been detected due to the absoption in air, photons above 50 keV might possibly penetrate the diaphragm walls and were supressed therefore. The visible outer ring is the inner surface of the vacuum chamber with a diameter of 60 mm. The plasma zone has a diameter of about 30 mm and is located in the center of the chamber. There exist three angular directions at the ends of the plasma zone in which the electrons are not properly confined, due to a destructive superposition of the axial field component of the hexapole and the solenoid coils. In these three directions an electron current out from the plasma zone is observed, visible by the characteristic and bremsstrahlung emission from the vacuum chamber wall. In the 12.6 keV region (Fig.**6.53**)

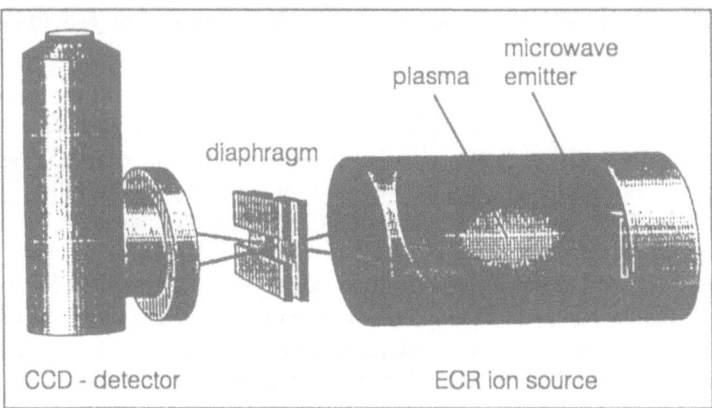

Figure 6.51   Experimental arrangement for measuring the X-ray emission from the ECR plasma.

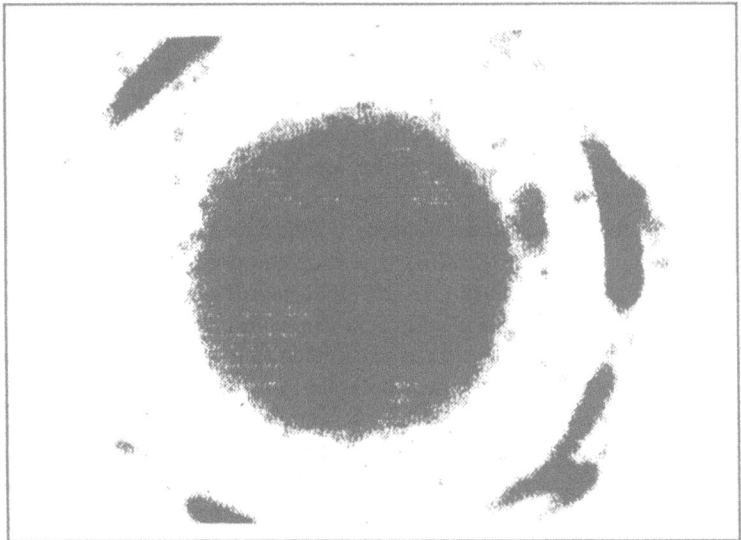

Figure 6.52   Integral axial X-ray emission.

mainly the plasma is emitting Krypton X-rays. Here an nearly homogeneous emission of Krypton X-rays over the plasma cross-section is observed. Nevertheless, it is neccessary to note, that the direction of observation is along the z-axis and the resulting picture corresponds to an integration over an baseball shaped plasma volume.

Fig.**6.54** shows the radial intensity distribution of the emitted radiation in three spectral ranges. It is clearly visible, that the high energetic bremsstrahlung radiation is emit-

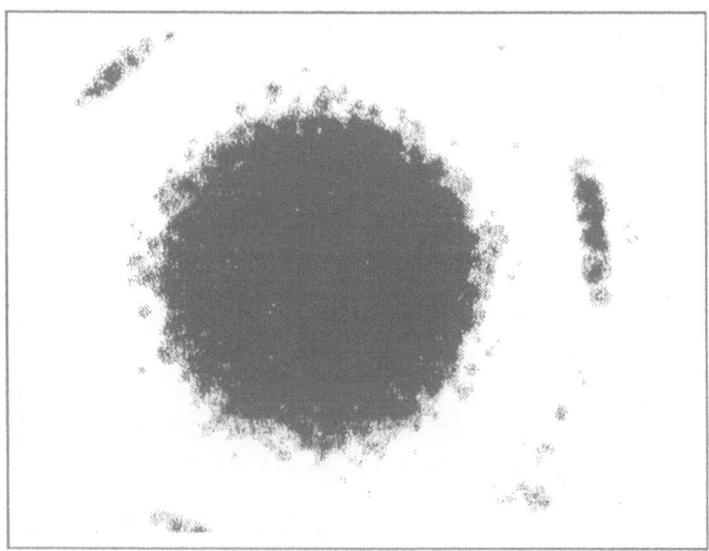

Figure 6.53   Axial X-ray emission of an ECR plasma in the spectral range of the characteristic Krypton K X-rays (12.3...12.9 keV).

ted from a smaller radial range than the characteristic Krypton X-rays and low energetic bremsstrahlung radiation. Additionally, it tends to peak in the center of the emitting plasma volume, which effect is much less significant for the low energetic radiation. It seems, that the highest energetic radiation is generated preferably in the endcaps of the plasma ellipsoid, but this assertion is not verifialbe without a radial view of the plasma.

The described method allows it, to measure geometric plasma profiles and radial intensity distributions of emitted X-rays in different spectral ranges. Plasma X-ray images give information on directions of electron losses in the vacuum chamber and after corresponding mathematical treatment also information on electron density profiles inside the plasma.

6.2.4.2.4   Wavelength Dispersive Measurements   The use of crystal diffraction X-ray spectrometers is an excellent tool to analyze small effects in the atomic subshells of atoms and ions and to identificate ionic charge states. The high resolution power of diffraction spectrometers allows investigations of ion charge state distributions, where other spectrometers are not able to give detailed information due to their resolution.

Fig.**6.55** shows a diffraction spectrum of the Krypton $K_{\alpha_{1,2}}$ doublett, measured in second order of diffraction with a bent in Johann geometry (radius of Rowland circle 200 mm) LiF(200) crystal at the 14.5 GHz ECR ion source of the TU Dresden [268]. The measured spectrum shows for the K_{α_1} as well as for the K_{α_2} transitions a clear asymmetric form arising from superimposing X-ray transitions caused by ions of different charge states. The measured spectrum can be approximated by Voigt profiles with contributions from the natural line width and from the geometry function.

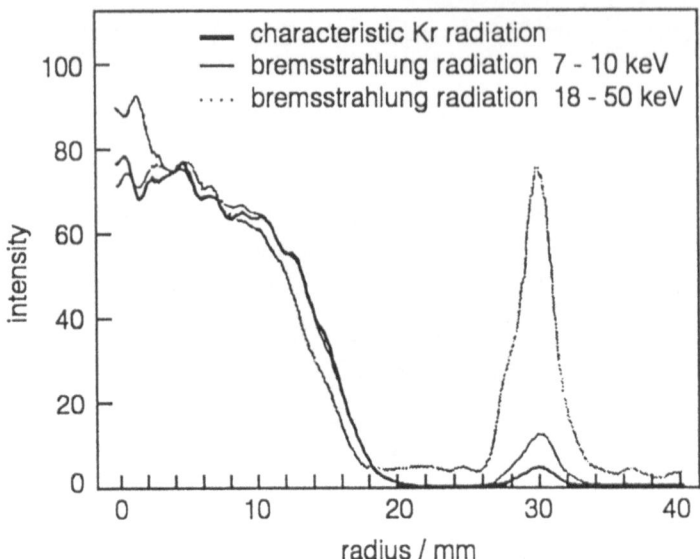

Figure 6.54   Radial intensity distribution of the emitted radiation in different spectral regions.

Analyzing the energy shifts of the peak components information on the ion charge state distribution inside the ECR plasma can be deduced. Thus three components were identified, where the first (most intense) one corresponds to ionization states from Kr^{0+} up to Kr^{18+}. This peak is representative for ionization of electrons from the M shell, e.g. of $3s^23p^63d^{10}$ electrons. Ionization of electrons from the M shell causes energy shifts for Krypton of less than 4 eV, so that only one superimposed line containing all contributions from ionization states up to Kr^{18+} is found. The two peak components with much lower intensity arise from ionization of one or two L shell electrons, where energy shifts are clearly stronger [10].

It must be noted, that the energy resolution obtained is not the ultimate resolution. In these experiments diaphragm widths of 250 μm are chosen to get a high registration efficiency. As the measurements have shown, it is also possible in a reliable time to measure with much smaller diaphragm widths, e.g. with better resolution.

In the case of poor magnetic confinement diffraction spectra are measured from contaminating elements sputtered from wall materials. So, an iron spectrum showing satellite structures from additional M and L vacancies was obtained in first order of diffraction by an LiF(200) crystal.

6.2.4.3   VUV Spectroscopy

6.2.4.3.1   VUV Spectroscopy on ECR Plasmas   ECR ion sources emit in a wide electromagnetic wavelength range. Line and continuous radiation are emitted beginning with the microwave region up to X-rays. These kinds of radiation characterize the plasma state and could be used to analyze the plasma without any disturbance of the plasma.

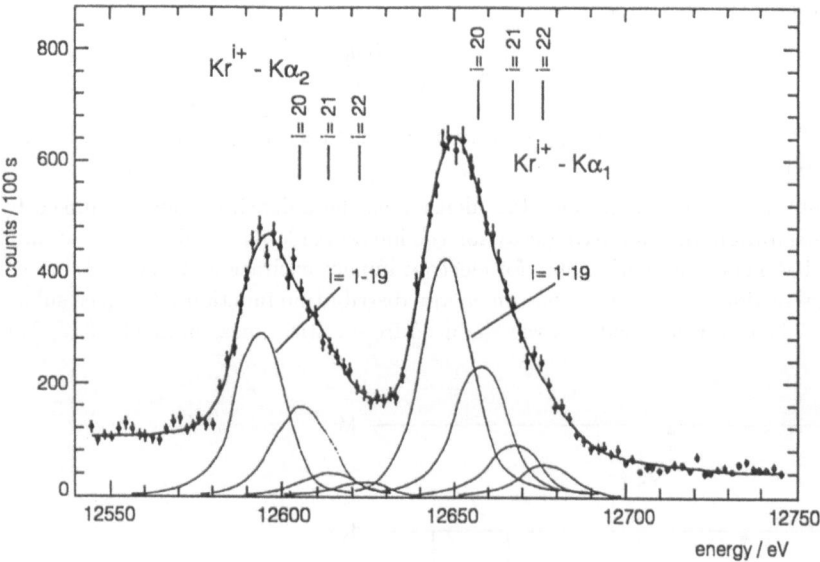

Figure 6.55   Diffraction spectrum of K X-rays from Krypton ions in the plasma of the 14.5 GHz ECR ion source of the TU Dresden.

Beside diagnostics of ECR plasmas, such methods are widely used for investigations of plasmas properties in astrophysics and in plasmas of thermonuclear devices.

Because of the specific plasma parameters of ECR ion sources fundamental standard approaches of classic spectroscopic diagnostics [312, 313, 314] are not applicable. This is the reason, that up to now only a few publications on spectroscopic diagnostics of ECR plasmas are known (see for instance [288, 315, 316]).

ECR ion sources are intense sources of electromagnetic radiation, so that spectroscopy in the visible, UV, VUV and X-ray ranges is possible. This fact will also be used to determine source parameters.

6.2.4.3.2   Determination of Ion Densities in an ECR Discharge

The spectroscopy of VUV transitions from ions of different charge states allows it to determine ion densities. In this paragraph we will illustrate the principal method following investigations on Nitrogen ions by Vinogradov et al. [288].

Vinogradov et al. [288] for VUV Spectroscopy used a symmetric normal incidence monochromator equipped with a holographic Pt grating of 3600 lines per mm. The spectrometer entrance slit had a width of $10\mu m$ to $30\mu m$ corresponding to a resolution of 0.01 nm to 0.03 nm. The absolute photon efficiency $\varepsilon(\lambda)$ was determined by using a diaphragm close to the resonator for defining the solid angle and excitation of lines with known excitation cross-sections by an electron beam as a secondary standard. The relation between the photon counting rate N_{jh} given as the integral over a definite line from the decay of an excited level j to level h and the population density n_j of the state j is given by

$$n_j \, A_{jh} = \frac{4\pi N_{jh}}{\Delta\Omega \, V_P \, \varepsilon(\lambda)} \tag{6.87}$$

where $\Delta\Omega$ is the solid angle, V_P the plasma volume seen by the spectrometer and A_{jh} the Einstein coefficient.

To obtain from n_j the ground state density n_0 these densities must be linked by balancing excitation and radiative decay for the ions considered (modified corona model). For this balancing rate coefficients for electron impact excitation are needed, i.e. excitation cross-sections σ_{0j} and the electron energy distribution function $F(E)$ (see subsection 6.2.4.1). The electron density n_e was deduced from optical measurements [317, 318].

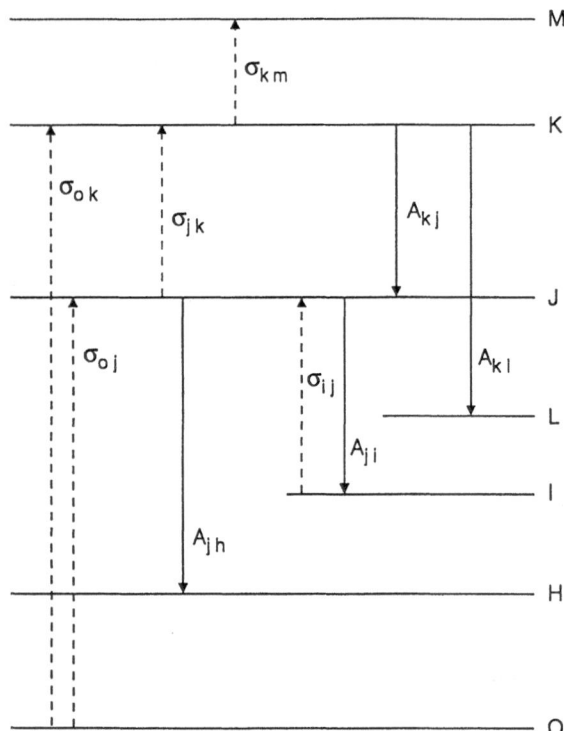

Figure 6.56 Level population and depopulation in the modified corona model.

The modified corona model considers population and depopulation of a certain level j (see Fig.**6.56**). At low gas pressures of the discharge (below 10^{-2} Pa) the plasma is optically thin and has low densities, i.e. photoionization, photoexcitation and recombination processes can be neglected. Diffusion losses can also be neglected for usual optical levels having life times of about 10 ns compared to confinement times of microseconds (note: this may be different for metastable and highly excited states). This population is mainly due to the excitation from lower lying levels and the radiative decay of higher levels. Depopulation is due to the radiative decay of levels j and electronic excitation to higher levels and ionization from level j. Assuming quasistationarity and taking into account only one step cascading the balance equation for n_j holds

$$n_e n_0 < \left[\sigma_{0j} + \sum_{k>j} \sigma_{0k} \frac{A_{kj}}{\sum_{l<k} A_{kl} + n_e \sum_{m>k} < \sigma_{km} v_e >} \right] v_e >$$ (6.88)

$$+ n_e \sum_{i<j} n_i < \sigma_{ij} v_e > -n_j \left[A_{jh} + \sum_{i<j} A_{ji} + n_e \sum_{k>j} < \sigma_{jk} v_e > \right] = 0 .$$

Here v_e is the electron velocity and the σ_{kl} are cross-sections for excitation from level k to level l. Excitation by charge exchange [318] was not taken into account. Equations (6.87) and (6.88) can be solved for the ground state density n_0

$$n_0 = \frac{4\pi N_{jh}}{A_{jh} \Delta\Omega V_P \, \varepsilon(\lambda)} \left[A_{jh} + \sum_{i<j} A_{ji} + n_e \sum_{k>j} < \sigma_{jk} v_e > \right]$$ (6.89)

$$\times \left(n_e < \left[\sigma_{0j} + \sum_{k>j} \sigma_{0k} \frac{A_{kj}}{\sum_{i<k} A_{ki} + n_e \sum_{m>k} < \sigma_{km} v_e >} \right] v_e > \right.$$

$$\left. + \frac{n_e}{n_0} \sum_{i<j} n_i < \sigma_{ij} v_e > \right)^{-1} .$$

For evaluating n_0 from measured line intensities all atomic data (A_{kl}, σ_{kl}) and plasma parameters like $F(E)$ and n_e must be known. As a rule, in the literature only a few published data for different transitions to obtain electron excitation cross-sections are known. For most transitions detailed calculations must be done.

In the case of Nitrogen in [288] for optically allowed transitions the semiempirical Van Regemorter formula [319] was used to calculate excitation cross-sections

$$\sigma_{ij} = 2hcR_H \frac{4\pi^2 f_{ij}}{k^2 \sqrt{3}(E_i - E_j)} g(k_i, k_j)$$ (6.90)

with $R_H = 0.529 \cdot 10^{-10}$ m. The Gaunt factor $g(k_i, k_j)$ is approximated by

$$g(k_i, k_j) = a + b \frac{1}{1 + \left(\frac{x}{c}\right)^d} .$$ (6.91)

Here $x = \sqrt{E/(E_i - E_j)}$ and f_{ij} is the oscillator strength of the transition i to j. For neutral atoms holds $a = 0.0074069$, $b = 1.571346$, $c = 4.20839$ and $d = -1.8304$. The fitting coefficients for positive ions are $a = 0.17912$, $b = 1.25915$, $c = 4.314785$ and $d = -2.41377$ [288].

For higher energies the Bethe approximation can be used

$$\sigma_{ij} = \frac{2hcR_H 4\pi f_{ij}}{k^2 (E_i - E_j)} \ln \frac{4E}{E_i - E_j} .$$ (6.92)

For evaluating n_0 by using (6.89) the n_i must be known, i.e. the population densities of the i-th level from which the excitation to level j may take place. Most important are metastable levels close to the ground state. To obtain these densities, a similar balancing is necessary

$$
n_e n_0 < \left[\sigma_{0i} + \sum_{k>i} \sigma_{0k} \frac{A_{ki}}{\sum_{l<k} A_{kl} + n_e \sum_{m>k} < \sigma_{km} v_e >} \right] v_e >
$$

$$
- n_i \left[A_{i0} + n_e \sum_{k>i} < \sigma_{ik} v_e > \right] = 0 . \tag{6.93}
$$

Thus, it follows

$$
n_i = \frac{n_e n_0 < \left[\sigma_{0i} + \sum_{k>i} \sigma_{0k} \dfrac{A_{ki}}{\sum_{i<k} A_{ki} + n_e \sum_{m>k} < \sigma_{km} v_e >} \right] v_e >}{A_{i0} + n_e \sum_{k>i} < \sigma_{ik} v_e >} . \tag{6.94}
$$

For optically forbidden transitions (i→ 0)

$$
A_{i0} \ll n_e \sum_{k>i} < \sigma_{ik} v_e >
$$

yields

$$
n_i = \frac{n_o < \left[\sigma_{0i} + \sum_{k>i} \sigma_{0k} \dfrac{A_{kj}}{\sum_{i<k} A_{ki} + n_e \sum_{m>k} < \sigma_{km} v_e >} \right] v_e >}{\sum_{k>i} < \sigma_{ik} v_e >} . \tag{6.95}
$$

Using this expression for the ground state density n_0 follows

$$
n_o = \frac{\dfrac{4\pi N_{jh}}{A_{jh} \Delta \Omega V_P \varepsilon(\lambda)} \left[A_{jh} + \sum_{i<j} A_{ji} + n_e \sum_{k>j} < \sigma_{jk} v_e > \right]}{n_e < \left[\sigma_{oj} + \sum_{k>j} \sigma_{ok} \dfrac{A_{kj}}{\sum_{i<k} A_{ki} + n_e \sum_{m>k} < \sigma_{km} v_e >} \right] v_e > + B} \tag{6.96}
$$

where

$$
B = < \sigma_{ij} v_e > \sum_{i<j} \frac{< \left[\sigma_{0i} + \sum_{k>i} \sigma_{0k} \dfrac{A_{ki}}{\sum_{i<k} A_{ki} + n_e \sum_{m>k} < \sigma_{km} v_e >} \right] v_e >}{\sum_{k>i} < \sigma_{ik} v_e >} . \tag{6.97}
$$

For further simplification the contributions of electronic excitation from different levels i were estimated. Except for energies close to the threshold the cross-sections for optically forbidden transitions are more than one order of magnitude smaller than those for optically allowed transitions. The same should hold for other optically allowed and forbidden transitions. Therefore, optically forbidden transitions can be neglected in first order and (6.96) can be simplified to

$$
n_o = \frac{\dfrac{4\pi N_{jh}}{A_{jk}\Delta\Omega V_P \varepsilon(\lambda)} \left[A_{jh} + \sum\limits_{i<j} A_{ji} + n_e \sum\limits_{k>j} <\sigma_{jk}v_e> \right]}{n_e < \left[\sigma_{0j} + \sum\limits_{k>j} \sigma_{0k} \dfrac{A_{kj}}{\sum\limits_{i<k} A_{ki} + n_e \sum\limits_{m>k} <\sigma_{km}v_e>} \right] v_e >} .
\tag{6.98}
$$

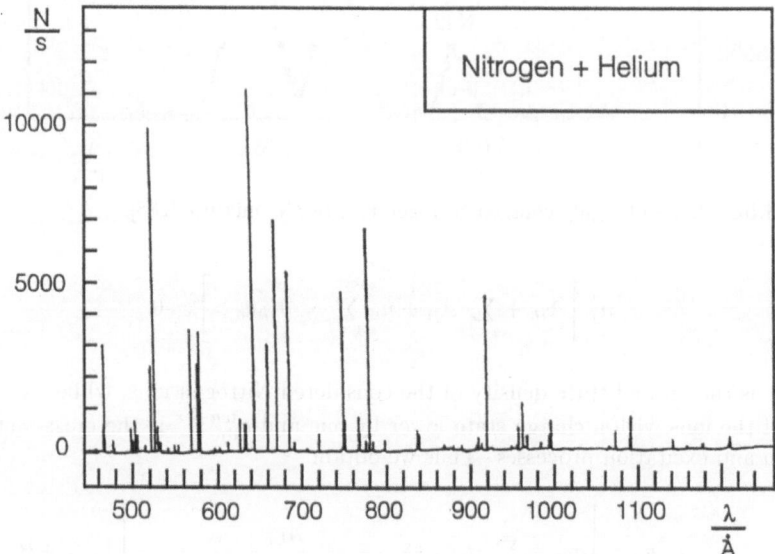

Figure 6.57 VUV spectrum of an ECR discharge in a He/N₂ plasma [288].

A spectrum obtained from an ECR discharge in a He/N$_2$ plasma is shown in Fig.**6.57**. The more intense lines are due to electronic transitions of NI, NII, NIII and HeI. Lines of more highly charged ions are difficult to detect. Corresponding results are shown in Fig.**6.58**.

Differencies due to the use of different data bases can occur and can not be excluded. There is also a disagreement when calculating n_0 from the intensities of different lines. These differences may be due to further excitation mechanisms neglected in the model discussed above. A process having different importance for different lines may be direct ionization plus excitation by producing a hole in a lower shell in the ionization process. To take these processes into account, the balance equation (6.88) has to be modified to

$$
n_e n_N^0 < \left[\sigma_{0j} + \sum\limits_{k>j} \frac{A_{kj}}{\sum\limits_{l<k} A_{kl} + n_e \sum\limits_{m>k} <\sigma_{km}v_e>} \right] v_e >
$$
$$
+ n_e \sum\limits_{i<j} n_N^i < \sigma_{ij}v_e > + n_e N_{N-1}^0 < \sigma_{0j}^{ion+ex}v_e >
\tag{6.99}
$$

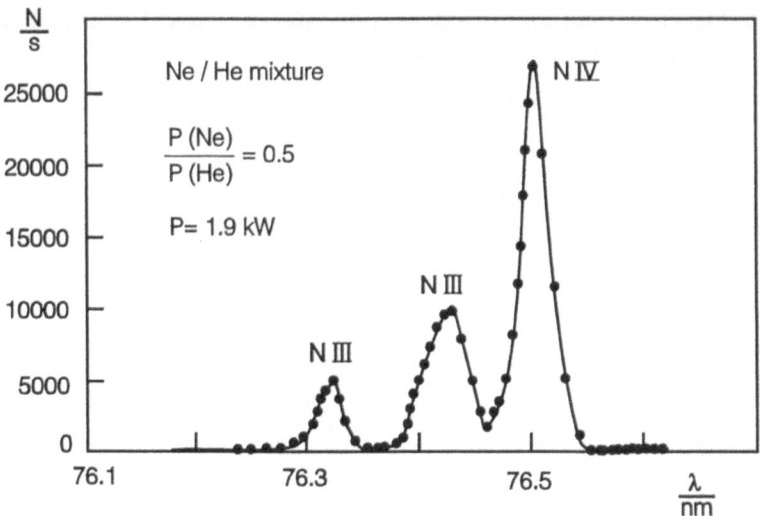

Figure 6.58   Lines of highly charged Nitrogen in a He/N$_2$ mixture [288].

$$- n_j \left[A_{jh} + \sum_{i<j} A_{ji} + n_e \sum_{k>j} < \sigma_{jk} v_e > \right] = 0$$

where n_N^0 is the ground state density of the considered Nitrogen ions, while n_{N-1}^0 is the density of the ions with a charge state lower by one and σ_{0j}^{ion+ex} is the cross-section for ionization and excitation processes. Thus we obtain

$$\left(n_N^0 \right)^{-1} = \frac{n_e < \left[\sigma_{0j} + \sum_{k>j} \sigma_{0k} \dfrac{A_{kj}}{\sum_{i<k} A_{ki} + n_e \sum_{m>k} < \sigma_{km} v_e >} \right] v_e > + B}{\dfrac{4\pi N_{jh}}{A_{jh} \Delta\Omega\, V_P\, \varepsilon(\lambda)} \left[A_{jh} + \sum_{i<j} A_{ji} + n_e \sum_{k>j} < \sigma_{jk} v_e > \right]}$$
$$+ \frac{\dfrac{n_e n_{N-1}^0}{n_N^0} < \sigma_{0j}^{ion+ex} v_e >}{\dfrac{4\pi N_{jh}}{A_{jh} \,\Delta\Omega\, V_P\, \varepsilon(\lambda)} \left[A_{jh} + \sum_{i<j} A_{ji} + n_e \sum_{k>j} < \sigma_{jk} v_e > \right]} \qquad (6.100)$$

where B is given by (6.97).

Special problems connected with the population of different levels are discussed in detail in [288], where also extensive tabulations of lines for different Nitrogen charge states are given. For all these lines densities can be calculated from measured intensities at least to a first approximation. Transition probabilities A_{ki} are not available for all these lines. In special cases for cascade transitions or branching transitions the Einstein coefficients are not known. For these cases detailed calculations must be done. An example for determining Nitrogen ion densities is shown in Fig.**6.59**.

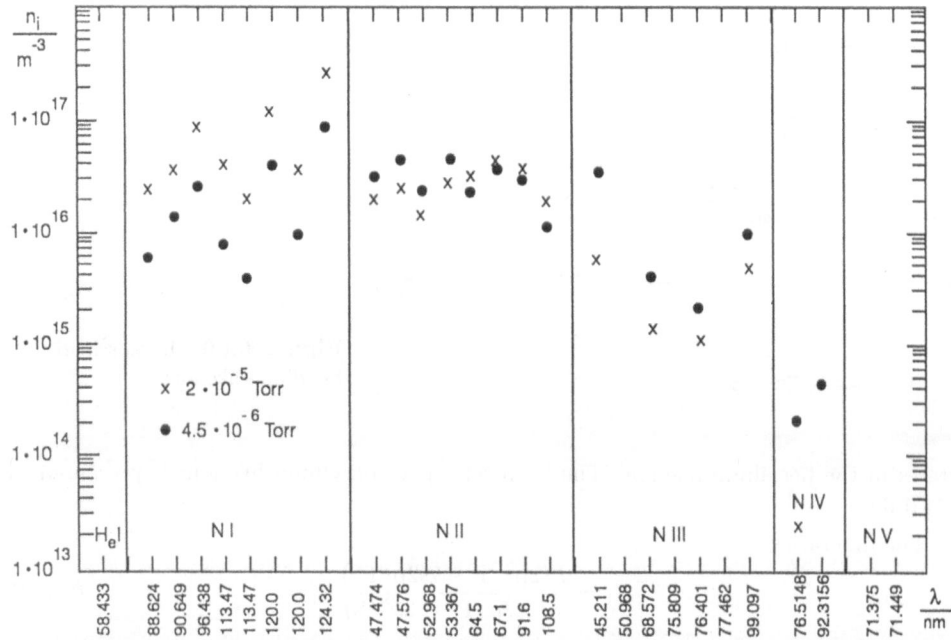

Figure 6.59 Density distribution of Nitrogen ions, derived from spectroscopic VUV measurements Vi93.

For molecular lines the photon flux can be calculated by the expression

$$\Phi(\lambda) = \varepsilon(\lambda)\, n_{\mathrm{mol}}\, n_e\, V_P\, \frac{\Delta\Omega}{4\pi}\, \frac{\int\limits_{E}^{\infty} Q(E)\sqrt{\frac{2E}{m_e}}\, F(E)\, dE}{\int\limits_{E}^{\infty} F(E)\, dE} \tag{6.101}$$

where $Q(E)$ is the electron impact excitation cross-section for the N_2 fluorescence. Values for $Q(E)$ from threshold up to 100 eV are given in Ref.[320].

6.2.4.3.3 Determination of the Electron Density The investigation of line intensity ratios between triplet and singulet transitions in beryllium-like ions [290] is a possible way for the determination of the electron density in ECR ion sources. Such intensity ratios have also been used to determine the electron density of the solar atmosphere [321]. Calculations of emission ratios of beryllium-like ions in solar and astrophysical spectra are known from [322, 323, 324] to determine the electron density and the electron temperature.

The determination of the electron density from the intensity ratio of triplet and singulet lines in beryllium-like ions is done on the basis of calculations of level population densities. A beryllium-like ion involves four electrons and can be described by six LS states of the two outer electrons in the L shell with the levels $2s^2\,^1S$, $2s2p\,^3P^o$, $^1P^o$ and $2p^2\,^1S$, $2s2p\,^3P^o$, $^1P^o$ and $2p^2\,^3P$, ^1D, ^1S where the triplet levels split due to the fine structure effects in three energetic states, respectively. Altogether there are ten energetic

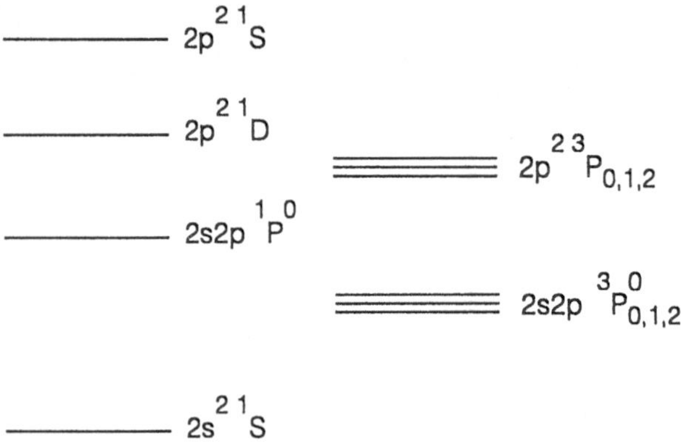

——————— $2p^2\ ^1S$

——————— $2p^2\ ^1D$

=========== $2p^2\ ^3P_{0,1,2}$

——————— $2s2p\ ^1P^o$

=========== $2s2p\ ^3P^o_{0,1,2}$

——————— $2s^2\ ^1S$

Figure 6.60 Level scheme of beryllium-like ions.

states in the beryllium-like ion. The term scheme of beryllium-like ions is presented in Fig.**6.60**.

The line ratios

$$R_1 = \frac{I\left(2p^2\ ^3P - 2s2p\ ^3P^o\right)}{I\left(2s2p\ ^1P^o - 2s^2\ ^1S\right)} \tag{6.102}$$

$$R_2 = \frac{I\left(2p^2\ ^3P - 2s2p\ ^3P^o\right)}{I\left(2p^2\ ^3P - 2s2p\ ^3P^o\right)} \tag{6.103}$$

can be used for electron density diagnostics in plasmas, because the intensity ratios R_1 and R_2 depends on the electron density [324, 321].

The line intensity ratio R_1 changes with the electron density, because the singlet 2s2p $^1P^o$ is populated by electron impact excitation of the ground state $2s^2\ ^1S^o$, while the triplet $2p^2\ ^3P$ is populated by electron collisions from the metastable state 2s2p $^3P^o$, whose population also depends on the electron density.

State populations for beryllium-like Carbon ions C III to Silicon ones Si XI were calculated for the ten ionic states with n = 2 [321, 322, 324]. For the calculation of population densities the knowledge of the Einstein emission coefficients A_{ik} for electron and photon excitation are necessary. For electron excitation rates R matrix calculations can be used, which are based on the assumption of a Maxwellian velocity distribution [325]. Although in the ECR plasma a Maxwellian velocity distribution evidently does not exist, it is possible to utilize these results for plasma diagnostics in the electron density region from $10^8 \ldots 10^{13}$ cm^{-3}, because the excitation coefficients of the transitions considered vary only very weakly with the electron temperature and have the same energy dependence [290]. Besides that the intensity ratios display only a slight electron energy dependence for triplet and singlet lines of beryllium-like ions, which is likewise useful for applications to electron density diagnostics.

From the calculated occupation densities N_k of individual ion states the intensity of a spectral line with the frequency ν_{ki} can be determined by

$$I_{ki} = \frac{1}{4\pi}\ h\nu_{ki}\ A_{ki}\ N_k\ . \tag{6.104}$$

Table 6.2   Atomic data for the intensity ratio R_1. The wavelengths λ were taken from [326] and the Einstein transition coefficients of the spontaneous emission A_{ik} from [322].

Ion		$2p^2\,^3P - 2s2p\,^3P^o$	$2s2p\,^1P^o - 2s^2\,^1S$
C III	λ/nm	117.49...117.65 117.57	97.70
	A_{ik}/s^{-1}	$1.33 \cdot 10^9$	$1.79 \cdot 10^9$
N IV	λ/nm	92.20...92.43 92.32	76.51
	A_{ik}/s^{-1}	$1.78 \cdot 10^9$	$2.38 \cdot 10^9$
O V	λ/nm	75.87...76.20 76.04	62.97
	A_{ik}/s^{-1}	$2.22 \cdot 10^9$	$2.86 \cdot 10^9$
Ne VII	λ/nm	55.86...56.45 56.16	46.52
	A_{ik}/s^{-1}	$3.11 \cdot 10^9$	$4.08 \cdot 10^9$

The intensity ratio R for two transitions with the frequencies ν_{ki} and ν_{lm}, with the Einstein transition coefficients A_{ki} and A_{lm} and the intensities I_{ki} and I_{lm} yields

$$R = \frac{I_{ki}}{I_{lm}} = \frac{n u_{ki} A_{ki} N_k}{\nu_{lm} A_{lm} N_l} = \frac{\lambda_{lm} A_{ki} N_k}{\lambda_{ki} A_{lm} N_l} \quad . \tag{6.105}$$

For the determination of the electron density from the intensity ratio of triplet and singulet lines beryllium-like ions C III, N IV, O V and Ne VII appear especially interestingly. Wavelengths and Einstein transition coefficients of the spontaneous emission for these ions are given in the Tables **6.2** and **6.3**. Numeric values for the Einstein transition coefficients were taken from [322, 323, 326, 327].

With the Einstein transition coefficients of the spontaneous emission and the wavelengths of dipole transitions from Tables **6.2** and **6.3** intensity ratios R_1 and R_2 can be calculated using calculated occupation densities for individual ion states [321, 322, 324] according to (6.105). For triplet transitions mean values are used as indicated in the tables.

Examples for the dependence of the intensity ratio R_1 as a function of the electron density for different ions are shown in Figs.**6.61** to **6.63**. An example for the intensity ratio R_2 is given in Fig.**6.64**. From the shown Figures it is obvious, that the electron density dependence of the individual intensity ratios varies significantly for different electron temperatures in different electron density regions.

Table 6.3   Atomic data for the intensity ratio R_2. The wavelengths λ were taken from [326] and the Einstein coefficients of the spontaneous emission A_{ik} from [322].

Ion		$2p^2\,^3P - 2s2p\,\,^3P^o$	$2p^2\,^1S - 2s2p\,^1P$
C III	λ/nm	117.49...117.64 117.57	124.74
	A_{ik}/s^{-1}	$1.33 \cdot 10^9$	$2.11 \cdot 10^9$
N IV	λ/nm	92.20...92.43 92.32	95.53
	A_{ik}/s^{-1}	$1.78 \cdot 10^9$	$2.92 \cdot 10^9$
O V	λ/nm	75.87...76.20 76.04	77.45
	A_{ik}/s^{-1}	$2.22 \cdot 10^9$	$3.87 \cdot 10^9$

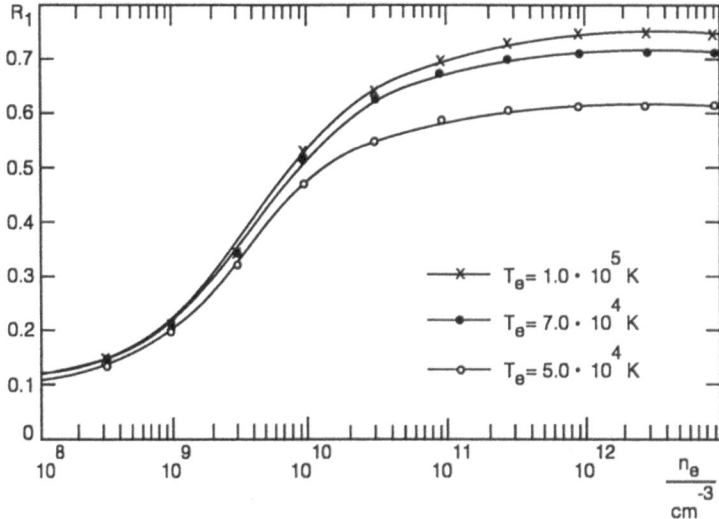

Figure 6.61   Intensity ratio R_1 as a function of the electron density n_e for C III.

Thereby, the emission lines of dipole transitions lie in the VUV region. Spectroscopy is accomplished under vacuum conditions to prevent radiation absorption by the Oxygen of the air. As dispersion elements in VUV spectrometers grids are used, whereby the spectrometers differ according to the analyzed wavelength region: Normal incidence spectrometers make it possible to measure in the wavelength region of about 30 nm to 200 nm. Grazing incidence spectrometers are used in the wavelength region below 30 nm.

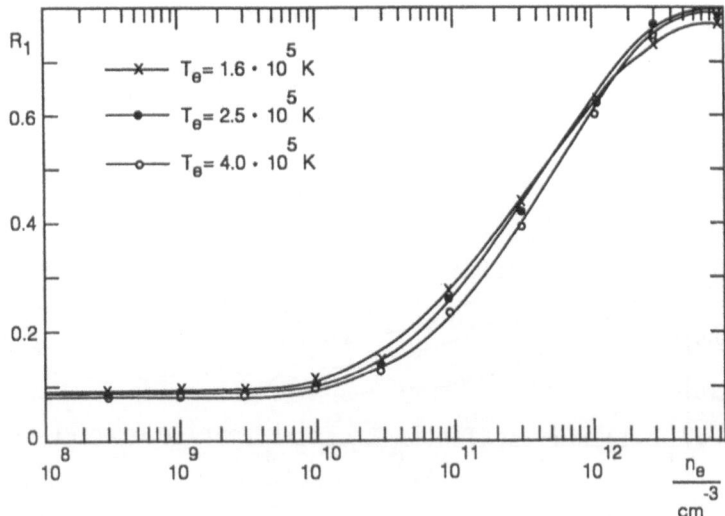

Figure 6.62   Intensity ratio R_1 as function of the electron density n_e for O V.

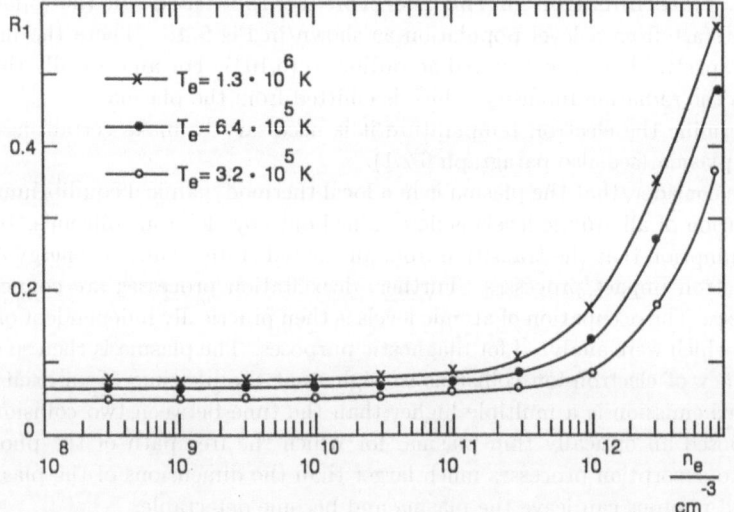

Figure 6.63   Intensity ratio R_1 as function of the electron density n_e for Ne VII.

Difficulties arrise for the determination of the registration efficiency of VUV spectrometers, since only a few light sources exist as intensity standards with quantitatively known emission values. An experimental determination of the registration efficiency of VUV spectrometers is experimentally possible by using synchrotrons, hollow cathode lamps, deuterium lamps and arc discharges. For instance, the determination of the absolute registration efficiency of VUV spectrometers is discussed in Ref.[328].

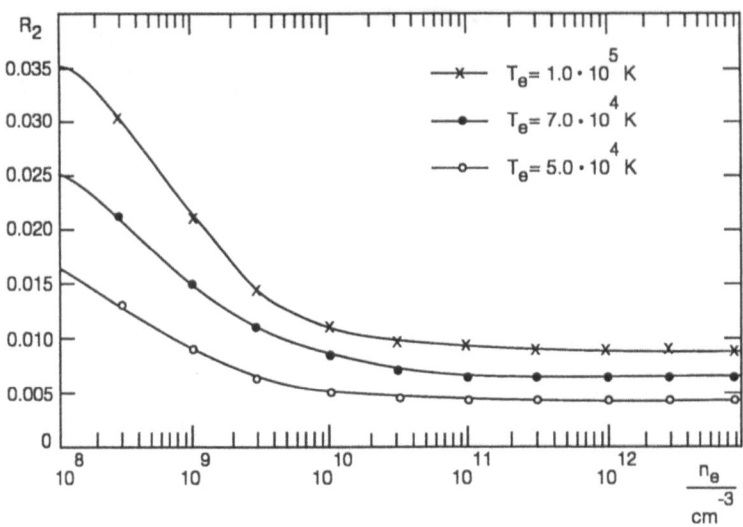

Figure 6.64   Intensity ratio R_2 as function of the electron density n_e for C III.

6.2.4.3.4 Determination of the Electron Temperature

In the following subsection we start from a level population as shown in Fig.**6.48**. There the intensity of individual spectral lines is calculated according to (6.104). For an optically thin plasma this is also the radiation intensity, which is emitted from the plasma.

To determine the electron temperature it is necessary to make certain assumptions about the plasma (see also paragraph 6.2.1).

First we consider, that the plasma is in a local thermodynamical equilibrium, i.e. that the occupation of all atomic levels is determined only by electron collisions. It proceeds on the assumption that the transition from an excited state results by energy deposition due to electron impact processes. Further, deexcitation processes are connected with energy losses. The occupation of atomic levels is then practically independent of radiative processes, which were analyzed for diagnostic purposes. The plasma is then so dense and the frequency of electron-ion collisions so high, that the life-time of an excited ion for spontaneous emission is a multiple higher than the time between two collisions. There is presupposed an optically thin plasma, for which the free path of the photons is in reference to absorption processes much larger than the dimensions of the plasma. Thus the emitted photons can leave the plasma and become detectable.

The *Boltzmann factor* for the ratio of the individual occupation densities has the form

$$\frac{N_k}{N_l} = \exp\left(-\frac{E_k - E_l}{kT_e}\right) \qquad (6.106)$$

with $E_{k,l}$ as energies of the individual atomic levels k and l. The ratio of the line intensity is as follows

$$\frac{I_{ki}}{I_{lm}} = \frac{\nu_{ki} A_{ki}}{\nu_{lm} A_{lm}} \exp\left(-\frac{E_k - E_l}{kT_e}\right) \qquad . \qquad (6.107)$$

In a series of works instead of the Einstein emission coefficients A_{ki} the oscillator strengths f_{ki} are used, which characterize the absorption probability for a certain transition. Both

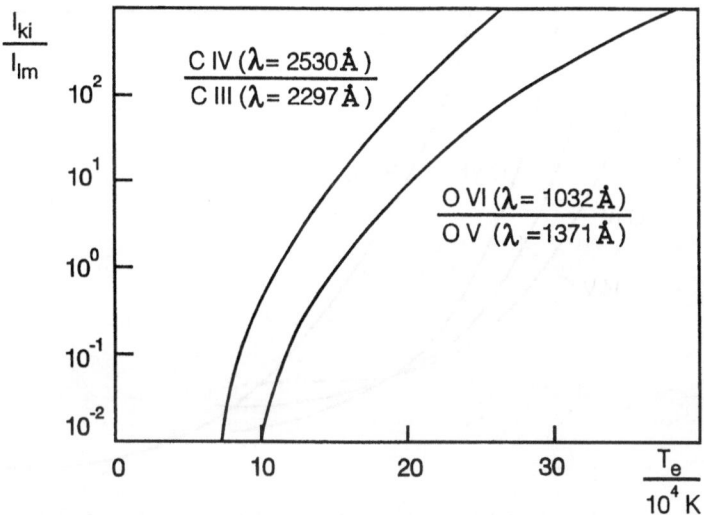

Figure 6.65   Dependence of the relative line ratios of ionized Carbon and Oxygen on the electron temperature T_e in the case of a local thermodynamic equilibrium [227].

quantities are interconnected by

$$A_{ki} = \text{const.} \frac{f_{ik}}{\lambda^2} \quad . \tag{6.108}$$

Thus results

$$\ln \frac{I_{ki}\lambda_{ki}^3 f_{ml}}{I_{lm}\lambda_{lm}^3 f_{ik}} = \frac{E_e - E_k}{kT_e} \tag{6.109}$$

and ultimately

$$kT_e = \frac{E_l - E_k}{\ln \dfrac{I_{ki}\lambda_{ki}^3 f_{ml}}{I_{lm}\lambda_{lm}^3 f_{ik}}} \quad . \tag{6.110}$$

For multiplet levels in (6.110) as correction the statistical weights of the individual lines are considered, i.e. (6.104) must be multiplied by the corresponding weights.

As it follows from the above mentioned equations, an exact measurement of T_e can result only for $kT_e \approx E_k - E_l = \Delta E_{kl}$. For lines in the visible and ultraviolet area, where ΔE_{kl} lies in the region of about 1 eV ... 10 eV, the method can be applied only for cold and dense plasmas. With growing electron temperature the sensitivity of the method decreases considerably. This can be partly compensated if there are known lines emitted by highly charged ions. There calculations are complicated by the fact, that for calculations of the atomic ground state occupations of different ionization stages the Saha equation (6.4) must be used.

As an example Fig.**6.65** shows the change of relative line intensities from Carbon and Oxygen ions [227]. The calculation of the intensity ratios were accomplished analogous to the above described procedure for a local thermodynamical equilibrium.

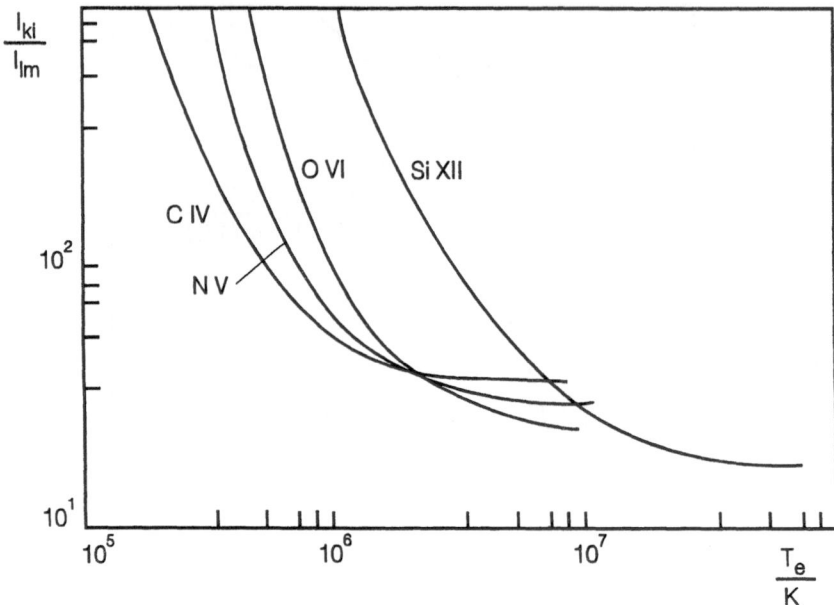

Figure 6.66   Calculated in the corona model relative line intensities for the transitions 2S –
$2P_{3/2}$ and $2S - 3P_{1/2,3/2}$ for selected elements.

In the following the case is considered where the conditions for a local thermodynamical equilibrium are not fulfilled and the level occupation does not follow a Boltzmann
distribution. Further it is assumed, that only electron impact excitation occurs, since
the plasma is optically thin and the radiation density is small, so that optical excitati-
ons can be neglected. Due to the small plasma density as the inverse process radiative
recombination dominates. Equilibrium conditions are reached in the plasma due to the
balance between electron impact excitation and radiative recombination. The balance
equation for the term k has then the form

$$n_e N_0 < \sigma_i(T_e) \, v_e >= N_1 \sum_i A_{ki} \qquad (6.111)$$

with $\sigma_i(T_e)$ as excitation cross-section into the level k from the ground state, N_1 as
occupation number of the excited state and N_0 as occupation number of the ground state.
The quantity $\sum_i A_{ki}$ describes the total probability for spontaneous radiating transitions
from the level k in all other lower lying states. Thus for the intensity of the line k → i
results

$$I_{ki} = \text{const.} \; < \sigma_1(T_e) v_e > \quad . \qquad (6.112)$$

With it for the intensity ratio of two lines follows

$$\frac{I_{ki}}{I_{lm}} = \text{const.} \; \frac{< \sigma_k(T_e) v_e >}{\sigma_l(T_e) v_e >} \quad . \qquad (6.113)$$

If for the atomic system considered above the excitation functions and the transition
probabilities are known, curves for the temperature dependence of the line intensity ratios

can be calculated and thus temperatures from experimentally line ratios I_{ki}/I_{lm} could be determined. As an example for practical interesting transitions in Fig. **6.66** relative line intensities as a function the electron temperature (energy) are presented. As seen from Fig.**6.66** the accessible temperature interval for measurements is essentially larger under the validity conditions of the assumed model. This model was applied first in astrophysics for the analysis the solar corona and is marked as *corona model* (see also paragraph 6.2.1.5).

Calculations in the framework of the *impact radiation equilibrium* (see also paragraph 6.2.1.4) were accomplished for instance in Refs.[329, 330].

6.2.4.3.5 Pressure and RF Power Dependencies of Spectral Lines   The dependences discussed above must be verified in experiments to study the relative intensities of some spectral lines in connection with the main parameters of an ECR ion source, because these parameters directly influence the electron temperature and the electron density in the ECR plasma. Thus, the Giessen group has studied the relative intensity of the $2s2p\ ^3P - 2p^2\ ^3P$ and the $2s^2\ ^1S - 2s2p\ ^2P$ transitions for C III, O V and Ne VII at different gas pressures [290]. The CAPRICE group reported VUV measurements performed with Oxygen to analyze the dependence of line intensities on RF power and gas pressure [289].

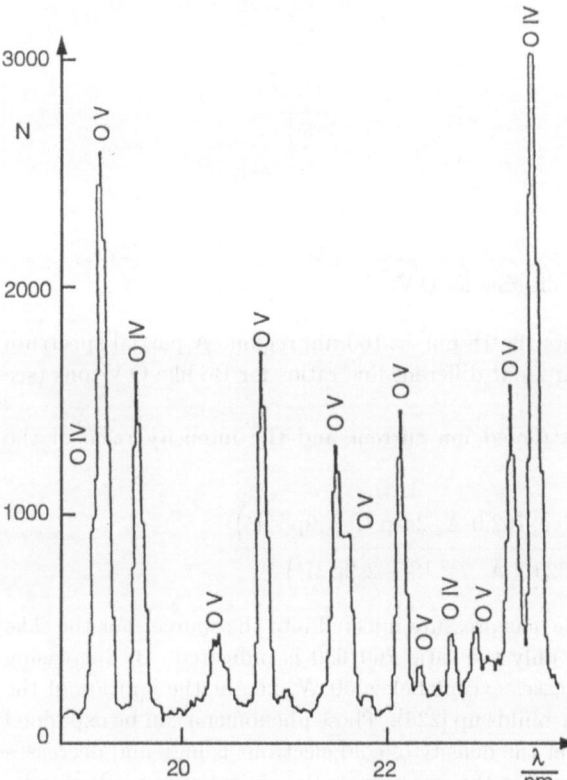

Figure 6.67 Partial Oxygen spectrum for on-axis spectroscopy on the plasma of a CAPRICE ECR ion source.

In the following, we present some results obtained by the CAPRICE group. The investigations at this source were performed on-axis with a 3 m grazing incidence spectrometer equipped with 300 lines/mm and 600 lines/mm gratings blazed at 51.2 nm and 25.6 nm respectively. Photons were detected by a channeltron. Thereby, the solid angle was $1.3 \cdot 10^{-5}$ sterad. Techniques of relative efficiency calibration are described in Ref.[331].

For on-axis spectroscopy the measured spectra contain an integration over all mechanisms which are in competition in different parts of the source. Nevertheless, the intensity ratios of some lines can give an estimation of the electron density.

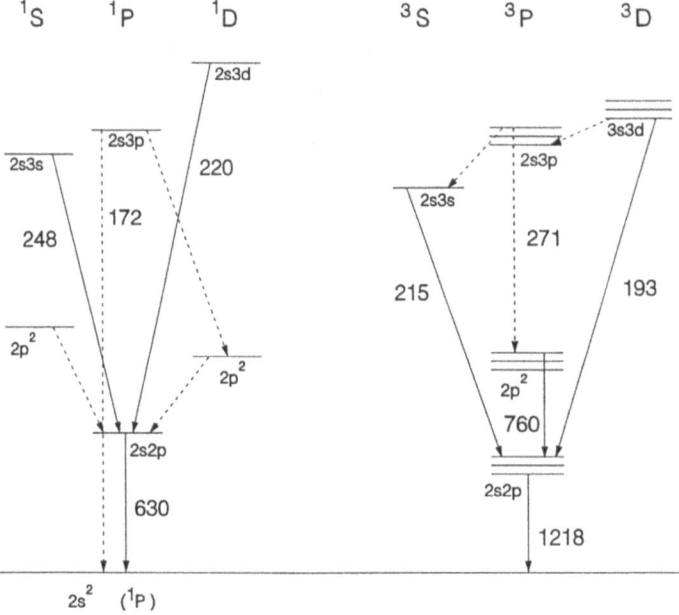

Figure 6.68   Schematic energy level diagram for O V.

The observed spectral range covers the 18 nm ... 160 nm region. A partial spectrum is shown in Fig.**6.67**. Investigated are four different line ratios for Be-like O V ions (see Fig.**6.68**).

Figs.**6.69** and **6.70** show the extracted ion current and the intensity ratio of the following two lines

$$\frac{760}{630} = \frac{I\left(758.7 \ldots 762.0 \text{ Å}; \, 2s2p \, ^3P - 2p^2 \, ^3P\right)}{I\left(629.7 \text{ Å}; \, 2s^2 \, ^1S - 2s2p \, ^1P\right)}$$

as a function of the RF power and the gas pressure injected into the source. For the sake of simplicity in Figs.**6.69** and **6.70** only the ratio 760/630 is indicated. By increasing the RF power, the ratio slowly decreases except below 50 W, i.e., at the ignition of the plasma when the hot electron density builds up [224]. These phenomena can be explained by the fact that at very low RF power the density of cold electrons is high and decreases slowly by increasing microwave power. For the oxygen ions the ionization is made step by

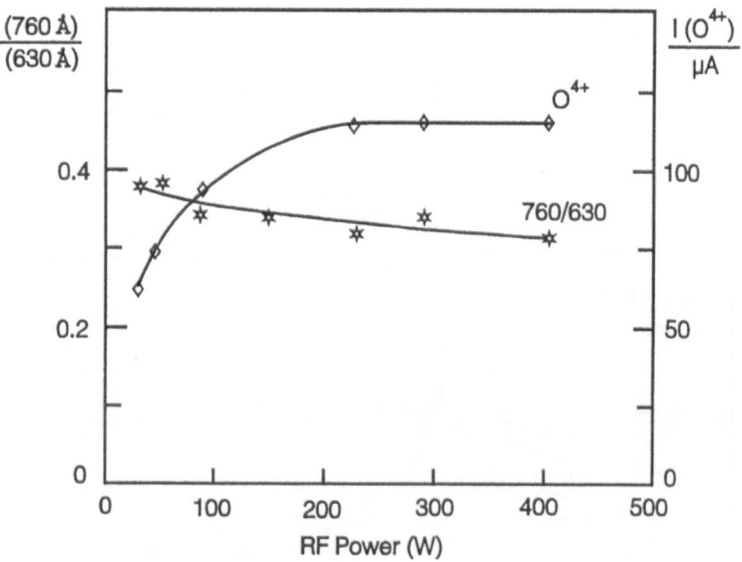

Figure 6.69 Variation of the O^{4+} intensity beam current and the 760/630 intensity ratio versus RF power injected in the ECR source.

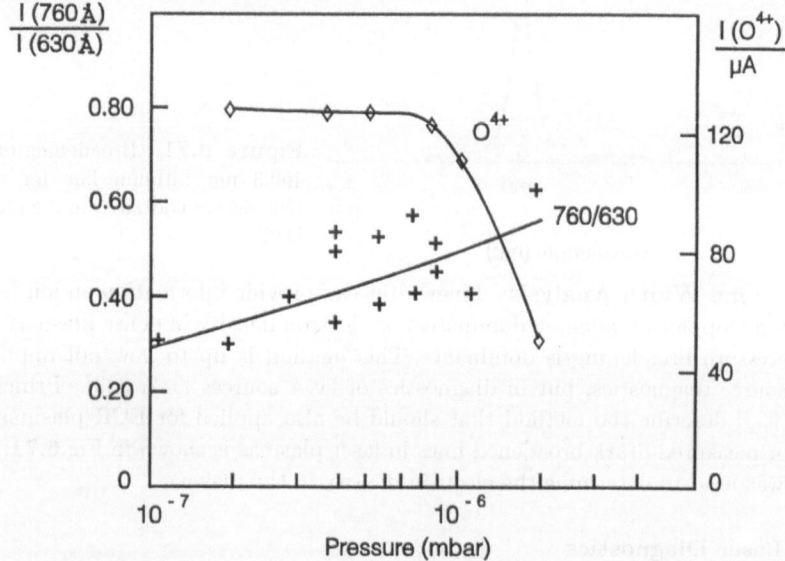

Figure 6.70 Variation of the O^{4+} intensity beam current and the 760/630 intensity ration versus pressure injected in the ECR source.

step, thus a certain RF power is required and the excess of power increases the collisonal processes. If the pressure in the source increases the intensity of the analyzed line ratio increases too as well as the cold electron density. In addition, the O^{4+} current decreases because of an increasing charge exchange and a reduced confinement time.

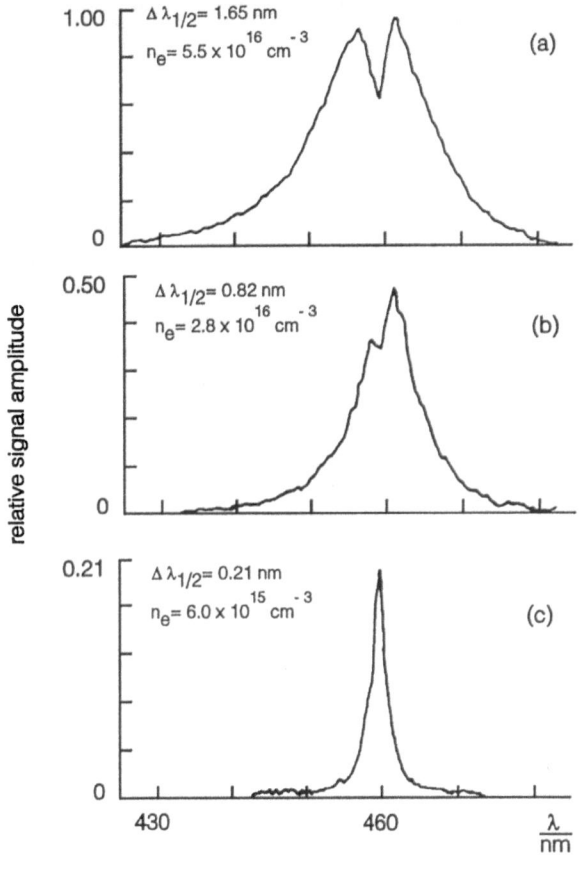

relative signal amplitude

wavelength (nm)

Figure 6.71   Broadening of the 460.3 nm Lithium line for different electron densities in the plasma [332].

6.2.4.3.6   Line Width Analysis   Linewidths can provide information on ion temperature (when Doppler broadening dominates) or electron density in dense plasmas when Stark or pressure broadening is dominant. This method is up to now not applied in ECR ion source diagnostics, but in diagnostics of laser sources [332, 333]. Principally Refs.[332, 333] describe the method that should be also applied for ECR plasmas. An example for measured Stark broadened lines in laser plasmas is shown in Fig.**6.71**. This spectrum was used to determine the electron density in the plasma.

6.2.4.4   Laser Diagnostics

Generally, active spectroscopic diagnostics use laser absorption or laser induced fluorescence. In this paragraph fundamental processes should be discussed, which are important for the description of the scattering of electromagnetic waves of free or weakly bound electrons.

A plane polarized wave of the frequency ω interacting with a free electron produces forced oscillations of the electron with the same frequency ω (see Fig.**6.72**). This dipole oscillation acts as a source of secondary scattered radiation. Thereby, the largest part of the radiation is emitted in the plane, which is localized vertical to the plane, which result the oscillations of the charge, i.e. vertically to the direction the electric vector $\vec{E}$ of the primary wave. Along the dipole axis the intensity of the scattered radiation vanishes.

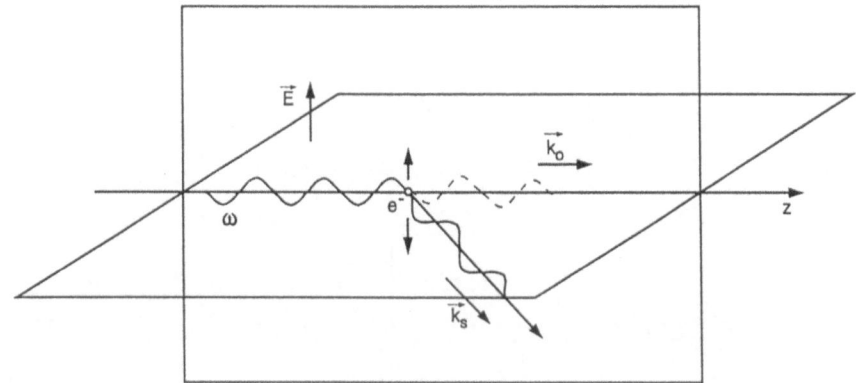

Figure 6.72   Scattering of electromagnetic radiation at a free electron.

For natural unpolarized light the intensity along the dipole axis totals 50% of the maximum intensity. The resulting intensity distribution is presented in the form of a polar diagram in Fig.**6.73**. The spatial intensity distribution results by rotation of the figure plane around the z-axis, i.e. around the direction of the primary beam. It counts

$$I = I_{\pi/2} \left(1 + \cos^2 \Theta \right) \tag{6.114}$$

with I as the radiation intensity registered under the angle Θ and $I_{\pi/2}$ as the intensity scattered under the angle $\pi/2$.

The absolute intensity of the scattered radiation is described by the total effective scattering cross-section σ_e, which is calculated from the relationship of the energy E_s scattered off an electron in the time interval in all directions to the current density E_0/A of the incident energy

$$E_s = \sigma_e \frac{E_0}{A} \tag{6.115}$$

with A as acting area. In the classic theory for an electron the effective scattering cross-section totals πr_0^2 with r_0 as classic electron radius. Practical investigations lead to the effective differential cross-section, which describes the scattering in a solid angle element $d\Omega$ under an angle ϕ in direction of the vector $\vec{E}$ of the impinging electromagnetic wave

$$d\sigma_e = r_0^2 \sin^2 \phi \, d\Omega \quad . \tag{6.116}$$

After integration the total effective cross-section for the scattering at an electron in 4π is

$$\sigma_e = \frac{8\pi}{3} r_0^2 \quad . \tag{6.117}$$

It is assumed, that a light beam gets through a plasma of the density n_e, of the power P_0 and of the cross-section A. The number of scattering centers along the way x is $n_e x A$. If it is assumed, that the total intensity of the scattered light is proportional to the

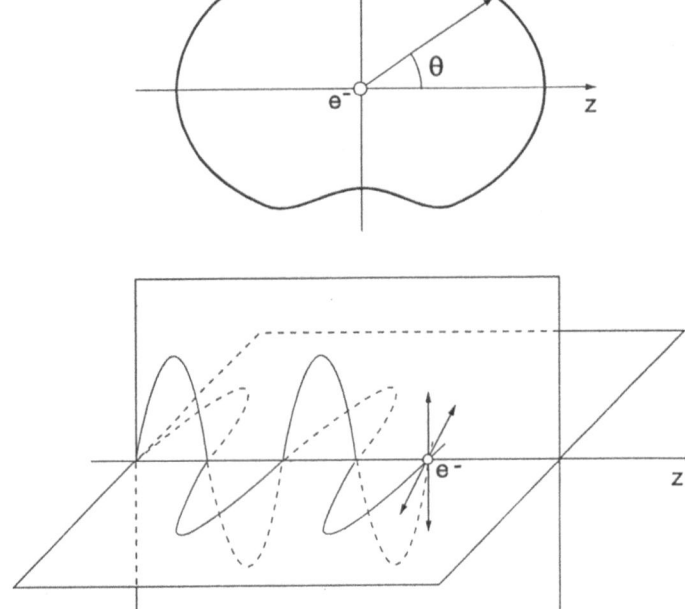

Figure 6.73 The intensity distribution of secondary scattered radiation.

number of scattering centres, then in the solid angle Ω under the angle ϕ to the direction of the vector $\vec{E}$ the power

$$dP = n_e x A \frac{P_0}{A} \, d\sigma_e = x n_e r_0^2 \sin^2 \phi \, P_0 \, d\Omega \qquad (6.118)$$

is scattered.

So the scattered power is very small and lies at typical experiment conditions ($\phi = 90^o$, $l = 1$ cm, $d\Omega \leq 10^{-3}$ sterad) for comparatively dense plasmas ($n_e \approx 10^{14}$ cm^{-3}) at $dP/P_0 \approx 10^{-14}$, so that with classic light sources scattering experiments at the plasma lead not to measurable results. This situation changed with the introduction of lasers in plasma diagnostics.

Besides continuously working lasers pulsed lasers with impulse lengths of milliseconds up to picoseconds are known. The energy transported by the laser pulses varies in a wide range of Joules up to tens of Gigawatts.

The principle of the laser diagnostics is presented in Fig.**6.74**. Laser diagnostics makes it possible, to get in extremely short time intervals local information about the plasma properties without changing the characteristics of the analyzed plasma.

First some assumptions should be met for the application of possible laser diagnostics methods. The plasma volume should be subdivided in elementary cells. The dimensions of this elementary cell should be so small, that the phases of the electric fields of the scattered waves coincide in a point far away from the elementary cell. If the number of elementary cells is sufficiently large, the scattered wave contains all possible phases

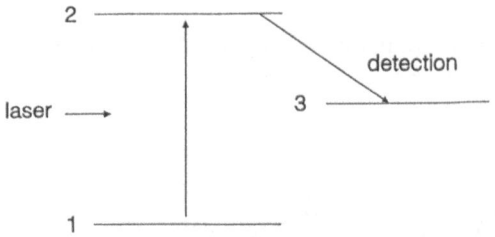

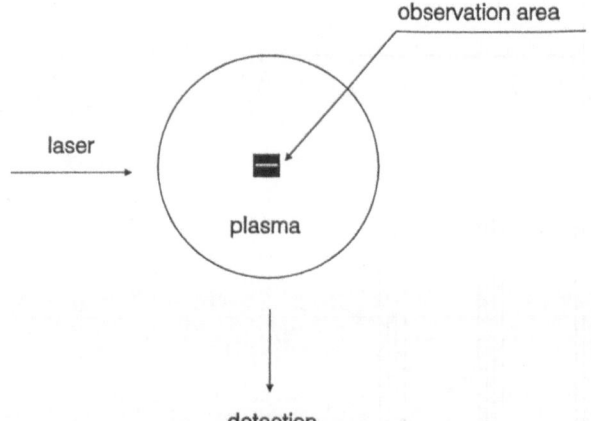

Figure 6.74 Principle of laser induced fluorescence and detection.

ϕ_j and the resulting intensity disappears, if the number of scattering centres is equal for each elementary cell, i.e. the condition for the appearance of a scattered wave are density fluctuations. If the mean electron number in the elementary cell is indicated with $\bar{n}$ and with δn_j the deviation from the mean value $\bar{n}$, for the resulting field it follows that

$$E_s \sim \sum_j \left(\bar{n} + \delta n_j \right) e^{i\phi_j} . \tag{6.119}$$

As already mentioned, the first term in (6.119) vanishes for a sufficiently large amount of elementary cells. Since the intensity is proportional to the square of the amplitude, it results in

$$I_s \sim \sum_{j,k} \delta n_j \bar{\delta} n_k \, e^{-i(\phi_k - \phi_j)} . \tag{6.120}$$

For the case of neglecting correlation, i.e. in case of independently treating each other fluctuations in the individual elementary cells, (6.120) simplifies to

$$I_s \sim \sum_j \left(\sigma_{\bar{n}_j} \right)^2 . \tag{6.121}$$

According to the Poisson statistics $\sigma_{n_j} = \sqrt{\bar{n}}$ follows

$$I_s \sim \sum_j \bar{n} \sim N . \tag{6.122}$$

Thus the net intensity of the scattered radiation is proportional to the number of scattering centers. Physically this characterizes the case of a thin and hot plasma.

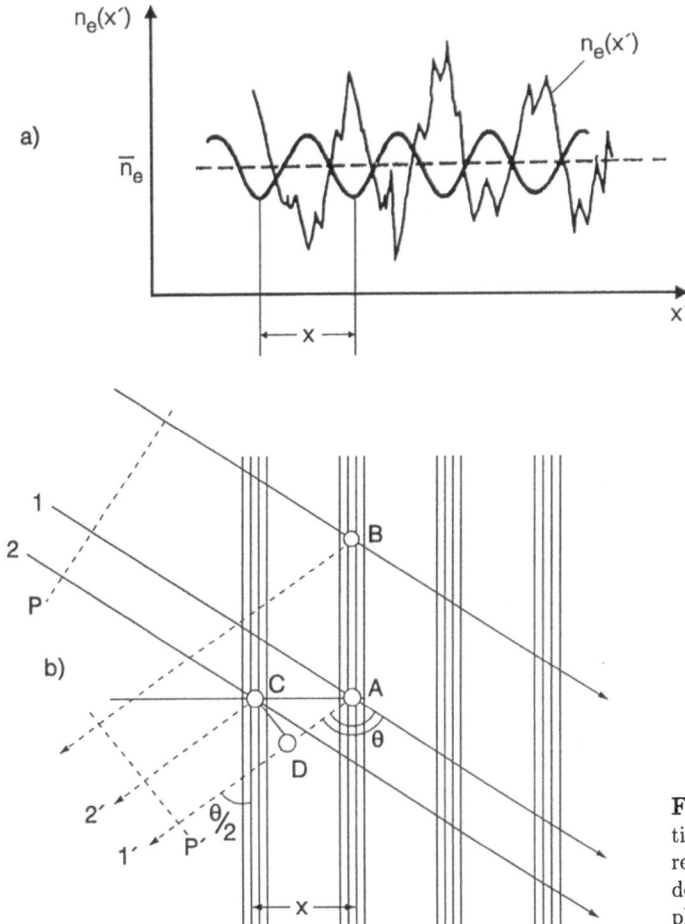

Figure 6.75 Formation of a current of scattered radiation as result of density fluctuations in the plasma.

Further the case of correlated density fluctuations is considered. It is assumed, that the plasma dimensions are small in comparison to the plasma – detector distance. The incident and scattered waves are plane waves and scattered under the angle Θ. Then the wave amplitudes in the scattering process off different electrons are approximately the same, but with different phases. For the appearance of a current of scattered quanta yields, that the electron density $n_e(x)$ is changeable, i.e. it fluctuates around $\bar{n}_e$ (see Fig.**6.75**a). In analogy to Fig.**6.75**b a point A in the plasma volume at the intersection of the incidenting beam and the beam scattered under an angle Θ is considered. An electron, which is found at an arbitrary point B on the angle halving the scattering angle Θ, emits a scattered wave in the direction Θ, which is found in phase with the radiation emitted at the point A, since the path difference between the planes P and P' is zero.

From Fig.**6.75** it is obvious, that radiation of equal phase arrive at the observation point from all plasma electrons, which are found in a plane parallel and so far to the chosen plane AB, that the path difference $\Delta = 2AD$ is equal to the wave length λ_0 of the incidenting radiation

$$\Delta = 2AD = 2x \sin \frac{\theta}{2} = \lambda_0 \qquad (6.123)$$

and with it

$$x = \frac{\lambda_0}{2 \sin \frac{\theta}{2}} \qquad . \qquad (6.124)$$

If the frequency of the scattered radiation coincides with the frequency of the the incidenting light (this counts for the case of resting scattering centers) or differs from it only unimportantly, so the coherence length x can be connected with the change of the wave number vector $\vec{k}$ and with that of the vector of the incidenting wave $\vec{k_0}$. With $\vec{k_s}$ as wave vector of the scattered wave counts

$$\vec{k} = \vec{k_0} - \vec{k_s}; \quad k_0 = k_s = \frac{2\pi}{\lambda_0}; \quad k = k_0 2 \sin \frac{\Theta}{2}$$

respectively

$$k = \frac{4\pi \sin \frac{\Theta}{2}}{\lambda_0} \qquad . \qquad (6.125)$$

Further, we explore the connections between plasma parameters and the fluctuation spectrum. The quasi-neutrality of the plasma and also the fluctuations of the electron density are caused by the thermal movement of the plasma particles. At roughly equal electron and ion temperatures T_e and T_i the oscillation frequency of the electrons in the plasma is clearly higher. Thus, the fluctuations of the electron component can be written in the form

$$\delta n_e(\vec{r}, t) = \delta n'_e(\vec{r}, t) + \delta n''_e(\vec{r}, t) \qquad . \qquad (6.126)$$

The first term corresponds to the high-frequency oscillations of free electrons, but the second one to the low-frequency electron oscillations, which follow to the dislocations of the ions in the plasma. Accordingly the spectrum of the scattered electromagnetic radiation contains two components: an electron and an ion component.

As a spatial scale for the change of the quasi-neutrality counts the *Debye-radius* δ. Within the Debye sphere the electric field of the test charge disappears due to the screening by the spatial plasma charge. The Debye length was introduced for a resting test charge. If fast electrons move, screening results within the Debye sphere by a reduction of the electron cloud. If a slow ion of the charge i moves, then its screening function divides in an ion cloud with the charge $+i/2$, which repulses the test charge, and in a electron cloud of the charge $-i/2$, which attracts the test charge.

The scattering of laser light in the plasma is characterized by [334, 335]

$$\alpha = \frac{1}{k\delta} = \frac{x}{2\pi\delta} = \frac{\lambda_0}{4\pi \sin \frac{\Theta}{2}} \qquad . \qquad (6.127)$$

The parameters α compares two quantities: the characteristic plasma size δ and the coherence length $x = 2\pi/k$, which is connected with the wave length of the laser radiation and the scattering angle.

The following cases are possible:

1. $\alpha \ll 1$.

For this case the quantity x is essentially smaller than δ and the scattering of light on plasma electrons is not joined with the charge screening effect of the Debye sphere. Fluctuations of the electron density are determined by the thermal movement of practically free electrons. This is the case of a hot and thin plasma. Thereby, the spectrum of the scattered radiation is determined by the *Doppler effect*.

In the case of a Maxwellian distribution of the electron velocities the spectral contour of the scattered radiation is a Gaussian distribution, whose full width at half maximum of the thermal electron movement is proportional to

$$\Delta \omega_e \approx \text{const.} \sqrt{\frac{kT_e}{m_e}} . \tag{6.128}$$

For a Ruby laser follows at $\Theta = 90^o$ [227]

$$\Delta \lambda_e \left[\text{Å} \right] = 0.3 \, T_e^{1/2} \, [\text{K}] . \tag{6.129}$$

Thus the appearance of line broadening already assumes at relatively small plasma temperatures sufficiently large values.

2. $\alpha \gg 1$.

Here the Debye length is smaller than x and the fluctuation spectrum is determined by oscillations of the screening charges. Fluctuations joined with the ion movement dominate. Oscillations of ions with the charge i are accompanied by collective oscillations of an electron cloud with an effective charge $-i/2$. These uncompensated electric charges determine the scattering of the electromagnetic waves, i.e. it acts only the second term from (6.126). Thus the full width at half maximum of the spectral contour of the scattered radiation is proportional to the thermal movement of the ions . It results in

$$\Delta \omega_i \approx \text{const.} \sqrt{\frac{kT_i}{M}} \tag{6.130}$$

with M as ion mass.

3. $\alpha \approx 1$.

This case corresponds to the situation, that both contributions, the high-frequency (electron component) as well as the low-frequency (ion component) appear.

It should be pointed out, that in the first case the scattering power because of the screening effect is twice as large as the second case

$$P_e \sim \sigma_e(\Theta)n_e \quad \text{for case 1;} \quad p_e \sim \frac{1}{2}\sigma_e(\Theta)n_e \quad \text{for case 2 .} \tag{6.131}$$

In Fig.**6.76** spectra of scattered laser radiation for all three treated cases are presented. At $\alpha \approx 1$ resonances appear that are also observed for $\alpha \gg 1$ and correspond to the electronic plasma frequency

$$\omega_P = \sqrt{\frac{4\pi n_e e^2}{M}} . \tag{6.132}$$

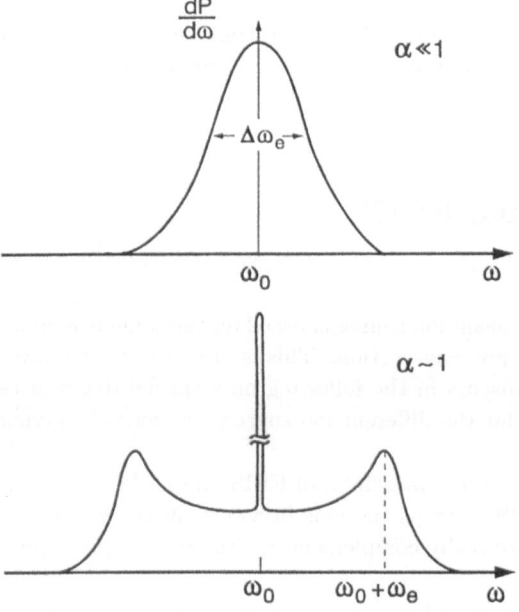

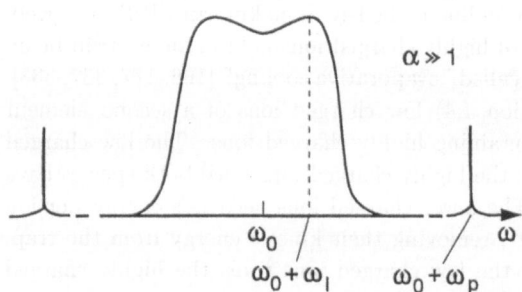

Figure 6.76 Spectra of scattered laser light for three different values of the parameter α.

An exact description leads to the result, that the frequency ω_1 of the electronic satellites relative to the laser frequency ω_0 oversteps the Langmuir frequency ω_P

$$\omega_1^2 = \omega_P^2 + 3v_e^2 k^2 \quad \text{or} \quad \omega_1^2 = \omega_P^2 + 3\left(\frac{v_e}{c}\right)^2 \left(4\pi \sin\frac{\Theta}{2}\right)^2 \qquad (6.133)$$

with v_e as thermal electron velocity. At small values of v_e/c the second term in (6.133) can be neglected.

Experiments, which are accomplished under the condition $\alpha \ll 1$ allow to determine the electron temperature of the plasma from the FWHM of different lines of the scattered laser radiation. If experiments are accomplished under the condition $\alpha \gg 1$, the ion temperature can be determined from the full width at half maximum of the ion peaks of the scattered laser radiation . From the absolute intensity of the ion peaks the electron concentration can be inferred. Additionally n_e can be estimated from the position of the electronic satellites.

So far well-diagnosed ion sources are still "black boxes" because only one value of density and temperature are available but spatial gradients would important to study. Particle and energy balances have rarely been carried out. Here laser fluorescent techniques can bring space resolution.

6.3   Diagnostics of EBIS and EBIT

6.3.1   Diagnostic Methods

In general, the diagnostics of an electron beam ion source is based on the same principles as discussed for ECR ion sources in the previous section. This is also true for for investigations of EBIT or ERIS. Thus, we discuss in the following only special diagnostics for the mentioned devices, typically for for the different ion sources or methods, giving characteristic results for the actual source.

As shown above (see chapter 4), the working principles of EBIS and EBIT are essentially the same: an EBIT is a short EBIS; it traps the ions in a very short and intense electron beam. Thus, EBIT and EBIS are really complementary: the former emits photons, the later emits ions.

6.3.2   Ion Cooling Diagnostics

As shown on the EBIT at LLNL [336] ion cooling is the key to making an EBIS work well. Both the number and trapping life-time of highly charged ions held in an electron beam have been greatly increased by a process called "evaporative cooling" [166, 177, 337, 338].

In evaporative cooling (see also section 4.4) low charged ions of a second element are continuously added to the beam containing highly charged ions. The low charged ions are heated by ion-ion collisions with the highly charged ions until both species have approximately the same temperature. The lower charged ions need less energy per ion to escape the beam, so they "evaporate", removing their kinetic energy from the trap. Thereby, the continuous evaporation of the low charged ions cools the highly charged ones.

Fig.**6.77** compares the n $= 3 \rightarrow 2$ transition lines of Gold ions without evaporative cooling to those with evaporative cooling for the case, that the EBIT was adjusted to maximize the formation of neon-like Gold ions (Au^{69+}). The narrower lines in the second case indicate a more highly ionized charge state distribution.

As a cooling medium Titanium ions were used. The techniques of titanium evaporation into the source is described in Ref.[166]. The light ions dramatically increase the intensity of heavy Gold ion X-rays emitted from the EBIT, as shown in Fig.**6.78**. This is assumed to result from increased density, which result from lower loss, presumably by heavy ion cooling. Cooled Gold ions can remain in the electron beam for hours. As longest $1/e$ life times about 4 hours were reported [166].

The density of highly charged Gold ions in the electron beam versus the density $n_0(Ti)$ of Titanium ions in the beam can be estimated from the intensity of the radiative recombination lines using calculated cross-sections [339]. Fig.**6.79**a shows the count rate of the n $3 \rightarrow 2$ lines in Gold ions as a function of time for different values of $n_0(Ti)$. The estimated density of Gold ions is plotted versus $n_0(Ti)$ in Fig.**6.79**b.

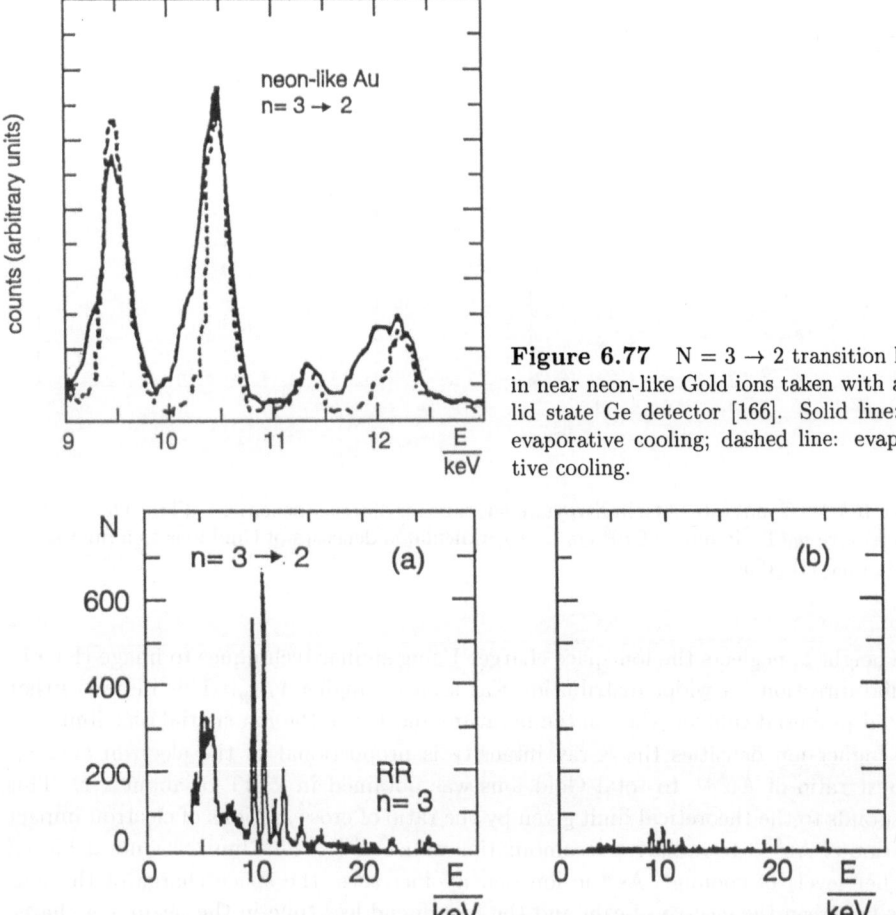

Figure 6.77 N = 3 → 2 transition lines in near neon-like Gold ions taken with a solid state Ge detector [166]. Solid line: no evaporative cooling; dashed line: evaporative cooling.

Figure 6.78 Time ruled Gold spectra, measured with a Ge detector. Shown are the n = 3 → 2 transition lines and the radiative recombination lines (RR) to n = 3. Titan shutter is kept open during (a) but kept closed during (b).

Measurements of the n = 3 → 2 lines [166] have shown for the limit, where the charge density of highly charged ions in the electron beam is less than 1% of the electron density, that the count rate is proportional to the square of the electron beam current, where the count rate is higher for higher axial potential V_A.

In an one-dimensional model the temperature should be proportional to QV_{trap}, i.e. the product of charge and trapping potential. A measurement of how the ions fill the trap should determine the temperature. Therefore in Ref.[166] a slit to axially image the ions onto a position sensitive detector gated to count only the n = 3 → 2 gold x-rays was used. The directly measured ion distribution can be characterized by its root mean square width.

To deduce the temperature it was assumed, that the density of highly charged ions in the beam is proportional to $\exp(-QV(z)/T)$, where $V(z)$, the axial electrostatic potential

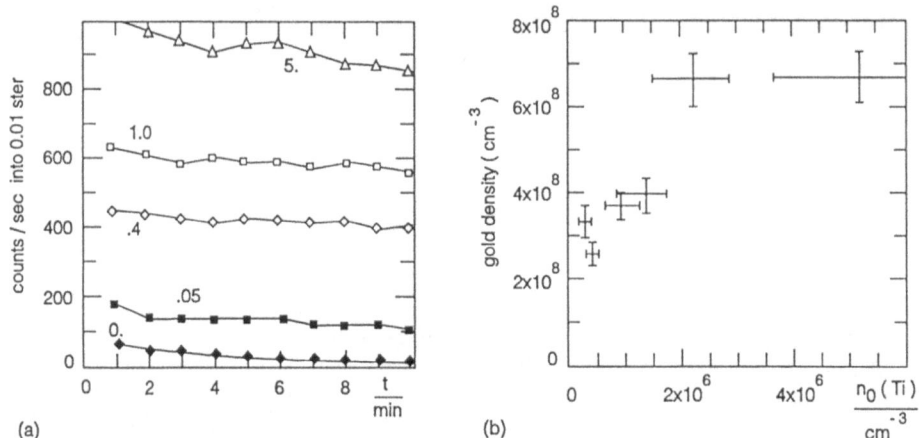

Figure 6.79   Count rate of Gold X-rays versus time for different densities $n_0(\text{Ti})$. The numbers in part a) give $n_0(\text{Ti})$ in units of 10^6 cm^{-3}. (b) Calculated densities of Gold ions as a function of estimated $n_0(\text{Ti})$ [166].

at the height z, neglects the ion space charge. Using similar techniques to image the ions in radial directions, a wider distribution was seen for higher V_{trap}. This indicates that the axial potential controls the ion temperature and hence the ion spatial distribution.

At higher ion densities the X-ray intensity is proportional to the electron current. The best ratio of Au^{69+} to total Gold ions was obtained in EBIT of about 1:2. This corresponds to the theoretical limit given by the ratio of cross-sections of electron impact ionization of Au^{68+} to radiative recombination onto Au^{69+}. This limit was not achieved at higher levels of cooling. As the ion density increases, the space charge of the ions begins to screen the electron beam and the ions spend less time in the beam, e.g charge exchange outside the beam destroys the charge states. In addition, charge exchange is expected to become important in the beam region at higher $n_0(T_i)$.

Levine et al. [338] report on evaporative cooling of Gold ions whereby low charged ions are supplied by ionization of the background gas in the beam. Thereby, an approximative equilibrium solution for the ion density can be found by assuming the charge states of low charged ions and negligible loss of highly charged ions. The heating rate is set equal to the energy loss through low charged ions and the rate of this ions should be equal to the ionization rate of the background gas with the density n_0. Thus

$$N_H = \alpha\, n_0 \qquad\qquad (6.134)$$

where N_H is the number of highly charged ions per cm in the electron beam and

$$\alpha \approx \frac{i(L)V_W \sigma_0 E_e A_H}{444.6\, L_C i^2(H)} \qquad\qquad (6.135)$$

The subscripts H and L denote high and low charged ions, L_C the Coulomb logarithm, σ_0 the ionization cross-section of the background gas, E_e the electron beam energy, V_W

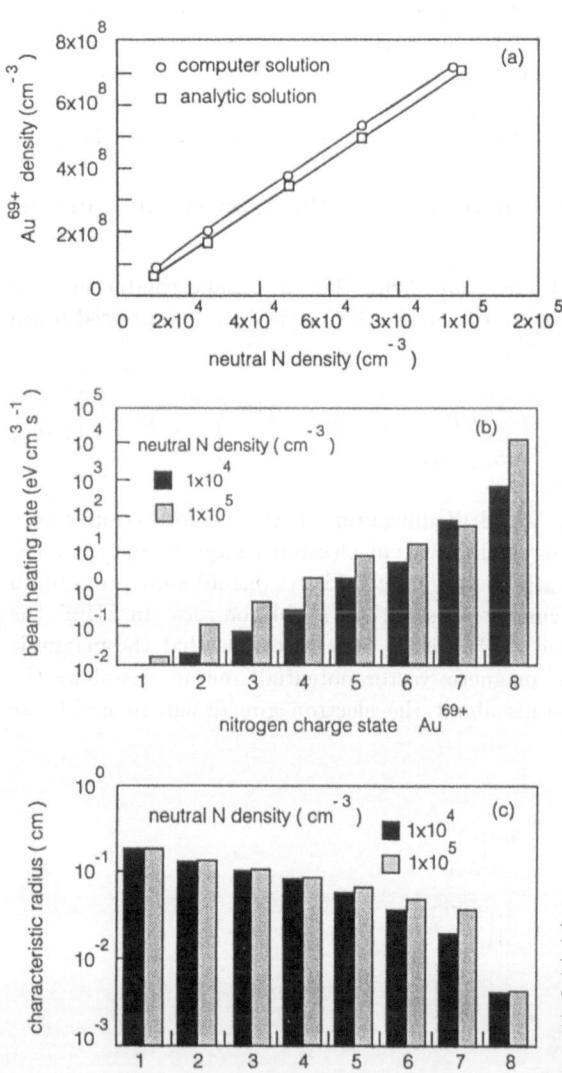

Figure 6.80 (a) Density of trapped highly charged Gold ions as a function of the Nitrogen coolant density for $V_W = 200$ V, $E = 18$ keV and $I = 100$ mA [338]. (b) Beam heating rate; (c) Ion density distribution for a 30 μm beam diameter in EBIT.

the wall potential and A_H the mass of the highly charged ions relative to the proton mass.

Results of computer simulation are shown in Fig.**6.80**. Fig.**6.80**a shows the dependence of the Au^{69+} ion density as a function of the density of Nitrogen atoms. Shown are results of the approximative analytical solution and results of computer simulation. Fig.**6.80**b shows the relative density of ions of different charge states on axis and Fig.**6.80**c the computer prediction of ion density distribution for a 30 μm beam diameter. The Nitrogen ions have a radius much larger than the beam and so the space charge due to coolant ions per cm is ten times larger than the charge due to Gold ions. When the neutral Nitrogen density is $1 \cdot 10^5$ cm^{-3}, the fractional neutralization of the electron beam is ≈ 20 %, but the Gold charge density is only $\approx 2\%$ of the electron beam charge density.

Numerical studies of the influence of different coolants are known from Ref.[177].

6.3.3   Electron Beam Geometry and Average Electron Beam Current Density in the EBIT

The EBIT beam is produced by a Pierce gun [338]. The 3 T superconducting coil compresses the beam further. According to Herrmann theory [340] the compressed beam radius r_0 (80% current) is given by

$$r_0 = r_B \left(\frac{1}{2} + \frac{1}{2} \left[1 + 4 \left(\frac{8kTr_e^2}{m_e\eta^2 r_B^2 B^2} + \frac{B_C^2 r_C^4}{B^2 r_B^4} \right) \right]^{1/2} \right)^{1/2} \tag{6.136}$$

where r_C is the cathode radius, r_B is the Brillouin radius, T the electron temperature at the cathode, m_e the electron mass, η the ratio of electron charge to mass, B the magnetic field intensity and B_C the cathode magnetic field. As (6.136) shows, to obtain a minimum beam radius (maximum current density), B_C should be zero. In EBIT, the cathode is mounted in a ferromagnetic shield with a bucking coil so that the magnetic field vanishes at the cathode and the magnetic vector potential contour lies along the spherical cathode contour [338]. Details about the electron gun design in EBIT are described in Refs.[338, 32].

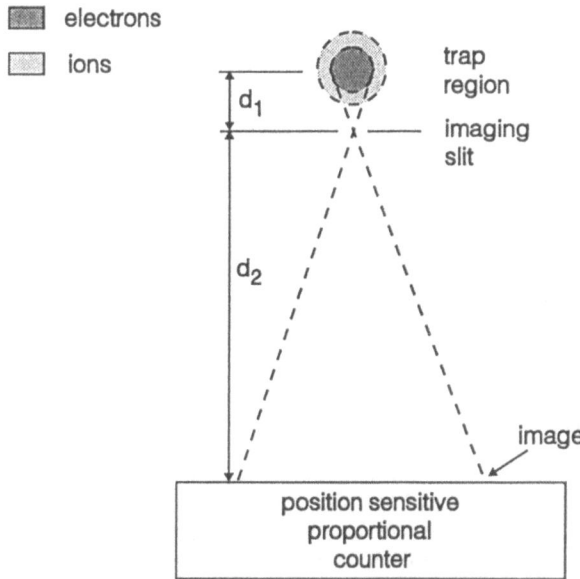

Figure 6.81 The experimental scheme for measuring the electron beam radius [338].

A measurement of the electron beam diameter allows to determine the current density. The scheme of the measuring geometry is shown in Fig.**6.81**.

For the diagnostic method used, the region of overlap of the electron beam and the trapped ion cloud is an X-ray source. A slit, aligned parallel to the EBIT axis, was located 1 cm from the beam and an X-ray image of the beam appears on a position sensitive proportional counter about 50 cm away. For analyzing the counter image a Gaussian radial density and an uniform ion distribution were assumed. For the ion density yields

$$\varrho_i = \varrho_0 \exp\left(-\frac{iV(r)}{kT_i}\right) \tag{6.137}$$

where $V(r)$ is the radial potential and T_i the ion temperature. To justify the approximation the ion temperature must be several times larger than the potential at the characteristic beam radius, i.e. $kT_i = \alpha i V(r_0)$, where α is large. If this condition is true, the distribution at the detector is given by

$$A(y') = A_0 \left(\text{erf}\left[\frac{1}{\sqrt{2}\sigma\mu}\right]\left[y' + \frac{w}{2}(1+\mu)\right] - \text{erf}\left[\frac{1}{\sqrt{2}\sigma\mu}\right]\left[y' - \frac{w}{2}(1+\mu)\right]\right) \tag{6.138}$$

where μ is the slit magnification, w the slit width, y' the distance on the image plane and σ is the variance of the beam radius.

In Ref.[338] the intensity of the X-radiation $A(y')$ from a mixture of neon-like, sodium-like and magnesium-like Gold ions was examined as the trap potential varied from 40 V to 340 V. The observed variance was almost constant from 140 V to 340 V, what indicates, that α was large.

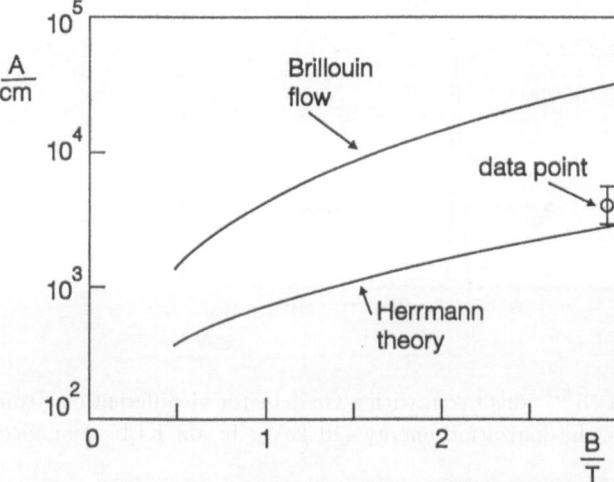

Figure 6.82 The current density for a measured electron beam radius in EBIT relative to the predicted by Brillouin flow and the Herrmann theory [338]. The theoretical curves are plotted assuming zero magnetic field and an electron temperature of 0.1 eV at the cathode.

These measurements indicated a beam radius of about $25 \ldots 30$ μm for 80% of the beam current. In Fig.**6.82** the current density for a measured beam radius is shown as a plot with results from the Brillouin and Herrmann theory [338].

Using (6.136), a 25 μm beam radius implies an effective cathode temperature of $T_C \approx 0.1$ eV. The radial compression from cathode to trap implies an electron temperature in the trap of $T_t \approx 300$ eV. If the electron flow from the cathode to the trap is laminar, the *variation of the electron energy across the beam* is only the difference in cathode to trap potential due to the beam space charge (about 15 V).

6.3.4   Electron Energy in the Centre-of-Mass System for Electron-Ion Collisions

The electron beam energy in the centre-of-mass system for electron-ion collisions can be measured analyzing electron-ion recombination. *Dielectronic recombination* (DR) is sensitive to the collisonal energy and the measurements can be done with a knowledge of several resonances and their relative strength. Dielectronic recombination is characterized by the resonant capture of an incident electron into a doubly excited state followed by X-ray emission

$$A^{i+} + e^- \to \left[A^{(i-1)+}\right]^{**} \to \left[A^{i+}\right]^* + e^- + X_{DR} \; . \tag{6.139}$$

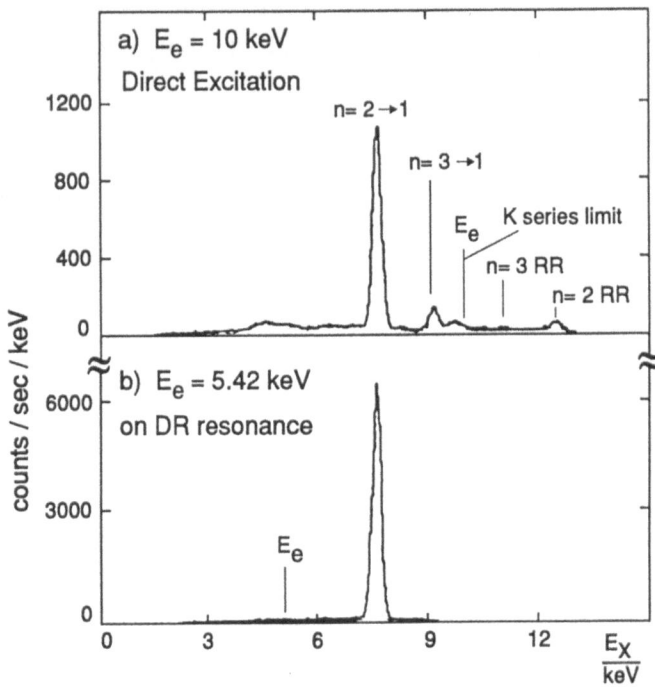

Figure 6.83 X-ray spectra of Ni^{26+} ions taken with a Ge detector at different electron beam energies E_e [341]. a: at the ionization energy (10 keV); b: on KLL resonance (5.42 keV).

The related process of resonant excitation (RE) occurs when the intermediate excited state decays by autoionization (Auger decay) to an excited state. The excited state decays by X-ray emission

$$A^{i+} + e^- \to \left[A^{(i-1)+}\right]^{**} \to \left[A^{i+}\right]^* + e^- \to A^{i+} + X_{RE} + e^- \; . \tag{6.140}$$

RE can only occur for an incident electron energy greater than the direct excitation threshold for the X-ray. The resonance strengths for nonoverlapping DR resonances can

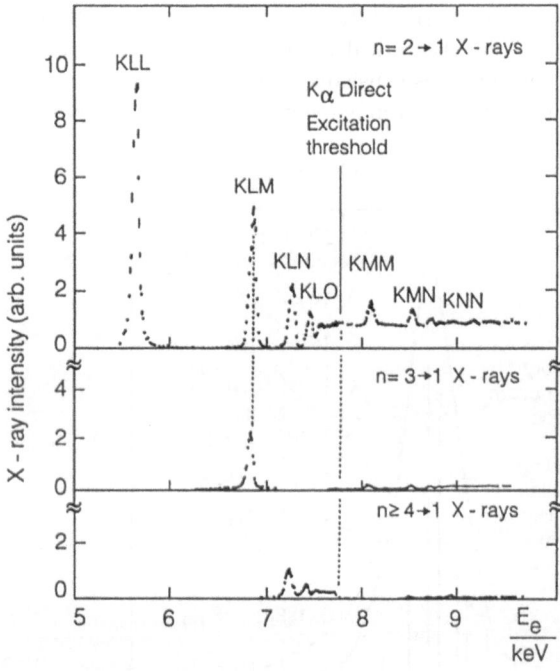

Figure 6.84 Dielectronic recombination excitation function of Ni^{26+} ions for separate X-ray bands [341].

be expressed in terms of the Auger and radiative widths of the resonance [341]

$$S(d,f) = \int\limits_{-\infty}^{\infty} \sigma(d,f,E)\, dE = \frac{2\pi^2}{k_i^2} \frac{g_d}{2g_i} \frac{A_r(d \to f)\, A_A(d \to i)}{\sum A_r + \sum A_A} \qquad (6.141)$$

where d and f refer to the resonant and final states, $\sum A_r$ and $\sum A_A$ are the total radiative and Auger widths, g_i and g_d are statistical weights for the initial and resonant states, and k_i is the incident electron wave number. For $\Delta n \neq 0$ transitions, $A_r \sim i^4$ for an isoelectronic sequence, while the Auger widths remain roughly constant as a function of ion charge i.

In Ref.[341] experiments with Ni^{26+} ions were described. The DR excitation function was measured by detecting the [Ni^{26+}]** K X-rays emitted at 90° to the electron beam direction with a planar Ge detector. For He-like target ions, exactly one K X-ray is produced per DR, so for a static ionization balance the number of K X-rays observed at a given beam energy is proportional to the DR cross-section at that energy. Details of the experimental techniques are described in Ref.[341].

Typical spectra taken at the ionization energy and on the energy of the KLL resonance are shown in Fig.6.83. The n = 2 → 1 and n = 3 → 1 X-rays are well separated from each other and from the higher members of the K series. Thus, in Fig.6.84 separate excitation functions for these energy bands are presented. The notation, e.g. the KLL resonances, for example, have an intermediate state in which an electron is captured into the L shell and another is excited from the K shell to the L shell.

In the case of Nickel ions, the electron beam energy distribution and the Nickel charge state distribution were estimated from a fit of the experimental data with the relative theoretical DR resonance strengths for each charge state. An electron energy offset of 77 eV was applied to the data to match the theoretical resonance energies, which can be calculated to very high accuracy. This offset is consistent with the estimated size of the space charge potential from ions in the trap and in agreement with the energy offset derived from the fit to the RR X-rays.

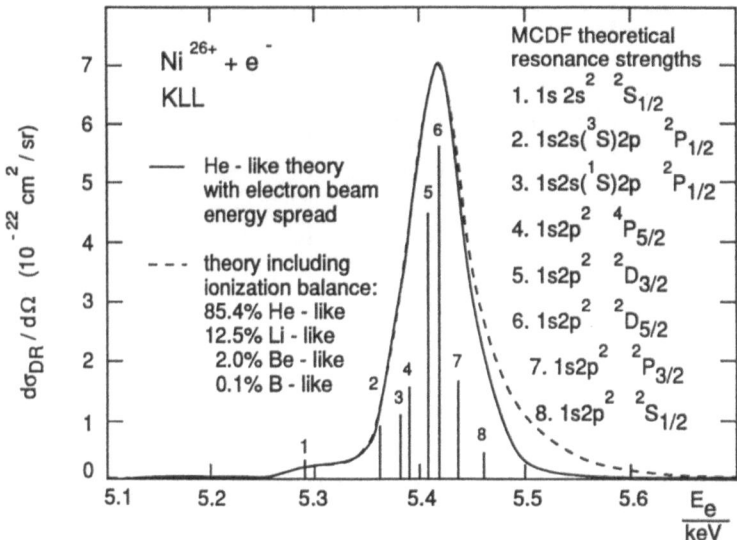

Figure 6.85 Comparison of experiment and theory for the KLL dielectronic recombination in Ni^{26+} [341]. The stick diagram shows the locations and relative amplitudes of the calculated resonances for He-like Nickel ions. The curves are theoretical resonance strengths folded with the electron beam energy distribution. Solid line: He-like ions only; dashed line: estimated charge state distribution. The systematic error in the data normalization was 11%.

A comparison of experimental and theoretical data for the KLL dielectronic recombination is shown in Fig.**6.85**. The behaviour of the resonance function allows to check the electron energy for the typical resonance energy. For other electron beam energies, other DR processes and other elements and ionization stages are possible probes for the determination of the electron energy.

Another example are measurements on neon-like systems as Au^{69+}. Fig.**6.86** shows an excitation spectrum for the near neon-like Gold. This is more complex than a helium-like system because there are many more resonant states.

U up to now a lot of different measurements of DR are known. First discussions are known from Massey and Bates [342] and first experiments were reported in 1983 for $\Delta n = 0$ [343, 344, 345]. Numerous subsequent measurements have now been performed. For instance at the LLNL EBIT beside X-rays due to DR of Ni^{26+} and Au^{69+} He-like Gold and F-like Xenon spectra [346] were measured. Briand et al. [347] reported on the observation of the KLL DR resonances on $Ar^{(12,13,14,15)+}$ and Ali et al. [348] on DR on He-like Argon on an EBIS.

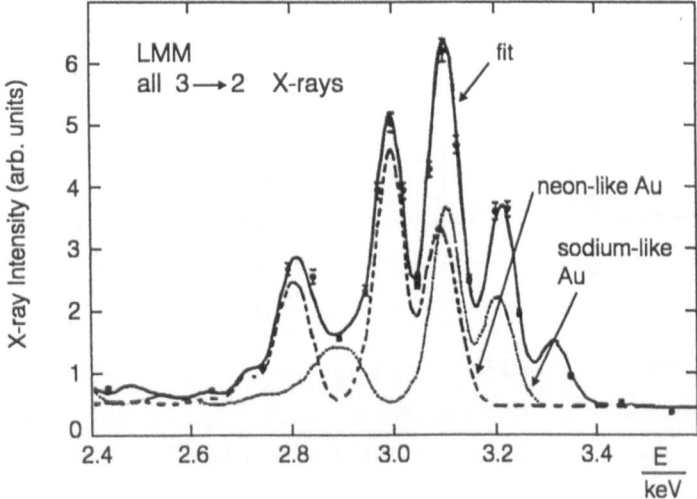

Figure 6.86 X-ray LMM spectrum from highly ionized Gold (mostly $i = 68+$ and 69+) obtained with a Ge detector [338].

6.3.5   Magnetic Measurements

The purpose of magnetic measurements are

- to evaluate the rectilinearity of the magnetic axes of superconducting solenoids and of bucking coils;
- to align the common magnetic axis with the source axis;
- to plot the curves $B_z = f(z)$ for various inductions.

Such measurements must be done in a carefully prepared non-magnetic environment (frameworks, removed of ion pumps ferrites, etc.).

The chosen method is, as shown for EBIS [349], the following: after the choice of a geometric axis on which the source is aligned chose the solenoid bore axis, a Hall probe measuring the B_r component rotates at a fixed abscissa [350]. From the modulation of $B(r) = f(\theta)$ the tilt and shift of the magnetic axis can be diagnosed.

Practically, this method is difficult to work with [349]:

- the sag and tilt of the probe holder have to be much lower than the defect being looked for;
- the measurements yield useful data only in the centre of the solenoid where the magnetic axis tilt can be deduced and in the field gradient where the axis translation can be obtained.

6.3.6   Electron Beam Geometry in EBIS

Magnetic field measurements must be followed by a series of tests of the electron beam rectilinearity and precise measurements of the current density. The drift tubes then must be finally fixed in the last stage and they are used to measure the electron beam transmission through the device with high precision. Methods of measuring the electron beam

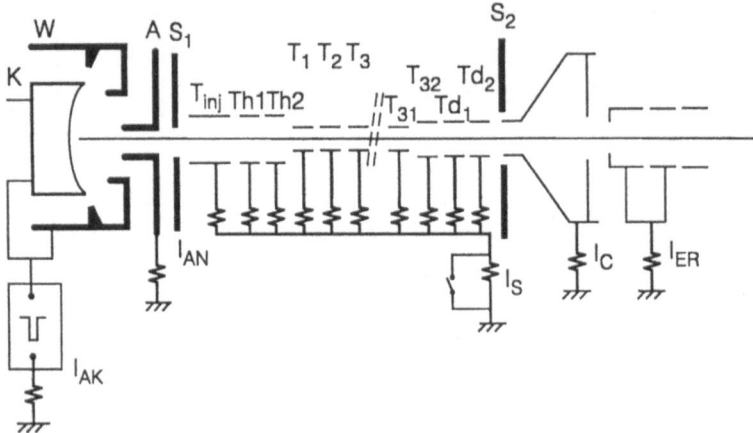

Figure 6.87   Electrical arrangement for electron beam transmission measurements [349].

diameter in EBIS after fixing the drift tubes are presented in [349]:

Movable Target. This method uses a thin sheet made of Tantalum or Tungsten (0.01 to 0.1 mm thick) movable along the source axis marked at the reticule centre of a telescope. The beam makes a bright spot on the target which is referenced to the reticule. By moving the target, with a bellow or a Wilson gasket, qualitative tests of the actions of different parameters can be performed very easily, together with accurate measurements of the rectilinearity [350]. The choice of the target is of particular importance to get a visible spot without crumpling. The visualisation of the beam image is made through the gun cathode with a Rank-taylor telescope giving the spot dimensions.

Drilling of holes in thin foils. This method uses a series of thin foils of Aluminium or Carbon regularly spaced the whole source long. The electron beam drills holes in the foils, allowing a better estimate of the magnetic compression than the first method. Here the electron beam parameters must be fixed during the drilling and each experiment requires much preparation and the entrance of air into the device between the experiments.

Beam Tester. Here a classical system of beam measurements is used [351]. The magneto-compression law may be determined with high accuracy as well as beam density profile, the ripple rate everywhere within the range of the longitudinal mobility of the beam tester.

All these methods change the electron beam characteristics [352] and the results underestimate the density. The final test remains the electron losses measurements on every electrode surrounding the beam from the catode to the collector as shown in Fig.**6.87**.

Correct transmission is achieved for

$$\frac{I_S}{I_C} + I_{ER} < 10^{-4} . \qquad (6.142)$$

Losses on the tubes at 4.2 K must be drastically limited ($I_S < 100$ nA) for a stable operation of the source. The transmission is improved either by moving the main coil or by shimming both ends as in a travelling wave tube. Obviously, the gun alignment and

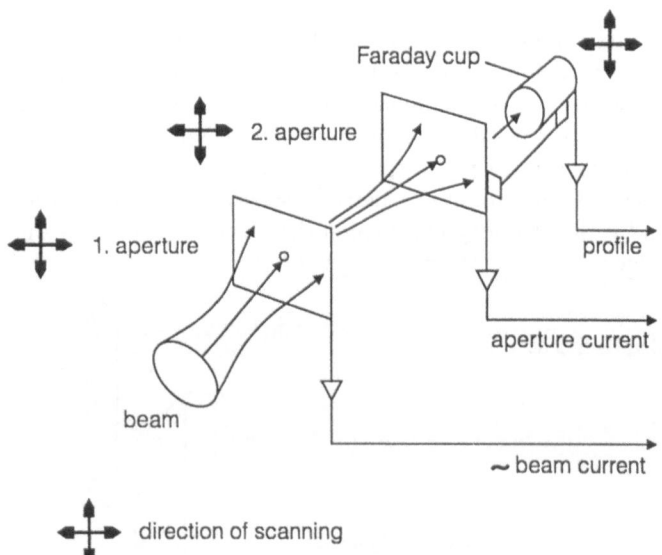

Figure **6.88** Scheme of an emittance measuring device [66].

bucking field must be perfectly adjusted to get

$$I_C + I_{ER} \cong I_{AK} \ .$$ (6.143)

This criterion is necessary for proper operation of the source but is not sufficient to get its total capacity. The electron beam axis within the collector has to be close to the ion beam line axis. This condition as a rule has never been fulfilled. Only a magnetic steering system around the electron collector would allow a correct alignment of the ion beam at the entrance of the lenses.

6.3.7   Diagnostics of Extracted Ion Beams from EBIS

At the source exit ions are accelerated according to the actual source potential and an electron repeller prevents primary electron transmission into the beam line. As a rule, the main ion beam diagnostic is the use of a Faraday cup (see paragraph 6.2.2.4.1) which is used either to measure the total current delivered by the source or as a time-of-flight detector.

Any ion beam diagnostics situated close to the source exit is disturbed by electrons flowing through the beam line (for example, an ion beam of ≈ 25 μA must be compared to some 100 μA of electrons) [349]. Thus, the choice of an electrostatic ion beam line and even the use of a coaxial collection-extraction is only valid if one takes care of the ion beam separation from the electron beam right after the electron repeller electrode, with either a magnetic or an electrostatic system.

The *ion beam diameter* and the *radial position* can be detected by using a special Faraday cup made with n coaxial rings, each divided into 4 sectors. From the $4*n$ pieces of information, it is possible to adjust the steering and the focussing of the ion beam line.

The *emittance* envelope for the transverse phase space can be measured by using systems with great diversity. Details for emittance measuring are given for instance by Strehl [246]. In most cases the principle can be derived from the scheme of Fig.**6.88**, assuming a low energy beam that can be stopped completely in an adequately designed aperture [66]. With respect to the definition of emittance the measuring technique as shown in Fig.**6.88** would be a nearly perfect technique, but also a very time consuming one. The most simplification of the scheme shown result from the assumption that there is no coupling between the two transverse phase spaces, leading to schemes, like the following: one horizontal or vertical slit in front of a detector sandwich or a profile grid, two horizontal or vertical slits with a Faraday cup behind the second slit, crossed slits with crossed scanning wires downstream, as shown in Fig.**6.89** etc.

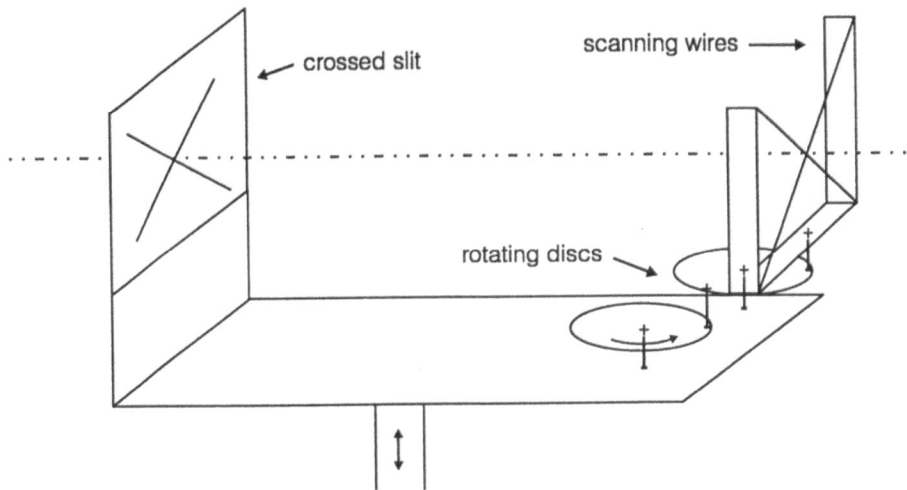

Figure 6.89   Scheme of an emittance measuring device with crossed slits and crossed scanning wires [66].

Further techniques for emittance measurements are also discussed for instance in Refs.[354, 355].

6.3.8   Diagnostics of the Ionization Process

Beside spectroscopic diagnostics, as described in subsection 6.2.4, other different diagnostics are applied [349]. *Time-of-flight* measurements allow to determine ions with different Energy E and so also with different charge states i according to $E = (i/A)U$ with U as accelerating voltage. The ion velocity is in the non-relativistic case given by

$$v = \sqrt{\frac{2E}{M}} \; . \tag{6.144}$$

If the energy E is known, the mass M can be directly determined from the velocity. Thus

$$M = \frac{2Et^2}{l^2} \tag{6.145}$$

where t is the time of flight measured in nanoseconds over a flight path l cm long. Considering the ion charge i, a determination of ion rates for species with different i

yields

$$t_i = \frac{Ml}{\sqrt{2Ui}} \qquad (6.146)$$

and

$$t_i = \frac{l}{c}\frac{E_0}{2E} \qquad (6.147)$$

with E_0 as ion rest mass. The mentioned relations are nonrelativistic assumptions, which are satisfied in the large majority of the cases studied.

A simple time-of-flight spectrometer consists of a chopper, a drift tube and a Faraday cup followed by a large bandwidth amplifier. In Ref.[349] the chopper is composed of two plates 1 cm long 2.5 cm spaced between two grounded rings of 2.5 cm diameter. It is located behind the focussing line. Without biasing the plates, the Faraday cup (2.2 m away from the chopper, 80 mm diameter) collects the total current. The analysis is achieved by biasing the plates at symmetrical potentials (± 300 V) to deflect the beam out of the detector and by cancelling the biases during short pulses (50-300 ns) to sample within the ionic peak. The sample may be selected anywhere within the peak by changing a modulator delay. The different charge states drift with different times of flight, so for instance typically 3 μs for light nuclei at $\cong$ 4 keV/charge. The resolution power in the for the described spectrometer in [349] was at about 40 and the calibration of the time-of-flight spectrometer was accomplished with Hydrogen lines and lines from single charged ions. The evolution of the charge state populations as a function of the ion confinement time allow the identification of a spectral line.

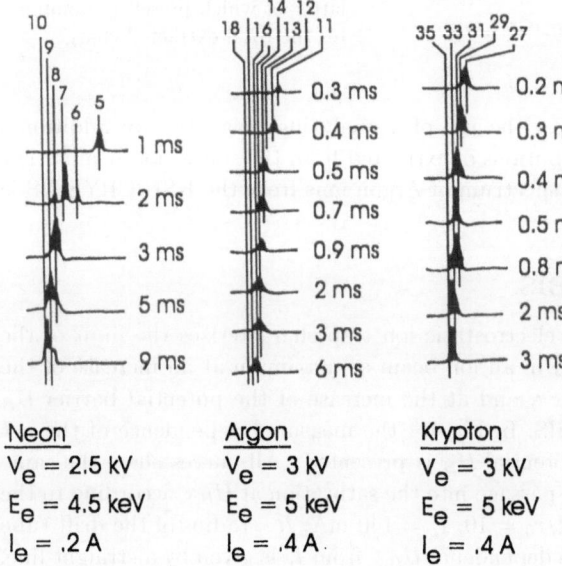

Neon
V_e = 2.5 kV
E_e = 4.5 keV
I_e = .2 A

Argon
V_e = 3 kV
E_e = 5 keV
I_e = .4 A

Krypton
V_e = 3 kV
E_e = 5 keV
I_e $\cong$ .4 A

Figure 6.90 Charge state distributions for different containment times with Neon, Argon and Krypton [356].

As an example, the evolution of the charge state population of Neon, Argon and Krypton are shown in Fig.**6.90** for different containment times. From the broadness of the time-of-flight lines it is possible to estimate the actual energy spread for any ion charge state.

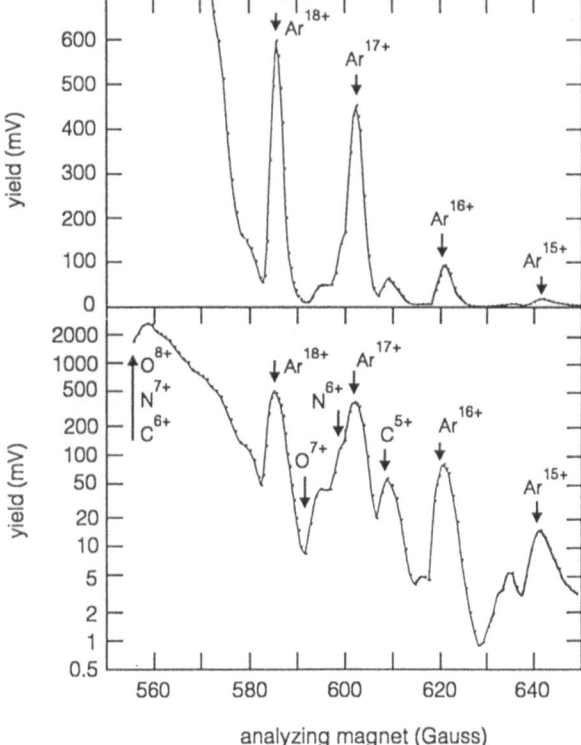

Figure 6.91 The ion charge state spectrum from the KSU CRYEBIS by varying the analyzing magnet current and measuring the analyzed ionic current with a channel plate detector behind four-jaw slits which are installed in the location of the exit focus [357]. The linear yield in the upper half shows how the Argon dominates the high-field region, with bare Argon ions being the most abundant Argon charge state. The lower logarithmic display indicates some peaks from the residual gas and the peak of the fully stipped residual gas, the latter of which presently dominates the total extracted charge.

As mentioned in paragraph 6.2.2.5 the use of a analyzing magnet is an additional tool to determine charge state distributions of extracted from the source ion beams. For example, in Fig.**6.91** a charge state spectrum of Argon ions from the KSU CRYEBIS is shown [357].

6.3.9   Ion Capacity of the EBIS

The capacity C^+ of an EBIS as an electrostatic ion trap characterizes the limit of the ion charge I^+, which can be reached in an ion beam as maximum at an increase of the gas current i_g, of the injection time τ_i and at the increase of the potential barrier U_B at the individual sections of the EBIS. In Fig.**4.4** the measured dependence of the net ion charge I^+ from the inclusion potential U_B is presented. All curves show the same behavior: a linear increase up to the passage into the saturation at $U_{B,S}$ according to the model the self-consistent field (for $R/r_0 = 10$, $I_e = 140$ mA; R – radius of the drift tube and r_0 – beam radius). Thereby, the dependence $U_{B,S}$ from I_e is given by a straight line.

In the model of the self-consistent field C^+ is equal to the number of the electrons in the EBIS and so proportional to I_e for $E_e =$ const. An entirety of measured curves $C^+ = f(I_e)$ is indicated in Fig.**6.92** [153]. A linear increasing of C^+ is observed only up to a certain value of C^+, from which appears a deviation from the linear growth of C^+. Then a reduction of C^+ follows at increasing I_e. This current is marked as critical current $I_{e,c}$. The dependence $I^+ = f(I_e)$ is presented in Fig.**6.92**b for $I^+ < C^+$. It becomes evident, that $I_{e,c}$ depends on the given boundaries not from the ion number in the trap.

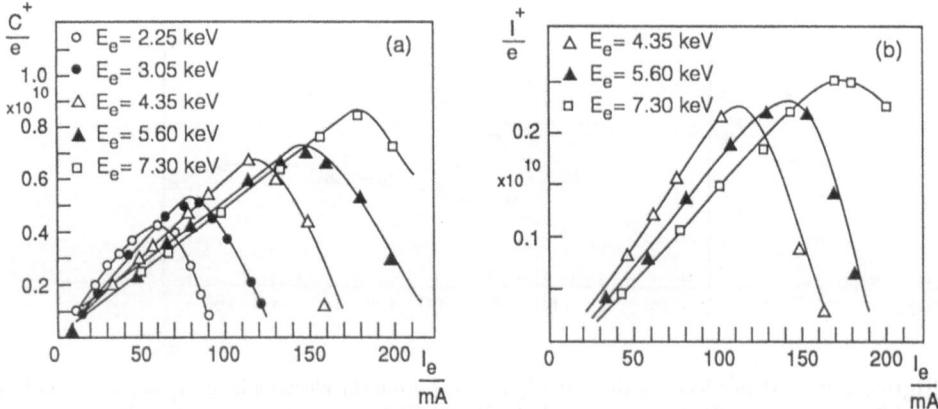

Figure 6.92 Dependence of the capacity C^+ (a) and the ion charge I^+(b) in the ion trap from the electron current I_e at different electron energies E_e [153].

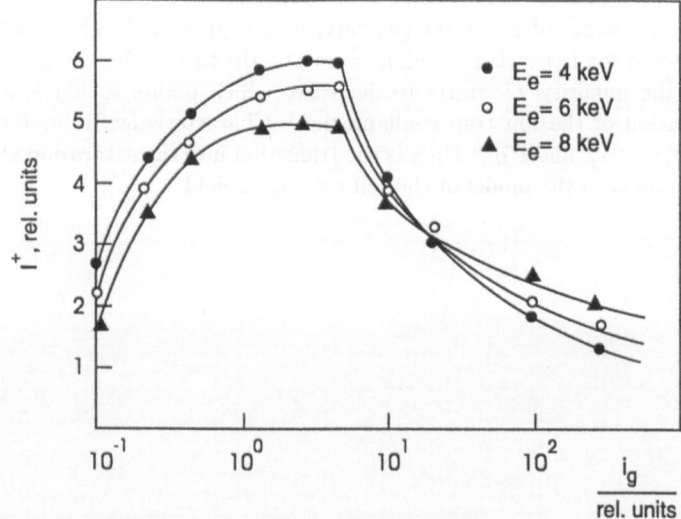

Figure 6.93 Dependence of the ion charge I^+ from the gas current in the source in relative units [153].

In the study of the dependence on $I_{e,c}$ from different parameters it was observed, that $I_{e,c}$ depends only from E_e, i.e., it increases linearly with the beam energy. This particularly points to the process which leads to ion losses and starts at a certain value of ϱ_e^l/v_e (ϱ_e^l – linear charge density of the electrons in the beam; v_e – electron velocity). This value is constant for all energies and currents of the electron beam. From the entirety

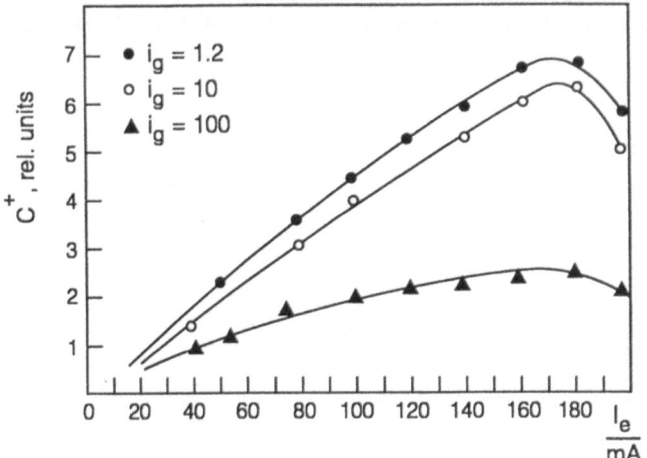

Figure 6.94   Dependence of the trap capacity C^+ from the electron beam I_e at $E_e = 7.43\,\mathrm{keV}$ for different values of the gas flow (in relative units) [153].

of the curves in Fig.**6.92** follows, that accordingly the model of the self-consistent field counts $C^+ \sim E^{-1/2}$ at $I_e = \mathrm{const.}$ and at $I_e < I_{e,c}$.

[6]. The dependence of I^+ on the gas current j_g is presented in Fig.**6.93**. The curves were constructed for $U_B > U_{B,S}$ and $\tau_i = \mathrm{const.}$ Up to C^+ the value of I^+ increases, at $i_g > i_g^{cr}$ the quantity I^+ starts to decrease, which points to ion loss processes or to a deformation of the ion trap configuration. Characteristically for the dependence $I^+ = f(\tau_i)$ at $i_g < i_g^{cr}$ and $U_B \geq U_{B,S}$ is the transition into the saturation with $I^+ = C^+$, which corresponds to the model of the self-consistent field.

Chapter 7

The Use of Electron Impact Ion Sources in Science and Technology

7.1 Introduction

The use of electron impact ion sources such as the ECR, EBIS and EBIT besides the work as injectors in ion accelerating devices allows it to accomplish different kinds of experiments with the ions trapped in the source or with extracted ions for external collision experiments. The prospects for the study of numerous highly charged ions led to a series of developments, experiments and techniques and to a further upgrade of the sources.

Basically, the research programs concentrate on the following major parts

- atomic structure investigations;
- investigations for thermonuclear fusion and astrophysics;
- ion-surface interaction studies;
- electron-ion interaction studies;
- beam diagnostics measurements;
- instrumental development.

Thereby, the character of investigations is different for activities on EBIS, EBIT and on ECR ion sources, because EBIS and EBIT devices have a monoenergetic, tunable electron beam, which can isolate spectra from individual charge states which is a great aid for understanding the complicated spectra from plasmas. The ability to quickly change the electron beam energy in EBIT made it possible to accomplish lifetime measurements of excited atomic states. Further more, EBIS and EBIT sources allow it to produce highest charge states, available for experiments inside the source configuration and after extraction also for collision studies. On the other hand, ECR ion sources are characterized by an electron energy distribution in the source plasma. This circumstance exludes different classes of experiments with ions inside the plasma (e.g. studies of resonant dielectronic recombination). In contrast to EBIS and EBIT sources it is possible for ECR ion sources to extract much higher currents of highly charged light ions and also of intermediate charged heavy ions. Thus, for different experiments one or the other ion source should be preferred.

It should be noted, that investigations accomplished at the named sources, give fundamental information about atomic physics processes and are also relevant for such fields as molecular and cluster physics, ion source physics, astrophysics, diagnostics of thermonuclear fusion plasmas, materials science and radiation physics.

In the following, we give some typical applications for electron beam ion sources, preferably for atomic physics. But it is important to note, that these are only examples to demonstrate different directions of investigations without claim to completeness. At present, the area of applications of electron impact ion sources is a dynamically developing subject, and in the present book it is simply impossible to characterize all scientific directions completely, connected with the development of the described ion sources. Thus, we give examples especially for the directions of investigations directly using the ions in the source. But we must note, that there is also a wide range of applications connected with ions extracted from the ion source. Many examples are found for instance in the corresponding publications of such groups as from Kansas State University, the Giessen group, the group from the KVI Groningen, the Grenoble group, the Berkeley group, the Bochum group and many others. Here we will give some examples of recent experiments provided at the Livermore EBIT as an advanced source for producing the highest ion charge states.

7.2   Application of Electron Impact Ion Sources to Heavy Ion Accelerators

The specific charge i/A is an important parameter to increase the final energy reached in a heavy ion accelerator. Here the specific charge dominates the final energy in an manner as shown in Table **7.1**.

Electron impact ion sources are able to produce relatively high currents of ions with high charge states. The development of this sources is not yet at its end, but it already influences the spectrum of heavy ion accelerators; especially the cyclotrons with the strong dependence on $(i/A)^2$ profit from these sources by external injection. So the use of ECR ion sources substantially increased the final energies and mass numbers of ions accelerated at cyclotrons.

The EBIS is limited to about 10^{11} charges per pulse, but offers the possibility to get even the heaviest ions with high charge states. EBIS are already used for injection into synchrophasotrons, were final energies of GeV/nucleon have been achieved. By increasing the product of the electron beam density times the containment time of the ions in the source values larger than 10^5 A/cm²s and of electron energy to about 300 keV U^{92+} can be produced.

Another method is to accelerate large ion currents with low charge states and shoot them through a dense electron target called a *stripper*. Fig.**7.1** compares the efficiency of stripper foils with ECR and EBIS ion sources.

Stripping at several MeV/nucleon corresponds to charge states produced by ECR ion sources. For the EBIS the corresponding stripper energy is above 10 MeV/nucleon. What the best solution is for an certain accelerator could not be answered generally.

Table 7.1   Final energies for different heavy ion accelerators. i – ion charge state; l – acceleration length; B – magnetic induction; R – radius of the circular accelerator; $\bar{E}$ – mean electric field; U – acceleration voltage; m_p – proton mass.

accelerator	final energy [MeV/amu]
van de Graff	$\frac{i}{A} U$ [MV]
linac	$\frac{i}{A} \Delta U = \frac{i}{a} \bar{E} l$
cyclotron	$k \left(\frac{i}{A}\right)^2 \qquad k = \frac{e^2}{2m_p} B^2$ [T] R^2 [m]
synchrotron	for $\quad v \ll c:\quad R^2 B^2 \frac{e^2}{2m_p} \left(\frac{i}{A}\right)^2$
	for $\quad v \approx c:\quad RBec\frac{i}{A}$

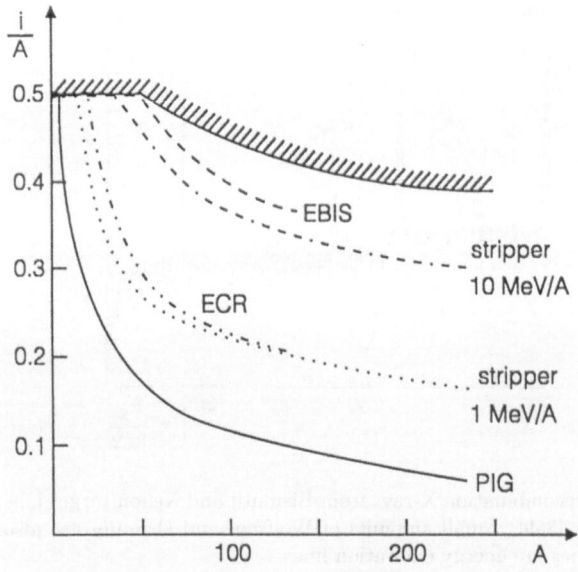

Figure 7.1   Maximum attained i/A by ECR, EBIS and Penning ion sources and by stripper foils [358].

So for instance, the stripping method is certainly the adequate solution for some linear accelerators, for cyclotrons ECR ion sources is a very good solution and for synchrotrons the pulsed EBIS has some merits [358].

7.3   Investigations on Highly Charged Ions from Electron Impact Ion Sources

7.3.1   Atomic Structure of Highly Charged Ions

7.3.1.1   Binding Energies of Hydrogen-Like and Helium-Like Ions

In the Super EBIT it is possible to produce hydrogen- and helium-like ions in such quantities, that spectroscopic measurements are possible. Especially of interest are here ions of the hydrogen-like isoelectronic sequence as the simplest atomic system in which multielectron interactions are absent. Helium-like ions are the next simplest system and a test system for two-electron interactions. Thus, at the Super EBIT of the LLNL a program was started to measure binding energies of 1s electrons in high-Z hydrogen-like and helium-like ions [359]. Ions with high Z are of special interest, because contributions from relativity and QED (lamb shift) both scale to atomic energy levels as Z^4.

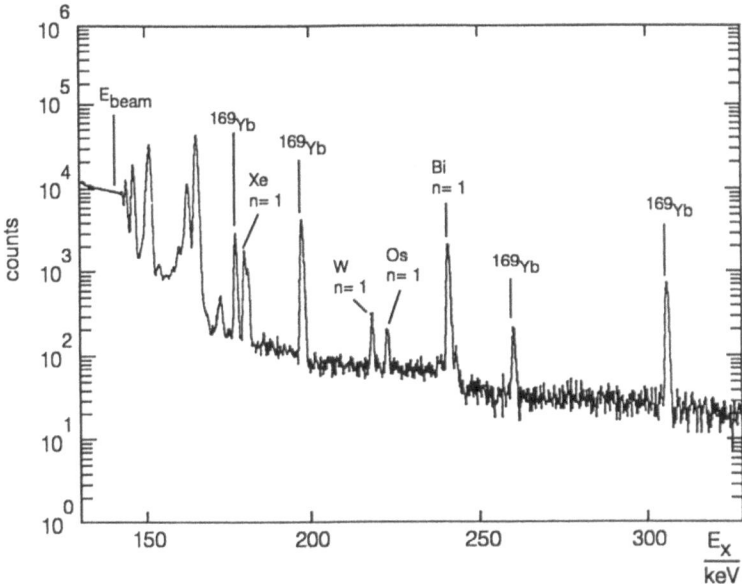

Figure 7.2   Spectrum of radiative recombination X-rays from Bismuth and Xenon target ions measured with a Germanium detector [359]. Small amounts of Wolfram and Osmium are also present in the trap. The Ytterbium lines are energy calbration lines.

A first result of Lamb shift measurements is shown in Fig.**7.2** for Bismuth and Xenon ions. Thereby, the energy from photons when electrons are captured into a 1s vacancy was measured with a Germanium detector calibrated with radioactive sources. For the determination of the 1s binding energies E_e and the binding energy E_B in the analyzed ion it yields for quanta from radiative recombination (RR)

$$E_{RR} = E_e + E_B(1s) \ . \tag{7.1}$$

Because the uncertainty in measuring the electron beam energy is much higher than the uncertainty from the binding energy measuring, it is necessary to measure binding energy differences to cancel out the electron beam energy.

7.3.1.2   Lamb-Shift in Lithium-Like through Neon-Like Uranium

Among the naturally occuring elements quantum electrodynamical effects are largest in Uranium because the effects increase as Z^4. Thus, high resolution X-ray measurements were done on Super EBIT for the $2s_{1/2}$-$2p_{3/2}$ transition in Uranium ions for U^{82+} through U^{89+} [360]. High resolution spectra from Ref.[361] are shown in Fig.7.3.

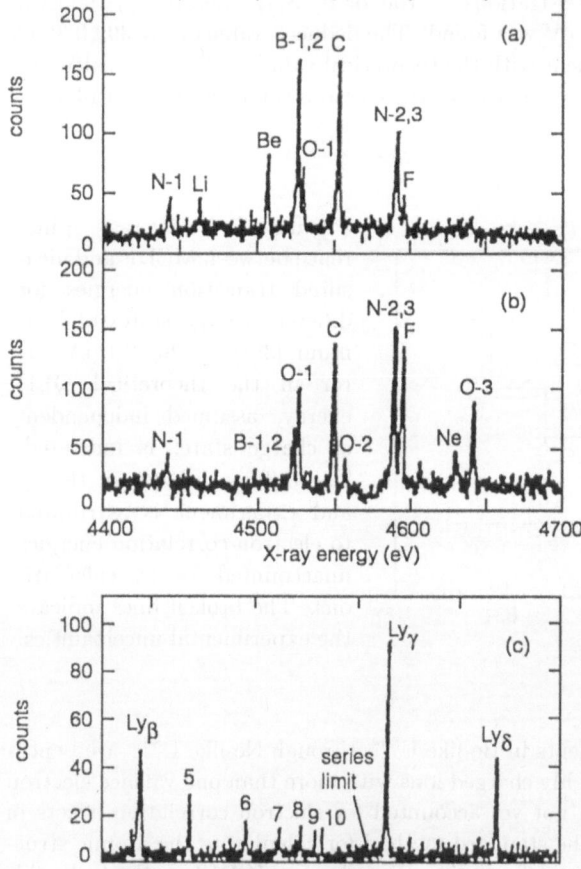

Figure 7.3 Spectra of the $2s_{1/2}$-$2p_{3/2}$ transitions in U^{82+} through U^{89+} [361]. The lines are labeled by the corresponding sequence. a: high charge state conditions, Li-like and Be-like lines are enhanced; b: lower charge states conditions, F-like and Ne-like lines are enhanced; c: calibration spectrum showing K shell X-rays from hydrogen- and helium-like Potassium.

To study the full range of charge states from U^{82+} to U^{89+} the ionization balance in the trap was varied by changing the electron energy, the neutral gas density and the effective current density. The beam energies used were above the 33 keV ionization potential of Li-like U^{89+} and below the 130 keV ionization potential of He-like U^{90+}. Further, the choice of the electron beam energy also ensures the absence of satellite transitions with a spectator electron in a high-n level that could shift the apparent line and that may be a possible source of error in beam-foil or plasma experiments.

The Uranium spectra were calibrated by recording of known reference lines for three separate arrangements of the X-ray crystal diffraction spectrometer. One calibration was provided by the Lyman series in H-like K^{18+} and He-like K^{17+} (see Fig.**7.3**). Other calbrations were provided by by the Lyman-α lines of Cu^{28+}, measured in second order Bragg reflection and by the $2p_{3/2}$-$3d_{5/2}$ and by the $2p_{1/2}$-$3s_{1/2}$ transition of Xe^{44+}. The corresponding energies are either theoretically well known or have been measured in earlier experiments.

As a result of the described investigations for the $1s^2 2s\ ^2S_{1/2}$-$1s\,2s\,2p\ ^2P_{3/2}$ splitting in U^{89+} an energy of 4459.37±0.35 eV was found. The deduced value of -47.39±0.35 eV for the 2s QED energy is in agreement with the theoretical value of -47.58 eV calculated by Blundell [362] and consisting of -62.57 eV for the electron self energy and +14.99 eV for the vacuum polarization.

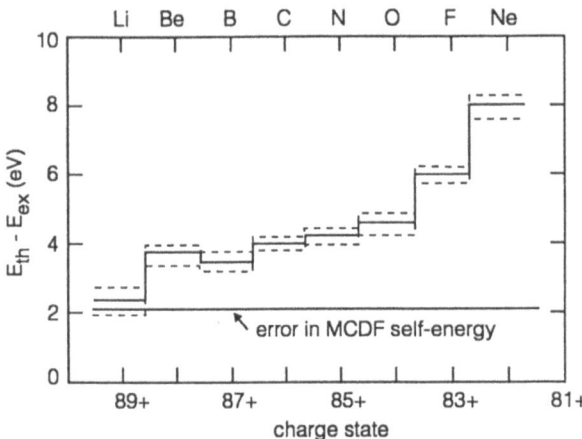

Figure 7.4 Average difference between MCDF and measured transition energies for different charge states of Uranium [361]. The 2.1 eV error in the theoretical QED energy, assumed independent of charge state, is indicated. The difference between theory and experiment is attributed to electron-correlation energies unaccounted for by calculations. The broken lines indicate the experimental uncertainties.

$2s_{1/2}$-$2p_{3/2}$ transition measurements in Be-like U^{88+} through Ne-like U^{82+} are benchmarks for testing calculations in highly charged ions with more than one valence electron because theoretical algorithms can not yet accounted for electron correlation effects in many-valence electron ions. Here the standard method for calculating the atomic structure of open-shell ions is the multi-configuration Dirac-Fock (MCDF) method. In this approach in Ref.[361] the energies for 2s-2p transitions were computed. A comparison between measured and computed energies shows large differences, shown in Fig.**7.4**. The difference of 2 eV arises from difficulties of calculating QED effects in many-electron atoms. The remainder results from the difficulty of calculating correlation energies. This difference increases with the number of valence electrons and thus with the complexity of the atomic structure.

7.3.1.3   The Self-Energy Contribution to the 2s-3p Transition in Neon-Like Ytterbium

The self-energy contribution to levels in hydrogen-like ions has been calculated with high precision from first principles [363, 364]. This is not the case for the self-energy contribution to levels in multielectron ions, where atomic structure calculations commonly use the so-called effective-charge approach [365, 366] to estimate the self-energy contributions.

Because no calculations of the self-energy contributions to transitions in multielectron ions with more than two electrons exist which are based on first principles, the accuracy of the effective-charge approach was tested by Beiersdorfer et al. by X-ray spectroscopy at an EBIT [171].

From tokamak plasmas measurements of the 2s-3p transition are known for Silver (Z=47), Xenon (Z=54) and Lanthanum (Z=57) with an experimental uncertainty of about 30...70 ppm [367, 368] and from beam-foil experiments for Gold (Z=79) and Bismuth (Z=83) with an uncertainty of about 170 ppm [369, 370]. Measurements on the EBIT are reported for neon-like Ytterbium (Z=70) with an uncertainty of 85 ppm [171].

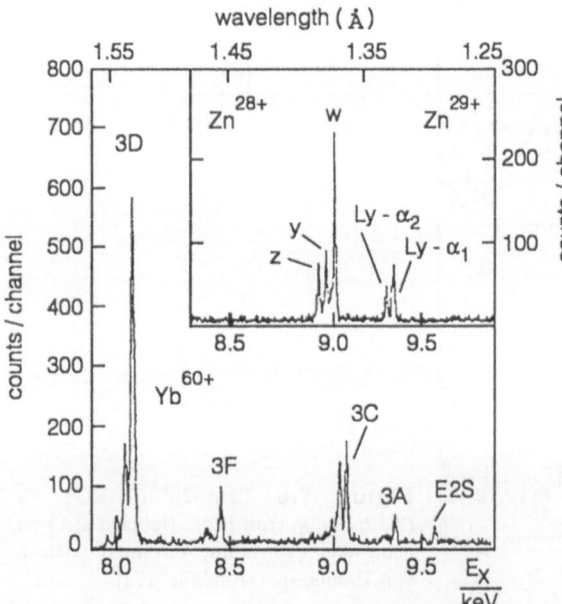

Figure 7.5   Spectrum of n=2→3 transitions in neon-like Ytterbium, obtained with a flat crystal diffraction spectrometer [171]. Unmarked features are transitions in lower charge states. The spectrum of hydrogen-like and helium-like Zinc used for calibration is shown in the inset.

A spectrum of the n=3 to n=2 transitions in Yb^{60+} is shown in Fig.**7.5** [171]. The observed lines are labeled in the notation of Loulerque and Nussbaumer [371] as 3C for $(2p^5_{1/2}\ 3d_{3/2})_{J=1} \rightarrow (2p^6)_{J=0}$, as 3D for $(2p^5_{3/2}\ 3d_{5/2})_{J=1} \rightarrow (2p^6)_{J=0}$ and as 3F for $(2p^5_{1/2}\ 3s_{1/2})_{J=1} \rightarrow (2p^6)_{J=0}$. The 2s-3p resonance line, labeled as 3A (see Fig.**7.6**), is situated on the high-energy side of line 3C and corresponds to the transition from $(2s_{1/2}\ 2p^6\ 3p_{3/2})_{J=1}$ to the ground state $(2s^2\ 2p^6)_{J=0}$. Its energy is strongly affected by QED effects because it involves a 2s core level. The spectrum also shows an electric quadrupole transition, labeled as E2S from the upper level $(2s_{1/2}\ 2p^6\ 3d_{3/2})_{J=2}$. This line is of interest in plasma diagnostics because of its density sensitivity [191].

To determine the wavelength of line 3A with high precision, the line has been measured with a von Hámos crystal diffraction spectrometer. A spectrum of line 3A was determined with respect to the theoretical wavelengts of the Ly-α_1 and Ly-α_2 lines of hydrogenic Zinc, because the wavelengths of the hydrogenic lines are very close to that of the neon-like line. The use of hydrogen-like lines as reference lines is possible because their wavelengths can be calculated with high precision [363, 364]. The reached precision is higher than the experimental precision for determining the transition energy of line 3A.

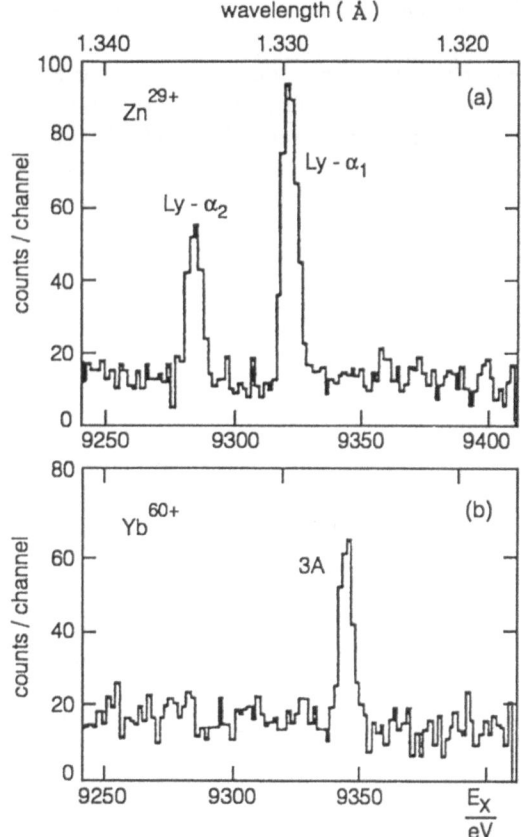

Figure 7.6 $(2s_{1/2}2p^6 3p_{3/2})_{J=1} \rightarrow (2s^2 2p^6)_{J=0}$ transition (labeled 3A) in neon-like Ytterbium, obtained with a von Hámos spectrometer [171].

The measured values of the transition energy of line 3A was compared to the value calculated with an MCDF code of Ref.[373]. The result shows, that the calculated transition energies are too large (2.97 eV for the measured with the von Hámos spectrometer and 3.4 eV with a flat crystal). This behaviour is the same as it was found in previous measurements of the neon-like Silver [374], Xenon and Lanthanum [368].

The self-energy contribution was obtained by subtracting the relativistic Coulomb-energy E_{Coul} and the contributions from the Breit interaction E_{Breit}, vacuum polarization E_{VP} and residual correlation energy E_{corr} from the transition energy of line 3A measured with the von Hámos spectrometer. Comparisons with values of the self-energy, calculated using the effective-charge approach and for hydrogenic wave functions and

Table 7.2 Self-energy contribution to the transition 3A in Yb^{60+} [171]. SE_{expt} is obtained by subtracting the calculated values for the relativistic Coulomb energy E_{Coul}, the transverse Breit energy E_{breit}, the vacuum polarization energy E_{VP} and the residual correlation energy E_{corr} from the measured transition energy E_{expt}. SE_{MCDF} is the self-energy computed using the effective-charge approach and SE_{hydro} is the self-energy computed assuming hydrogenic wave functions and a bare ytterbium nucleus. All values are in eV.

E_{expt}	E_{Coul}	E_{Breit}	E_{VP}	E_{corr}	SE_{expt}	SE_{MCDF}	SE_{hydro}
9341.86±0.80	9373.63	-15.40	3.09	-1.09	-18.37±0.80	-16.49	-19.91

a bare Ytterbium nucleus differ significantly from the measured value. The numerical results for the described analyse are shown in Table **7.2**.

7.3.1.4 Measurements of Radiative Branching Ratios in Lithium-Like Ions

Measurements of radiative transition rates provide fundamental tests of atomic structure calculations for highly charged ions. Although radiative rates have been measured for ions and various transitions with the beam-foil technique [375, 376, 377, 378, 178], radiative branching ratios of lithium-like ions Cr^{21+}, Mn^{22+}, Fe^{23+}, Ni^{25+} and Ge^{29+} were measured at the EBIT [380], studying the radiative decay of the level $1s\,2s^2$.

In single-configuration calculations the $1s\,2s^2$ level is electric-dipole forbidden to decay to the $1s^2\,2l$ lithium-like ground state configurations and can only decay by autoionization to the $1s^2$ helium-like ground state. Configuration interaction with the $1s\,2p^2$ configuration provides a way for this level to decay radiatively to two different lower levels

$$1s\,2s^2 \rightarrow 1s^2\,2p_{3/2} + h\nu_1 \qquad \text{(labeled by o)}$$

and

$$1s\,2s^2 \rightarrow 1s^2\,2p_{1/2} + h\nu_2 \qquad \text{(labeled by p)}$$

These two decay rates were calculated by many authors because of their importance for modelling K-shell tokamak spectra [381, 382, 383, 384] and spectra from the solar corona [385, 386]. Different calculations scatter about 5...15% [380].

Corresponding experimental investigations are made with a high resolution von Hamós spectrometer [387]. For instance, a spectrum of lithium-like Germanium is shown in Fig.**7.7**. The dielectronic resonance energy for populating the lithium-like level $1s\,2s^2$ from the helium-like ground state is 6941 eV, close to the resonance energy for the excitation of transitions u and r, i.e., the transitions $1s\,2p^2\,^4P_{3/2} \rightarrow 1s^2\,2p$ at 6964 eV and $1s\,2s\,2p\,^2P_{1/2} \rightarrow 1s^2\,2s$ at 7008 eV, respectively. Because the energy spread in electron-ion collisions is about 60 eV, these resonances are excited as well, and the transitions are seen in the spectrum in Fig.**7.7**.

Results for the measured branching ratios I_p/I_o are given in Table t7-3. The experimental branching ratio I_p/I_o rises from 0.88 to 1.57 as Z increases from 27 to 32. On the

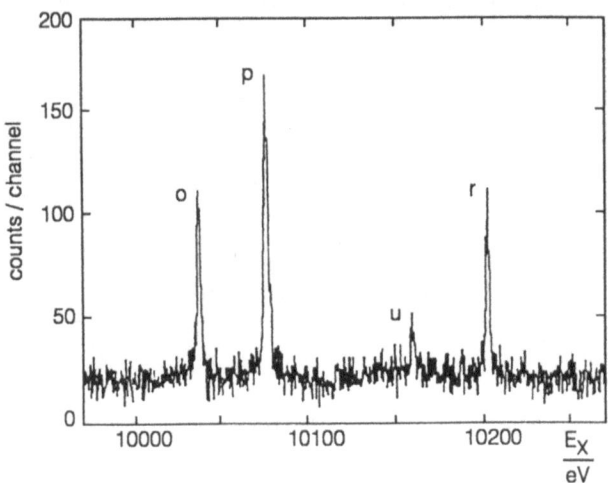

Figure 7.7   Spectrum of dielectronic satellite transitions in lithium-like Germanium [380] at
an EBIT beam energy of 6.96 keV.

Table 7.3   Measured bramching ratios I_p/I_o [380]. Besides the measured element the
used crystal is indicated. a: measurements employs a LiF(220) crystal; b: measurements
employs a Si(220) crystal.

Cr (a)	Mn (a)	Fe (b)	Fe (a)	Ni (a)	Ge (a)
0.88 ± 0.09	0.98 ± 0.03	1.00 ± 0.05	1.02 ± 0.04	1.15 ± 0.05	1.57 ± 0.08

other hand, a nonrelativistic LS-coupling calculation gives a constant ratio of 0.50. The
continued rise in the branching ratio reflects the increasing importance of the spin-orbit
interaction in the determination of the level structure and the mixing coefficients in the
configuration-interaction calculations.

A comparison of the measured relative intensities and the theoretical branching ratios
of I_p/I_o is given in Fig.**7.8**. A good agreement was found with MCDF calculations and
branching ratios calculated in the Z-expansion technique, not shown in Fig.**7.8** and falling
between the two types of MCDF calculations.

In Ref.[380] it was shown, that an ion source can be used to measure branching ratios of
X-ray transitions. Measurements of the branching ratios are less sensitive to the approach
employed in different calculations of dipole matrix elements than direct measurements
of the radiative transition ratios. Therefore, measurements of the radiative rates are
very desirable. The intensity of those dielectronic satellite transitions that possess a
radiative rate much smaller than the competing rate for Auger decay of the autoionizing

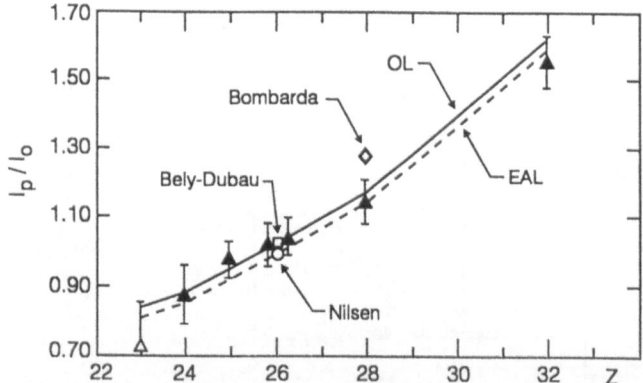

Figure 7.8 Comparision of measured and theoretical branching ratios I_p/I_o [380]. Solid triangles presents measurements. Open triangles represents data from Ref.[388]. The dashed and solid lines connect theoretical values obtained with MCDF calculations using the extended average level (EAL) and optimum level (OL) schemes, respectively. Theoretical branching ratios reported by Nilson [389], Bely-Dubau et al [381] and Bombarda et al. [384] are represented by an open circle, square and diamond, respectively. Branching ratios calculated by Vainshtein and Safronova [390] fall in between the two MCDF calculations and are not shown.

level (such as o and p in Fe^{23+}) is determined in first approximation by the radiative decay rate. Level specific measurements of the dielectronic recombination cross-section could, therefore, provide a precise tool for acessing the accuracies of different theoretical approaches employed in the calculations of dipole matrix elements. In [380] it was pointed out, that by appropriate normalization of the measured intensity of lines excited by dielectronic recombination one could expect to measure the strength of individual dielectronic resonances and to test theoretical models for weak transitions in various highly charged ions.

7.3.1.5 Observation of Magnetic Octupole Decay in Atomic Spectra

Observations of forbidden transitions (transitions that do not proceed via electric dipole E1 decay) provide fundamental tests of atomic structure theory as well as checks of level population calculations. Because the intensity of forbidden transitions is sensitive to the electron density, the lines are used as a diagnostic tool in the spectroscopy of solar, astrophysical and laser-produced plasmas [401].

Forbidden transitions with magnetic dipole M1, magnetic quadrupole M2 or electric quadrupole E2 multipolarity have been observed in the X-ray spectra of highly charged ions produced in high temperature low-density plasma sources such as tokamaks [76, 382, 391] and the Sun [245, 394, 395]. Electric quadrupole transitions have been observed in high-density laser-produced plasmas [396, 397] and in beam-foil interactions [398, 399]. The latter technique has also enabled the observation of the two-photon decay of atomic levels in highly charged ions [378, 400].

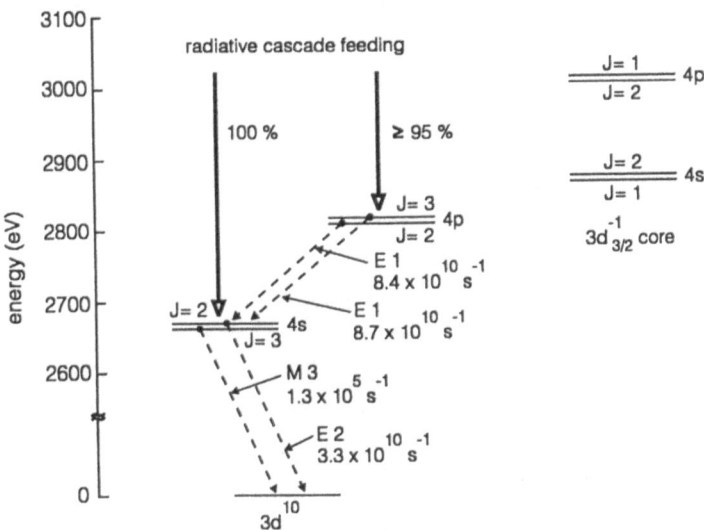

Figure 7.9   Level diagram of nickel-like U^{64+} for the lowest eight excited levels [401].

The first observation of magnetic octupole decay M3 in the X-ray spectrum of a highly charged ion was reported by Beiersdorfer et al. [401]. Here we give a short summary of this observations following [401]. The investigations were accomplished on the nickel-like systems Th^{62+} and U^{64+}. Nickel-like ions have in the ground state a closed-shell configuration $1s^2 \, 2s^2 \, 2p^6 \, 3s^2 \, 3p^6 \, 3d^{10}$. A diagram of the lowest eight excited levels in U^{64+} is shown in Fig.**7.9**. Here, dipole decay to the ground state of the lowest seven levels is forbidden because of selection rules for E1 decay require a change in total angular momentum $\Delta J = \pm 1$ as well as a change in parity. The $3d^{-1} \, 4p$ levels at 2836 eV and 2837 eV, for example, decay instead via allowed $\Delta n = 0$ E1 transition to the $3d^{-1} \, 4s$ levels. This option is unavailable to the lowest two excited $(3d^{-1}_{5/2} \, 4s)_{J=2}$ and $(3d^{-1}_{5/2} \, 4s)_{J=3}$. In the absence of collisions, these levels must decay to the ground state and must do so via an E2 and M3 transition, respectively.

A high-resolution spectrum, which clearly resolves the M3 transition from the neighbouring E2 transition is shown in Fig.**7.10**. The transition on the low-energy side of the nickel-like transitions is due to the radiative decay of the $3d^{-1}_{5/2} \, 4s^2$ level in copper-like U^{63+}. A detailed analysis of the low-energy region of the n=3→4 spectra of the nickel-like Th^{62+} and U^{64+} shows an intense line at 2558.7 ± 0.2 eV and 2689.3 ± 0.2 eV, respectively, that is attributed to the transition $(3d^9 \, 4s)_{J=3} \rightarrow (3d^{10})_{J=0}$. Calculations thereby indicate that the line is almost exclusively excited by indirect processes such as radiative cascades and electron capture.

A comparison of measured and calculated with the HULLAC-code [402] energies for E2 and M3 transitions in Th^{62+} and U^{64+} is given in Table **7.4**.

It was pointed out, that the lowest excited level in a highly charged helium-like ion (filled K-shell) decays by an M1 transition, that in a neon-like ion (filled L shell) by an M2 transition an that of a nickel-like transition by an M3 transition. Thus, one can

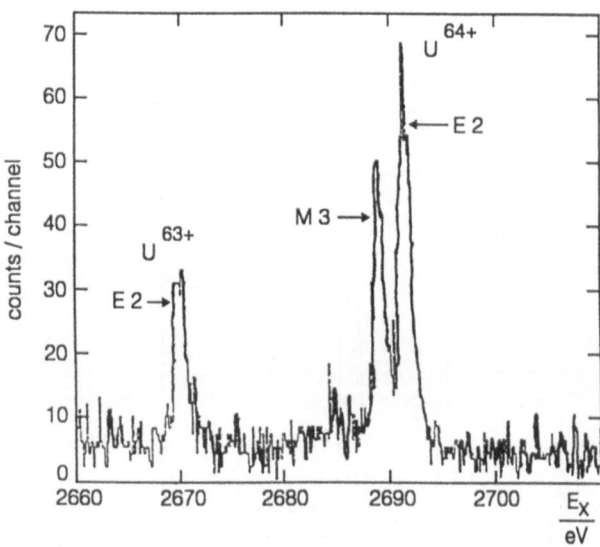

Figure 7.10   High-resolution spectrum of nickel-like U^{64+} recorded at an electron beam energy of 7.2 keV [401].

Table 7.4   Nickel-like measured and calculated transitions in Thorium and Uranium [401]. All values are in eV.

Transition	Type	E_{exp}	E_{th}	E_{exp}	E_{th}
		\multicolumn{2}{c}{Th^{62+}}	\multicolumn{2}{c}{U^{64+}}	
$(3d_{5/2}^{-1} 4s)_{J=3} \to {}^1S_{J=0}$	M3	2558.7 ± 0.2	2557.4	2689.3 ± 0.2	2688.4
$(3d_{5/2}^{-1} 4s(_{J=2} \to {}^1S_{J=0}$	E2	2561.3 ± 0.2	2560.0	2691.9 ± 0.2	2690.9

expected that the lowest excited level in a neodymium-like ion (filled N shell) decays by a magnetic hexadecapole M4 transition. Calculation for U^{32+} show transition rates of $3.3 \cdot 10^{-7}$ s^{-1} and $1.8 \cdot 10^1$ s^{-1} for the M4 and E3 transitions, respectively. Considering this values one can not expect to observe these transitions in any currently available device.

7.3.1.6 U^{q+} M-Shell X-Ray Transition Spectra

Although different trap studies have focussed on L-shell X-ray transitions on neon-like ions [168, 403, 173] and K-shell transitions for helium-like ions [404, 172] in [405] measurements of the 4f – 3d transition energies of a series of near nickel-like Uranium ions

(U^{60+} to U^{69+}) were accomplished. Previous studies of n=4 to n=3 X-ray transitions for nickel-like ions have been conducted using tokamaks, vacuum sparks and laser plasma sources [397, 406, 407]. In connection with the measured spectra the ionization balance for trapped ions within the EBIT was conducted, which can be compared to the ion charge state distribution reported for the extracted ions [174].

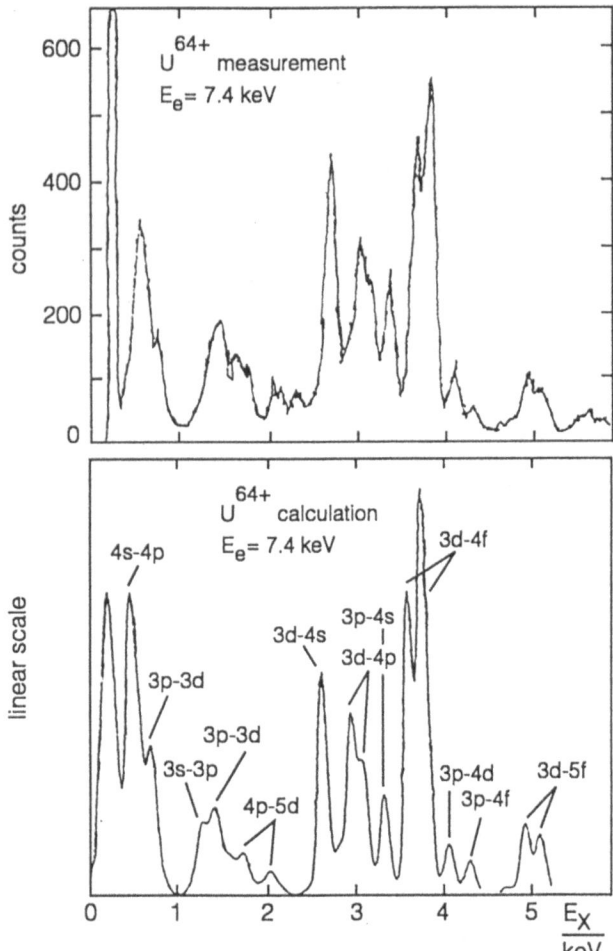

Figure 7.11 Comparison of a X-ray spectrum measured on EBIT with a Si(Li) detector at a beam energy of 7.4 keV with calculated X-ray intensities [405].

Measurements of X-ray spectra using a windowless Si(Li) detector at one of the beam ports of the EBIT, a Ge detector at a second port and a Bragg crystal diffraction spectrometer with a flat Ge crystal and a Xe-CH_4 filled position-sensitive proportional counter at a third port are described in Ref.[405]. The spectrometer were calibrated (energy and centroid dispersion) using theoretical values for nickel-like Uranium lines, e.g., the $3d_{5/2}$ - $4f_{7/2}$ (3705.5 eV), the $3d_{3/2}$ - $4f_{5/2}$ (3873.5 eV) and the $3p_{3/3}$ - $4d_{5/2}$ (4152.0 eV) lines, calculated with the GRASP code [373].

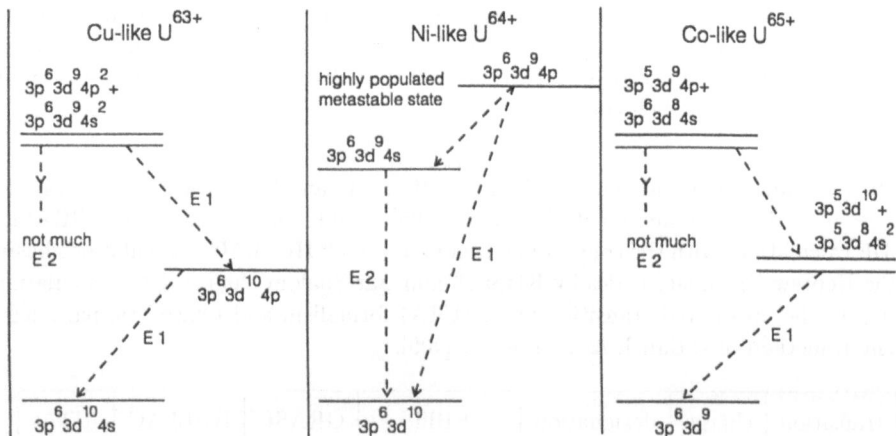

Figure 7.12   Decay schemes for copper-like, nickel-like and cobalt-like Uranium ions [405].

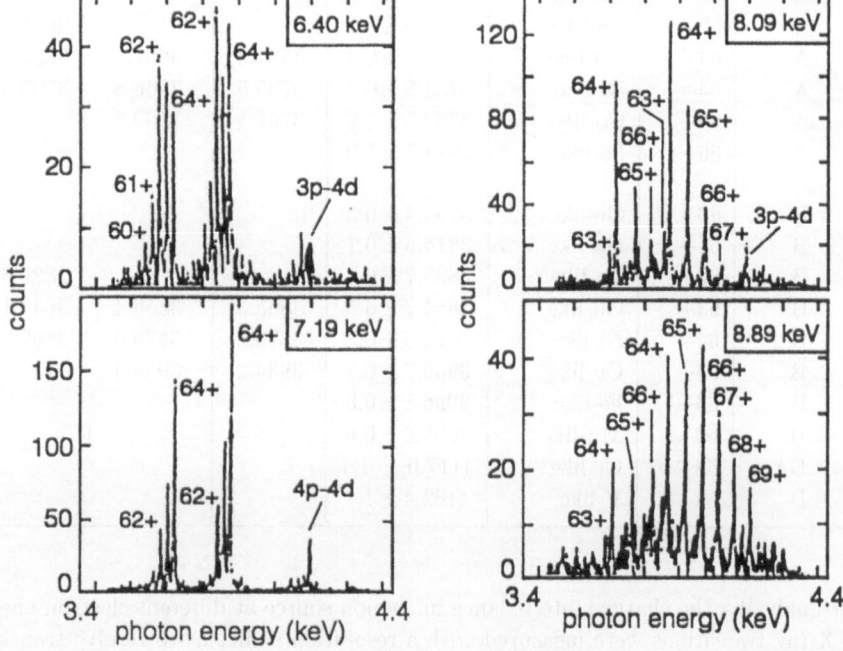

Figure 7.13   High resolution spectra for 3d - 4f X-ray lines for U^{60+} up to U^{69+} produced at electron excitation energies from 6.40 keV to 8.89 keV [405].

In addition, ions were extracted from the trap and were mass and charge analyzed for comparison with the results obtained with X-ray spectroscopy [174].

The dominantly nickel-like Uranium (4→3) transitions measured with a Si(Li)-detector are shown in Fig.7.11. A comparison of the measured spectrum with theoretical simulations gave an excellent qualitative agreement.

The strong electric quadrupole (E2) 4s-3d transition is an unique future in nickel-like ions. It results from the decay of the second excited level in U^{64+} and is mostly feed by radiative cascades as indicated in Fig.**7.12**. Corresponding excited levels in U^{63+} and in U^{65+} decay by E1 intrashell transitions.

Table 7.5   Measured and calculated $3d_{5/2}$ - $4f_{7/2}$ (A) and $3d_{3/2}$ - $4f_{5/2}$ (B) X-ray line energies in eV for Uranium ions U^{60+} – U^{69+} [405]. EBIT – measured results; GRASP – MCDF calculations with the code from Grant et al. [373]; HULLAC – calculations basing on the Hebrew University codes by Klapisch and Bar-Shalom [408]; UTA – calculations basing on the unresolved transition array (UTA) formalism and relativistic parametric potential method of C.Bauch-Arnoult et al. [409].

transition	charge	designation	EBIT	GRASP	HULLAC	UTA
A	60+	Ge-like	3627.6 ± 0.5			
A	61+	Ga-like	3647.7 ± 0.8			
A	62+	Zn-like	3667.6 ± 0.7			3671.3
A	63+	Cu-like	3686.9 ± 0.4	3687.3	3687.3	3692.1
A	64+	Ni-like	3705.5 ± 0.4	3705.5	3706.8	3709.8
A	65+	Co-like	3762.5 ± 0.6	3761.1	3760.8	
A	66+	Fe-like	3814.7 ± 1.0			
B	60+	Ge-like	3794.4 ± 0.3			
B	61+	Ga-like	3814.6 ± 0.7			
B	62+	Zn-like	3835.7 ± 0.3			3827.8
B	63+	Cu-like	3854.7 ± 0.5	3855.3	3856.4	3849.1
B	64+	Ni-like	3873.5 ± 0.2	3873.5	3876.0	3867.2
B	65+	Co-like	3935.7 ± 0.4	3934.2	3936.8	
B	66+	Fe-like	3996.1 ± 0.5			
B	67+	Mn-like	4055.8 ± 0.6			
B	68+	Cr-like	4117.0 ± 1.1			
B	69+	V-like	4182.4 ± 1.2			

For analyzing the charge state balance in the ion source at different electron energies 3d-4f X-ray transitions were measured with a resolution sufficient to resolve transitions from different charge states. Thus, the relative ion abundances could deduced from the relative X-ray line intensities.

The spectra shown in Fig.**7.13** resolve U^{60+}-U^{64+} ions at 6.40 keV, U^{61+}-U^{64+} ions at 7.19 keV, U^{63+}-U^{67+} ions at 8.09 keV and U^{63+}-U^{69+} ions at 8.89 keV electron excitation energies. In Fig.**7.14** an X-ray spectrum measured at $E_e = 8.9$ keV is compared with the charge state distribution of Uranium ions extracted from the EBIT. Thus, there arises a possibility to analyze the differences between the ion charge distributions inside of the source plasma and that one of the extracted ions. This may be an interesting tool for better understanding the process of ion extraction from the plasma meniscus.

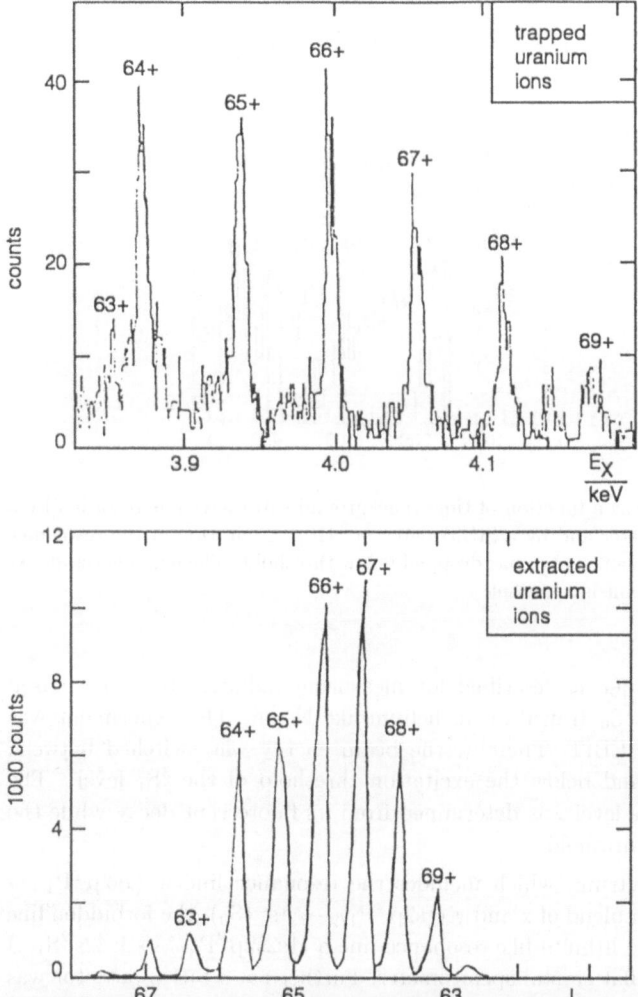

Figure 7.14 Comparison of X-ray line intensities from near nickel-like Uranium ions trapped in EBIT (as shown in Fig.7.13 with a charge distribution of extracted ions [405].

Quantitative results for the comparison between measured and calculated $3d_{5/2}$ - $4f_{7/2}$ and $3d_{3/2}$ - $4f_{5/2}$ X-ray energies are shown in Table **7.5**. The GRASP and HULLAC calculations represent in Table **7.5** a weighted average based on the respective excitation rates. The UTA calculations represent a centroid of all the transitions that would be present in a dense plasma and include transitions not observed in EBIT.

7.3.1.7   Radiative Lifetime of the $1s2s\,^3S_1$ State in Helium-Like Neon

Radiative transition rates provide information of the long-range behavior of atomic wave functions. Thus, measurements of the radiative transition rates are complementary to energy level measurements and test atomic structure calculations in a different way.

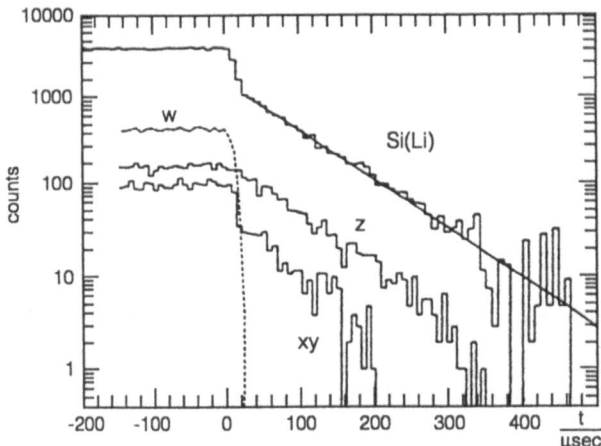

Figure 7.15   Line intensities as a function of time (background subtracted) as recorded by a flat crysral diffraction spectrometer and by a Si(Li) detector [410]. Excitation of the resonance line w ceases after the electron beam energy has dropped below threshold. The exponential decay of the other lines beyond this point is apparent.

In Ref.[410] a new technique is described for measuring radiative life times, used to study the $1s2s\,^3S_1 \rightarrow 1s^2\,^1S_0$ transition in helium-like Neon. The experiment was performed on the Livermore EBIT. Thereby, the beam energy was switched between 960 eV and 750 eV, above and below the excitation threshold of the $^3S_1$ level. The lifetime of the metastable $^3S_1$ level was determined from its fluorescent decay while the beam energy was below the threshold.

The resulting $2 \rightarrow 1$ spectrum, which includes the resonance line w ($1s2p\,^1P_1 \rightarrow 1s^2\,^1S_0$), the intercombination blend of x and y ($1s2p\,^3P_{1/2} \rightarrow 1s^2\,^1S_0$), the forbidden line z ($1s2s\,^1S_1 \rightarrow 1s^2\,^1S_0$) and the lithium-like resonance line q ($1s2s2p\,^2P_{3/2} \rightarrow 1s^22s\,^2S_{1/2}$) was recorded with a vacuum flat crystal spectrometer. Further on, a Si(Li) detector was used, which provide a much higher count rate with a much lower spectral resolution.

The temporal behavior of the individual lines of the recorded spectrum (time resolution 1.6 μsec) is shown in Fig.**7.15**. The origin of the time axis is the point at which the electron beam energy is switched from 960 eV to 750 eV. Line z and the Si(Li) ($2 \rightarrow 1$) show prominent exponential tails. A weaker tail also appears on the xy blend because of collisional transfer from the $^1S_1$ state to the ^3P states. From the temporal decay of the line emission a lifetime of 90.5 ± 1.5 μsec was obtained in good agreement with theoretical predictions.

The technique described can be used over eight orders of magnitude in time, from a few nsec to hundreds of msec [411]. Thus, with this technique it is possible to accomplish investigations of properties of highly charged ions in a regime unchartered by earlier experiments, because accelerator beam-foil experiments can not be made beyond a few hundred nsec (because of the long flight paths required) and laser fluorescence techniques can only be used in neutral or low charged ions because of photon energy limitations [411].

7.3.1.8   Proposed Spectroscopic Experiments at the Oxford EBIT

An EBIT has just been completed in the Clarendon Laboratory in Oxford [412]. The design is similar to the devices installed at Lawrence Livermore National Laboratory. It is intended that the Oxford EBIT will be used for X-ray and UV spectroscopy of hydrogenic and helium-like ions, laser resonance spectroscopy of hydrogenic ions and measurements of dielectronic recombination cross-sections in order to test the current understanding of highly charged ions.

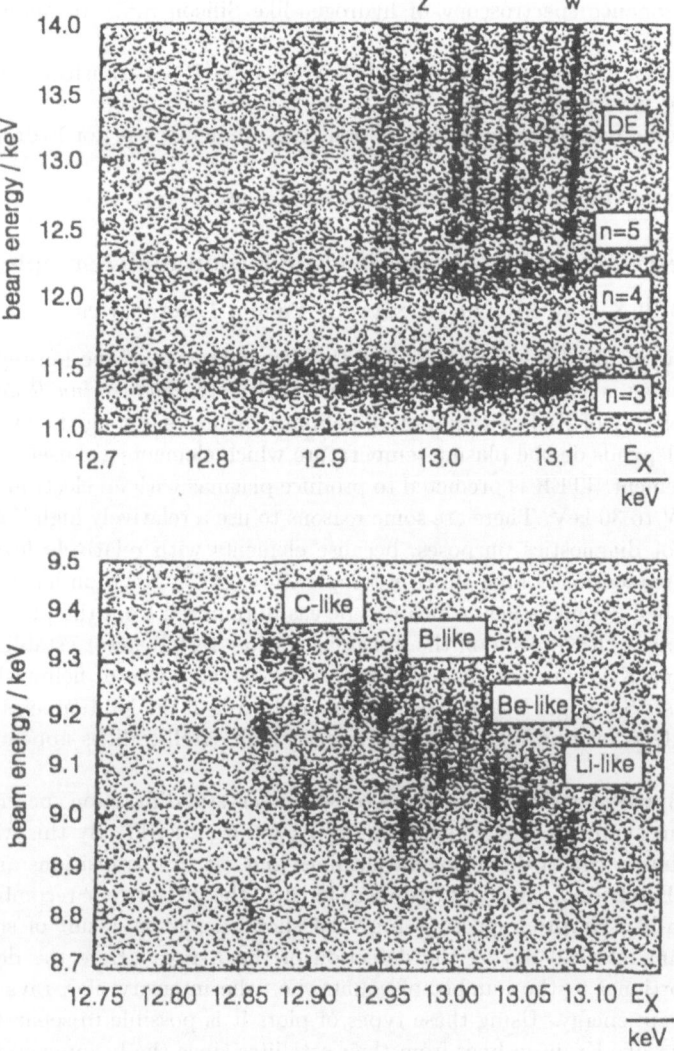

Figure 7.16   Detected photons in the region of helium-like Krypton resonance lines plotted as function of their X-ray energy and of the electron beam energy [413]. Left side: Energy of the electron beam sweeps above threshold and down to N=3 (KLM) satellites. Right side: Variation of the electron energies just covers the range of n=2 (KLL) satellite lines.

From the spectroscopic point of few the Oxford EBIT has six observation ports, although currently only four have windows fitted: two Beryllium windows, which are recessed to maximize the solid angle and transmission of low-energy X-rays, a Magnesium Fluoride window for UV spectroscopy and a quartz window for visible spectroscopy.

In the future, various experiments are planned for the Oxford EBIT. In Ref.[412] there are named

- measurement of the 1s Lamb shift in hydrogenic Nickel Ni^{27+} by high resolution X-ray spectroscopy;
- laser resonance spectroscopy of hydrogen-like Silicon Si^{13+} to determine the 2s Lamb shift;
- VUV emission spectroscopy of the $1s2s\,^3S_1 - 1s2p\,^3P_{2,1,0}$ transitions in helium-like ions, e.g. Ne^{8+};
- dielectronic recombination cross-section measurements, e.g. for hydrogen-like Iron Fe^{25+};
- investigation of visible transitions in lower-Z ions.

7.3.2   Investigations for Thermonuclear Fusion and Astrophysics

7.3.2.1   EBIT X-Ray Measurements for Fusion Diagnostics

Here we follow results from Ref.[413]. For the determination of the ion temperature of tokamak plasmas, such as in the *International Tokamak Engineering Reactor* (ITER), X-ray spectroscopy of K-shell radiation of helium-like ions is among the most promising methods. It depends on the plasma temperature which elements are useful for spectroscopic measurement. ITER is predicted to produce plasmas with an electron temperature around 10 keV to 30 keV. There are some reasons to use a relatively high-Z element such as Krypton for diagnostics purposes, because elements with relatively low Z are fully stripped at the mentioned high electron temperatures [413]. Krypton has the additional advantage of being a noble gas, which can be easily removed from the plasma chamber.

Experiments with Krypton on the Tokamak Fusion Test Reactor established the feasibility of using Krypton and suggested the the so-called z-line of helium-like Krypton $(1s2s\,^3S_1 \rightarrow 1s^2\,^1S_0)$ might be the most reliable indicator of the ion temperature, because it was the only transition for which interference with satellite lines appeared nefligible [414, 415].

In Ref.[413] investigations with a high resolution crystal diffraction spectrometer were described to measure the K-shell spectrum of Krypton on EBIT. By this way, positions and relative intensities of direct excitation lines and satellite transitions are measured, respectively. Furthermore, it was possible to identify the dielectronic recombination lines in the helium-like Krypton spectra. Fig.**7.16** shows a plot as result of scanning over a certain beam intervall during the detection of the X-rays. Here the density of the dots is proportional to the number of counts, i.e. the intensity of X-rays at a certain X-ray and beam energy. Using these types of plots it is possible to separate the direct excited helium-like Krypton lines from their satellites since the beam energy required is different in both cases. An example is given in Fig.**7.17** showing lineouts over different beam energy intervals. These plots are direct results from Fig.**7.16**a. The measurements show that z is indeed least blended by satellites and thus is indeed the best choice for determining the ion temperature.

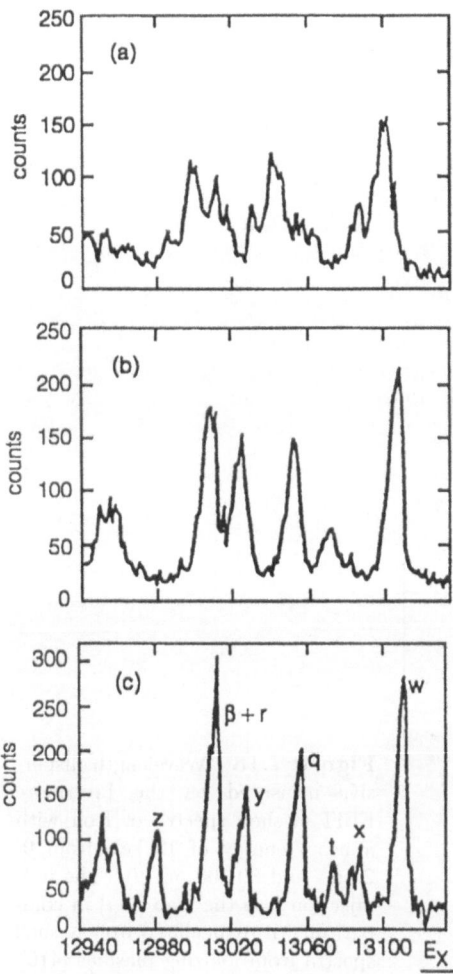

Figure 7.17   K-shell spectra of helium-like Krypton [413]. a: n=3 (KLM) satellites lines; b: dielectronic satellites with $n \geq 4$; c: spectrum having beam energies above thereshold.

7.3.2.2   Transient Ionisation Phenomena in Iron K-Shell Spectra

Spectra emitted by ionizing plasmas, such as those found in the shock-front-heated regions of a supernova remnant as well in solar flares or tokamaks with rapidly changing plasma parameters have a characteristic signature [416]. Here, the establishment of an ionization balance tends to lag behind the temperature changes and dramatically affects the emitted X-ray spectra.

To simulate transient ionization phenomena in a controlled environment and to identify and measure the lines produced under certain conditions, a technique at the LLNL EBIT was developed basing on recording a series of spectra shortly after the ion injection in the trap [417, 346]. This technique allows it to study in detail ion charge states of carbon-like up to chlorine-like ions, the line emission of which in most cases has never been analyzed before. The method allows also to monitor the time-dependent evolution of the charge balance and compare the measured spectra with the theory.

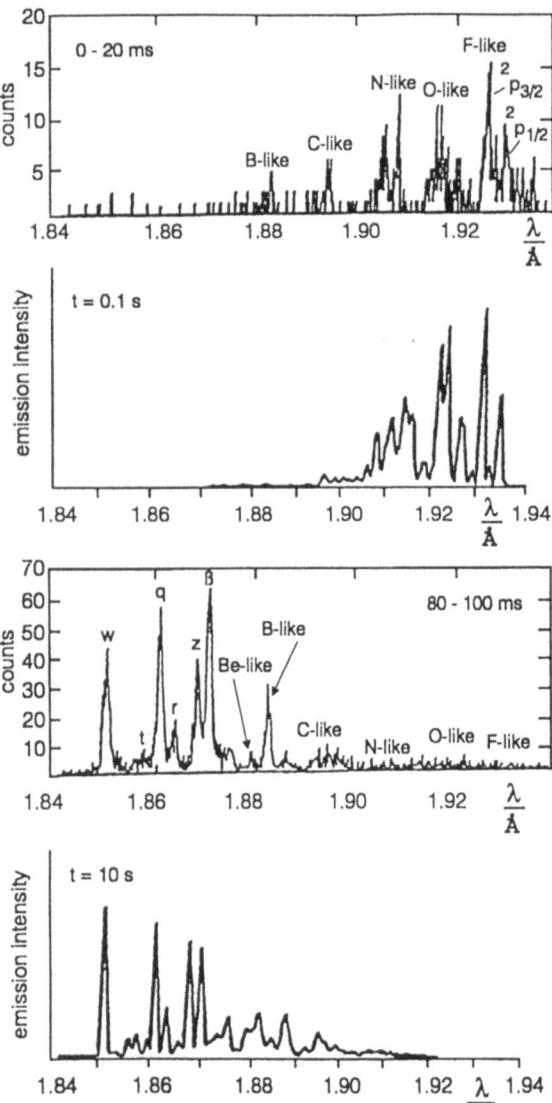

Figure 7.18 Wavelength dispersive measured on the Livermore EBIT K-shell spectra of Iron with a beam energy of 12 keV from 0-20 ms and 80-100 ms after the iron injection into the trap [417] in comparison with calculated iron K-shell spectra from ionizing plasmas [416], assuming a temperature jump from $5 \cdot 10^6$ K to $50 \cdot 10^6$ K at time t=0. The electron density was set to 10^{10} cm^{-3}.

As an example, in Fig.**7.18** wavelength dispersive spectra recorded at a beam energy of 12 keV are shown. The spectra cover the energy range from 6.39 keV to 6.73 keV and are presented as a function of time to show the sequential ionization process. The first spectrum, recorded between 0-20 ms after the injection of the Iron in the trap is dominated B-like to F-like charge states, whereas spectra recorded after 80 ms are closer to typical steady-state data. The measured spectra can be compared directly to theoretical predictions of shock-excited astrophysical plasmas as shown on the right side of Fig.**7.18**.

7.3.3   Ion-Surface Interactions

7.3.3.1   Ion-Surface Collisions Experiments with Slow Highly Charged Ions

In Ref.[419] ion-collision experiments with slow, highly charged Argon (up to Ar^{18+}) and Xenon (up to Xe^{48+}) ions from the Livermore EBIT were reported. Thereby, for measuring X-rays emitted in the neutralization process of highly charged Argon and Xenon ions on solid surfaces, a Copper target was inserted viewed by a Si(Li) detector with a solid angle of $3 \cdot 10^{-3}$ sr.

Argon K-spectrum. Ar^{17+} and Ar^{18+} ions produced in the trap have been used to bombard a solid Copper surface oriented at $45°$ relative to the beam axis. The Argon K x-ray emission following the decay of helium- and hydrogen-like excited Argon ions was observed with a Si(Li) detector located perpendicular to the in-beam axis. The observed X-ray spectra are shown in Fig.**7.19**.

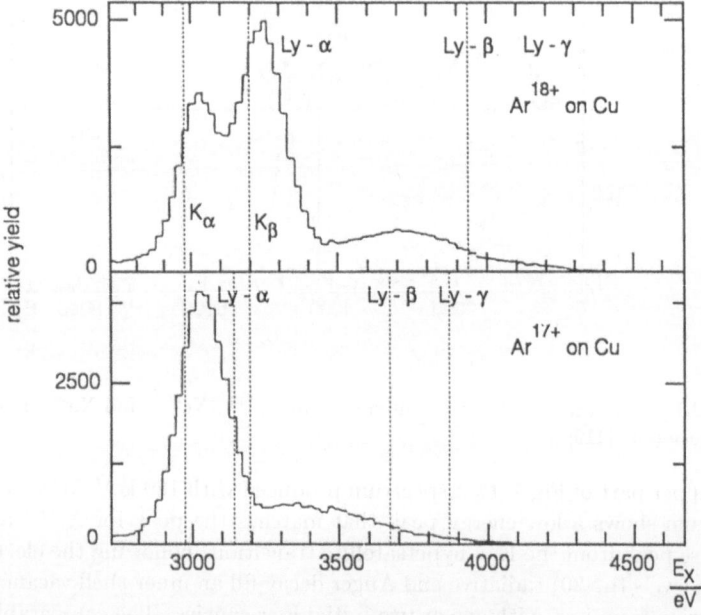

Figure 7.19   Argon K X-ray spectrum following 136 keV Ar^{17+} and 144 keV Ar^{18+} ion impact on a Copper surface ($45°$) [419]. The spectra include K_α and K_β satellite and $K_{\alpha''}$ and $K_{\beta''}$ hypersatellite X-ray emission lines.

The K X-ray spectrum from Ar^{17+} impact consists of the dominant intensity due to the shifted K_α X-ray emission (2p $\rightarrow$ 1s) at 3020 eV and due to the shifted K_β X-ray emission (3p $\rightarrow$ 1s) and intensity due to np $\rightarrow$ 1s transitions. The observed energy shifts with respect to the K_α line of Argon are about 70 eV for the 2p $\rightarrow$ 1s transition and about 220 eV for the 3p $\rightarrow$ 1s component with respect to the K_β line. The K_α and K_β lines are shifted due to multiple L- and M-shell vacancies, consistent with earlier reported results by Donets et al. [420], who used slow 17 keV Ar^{17+} ions. The K X-ray peak agrees with K_α and K_β lines for Argon atoms carrying one K, several L and several M vacancies. This

confirms the picture of a "hollow" atom, whereby electron capture into empty n levels of
the ion from the surface occurs resonantly up to neutralization when the ion approaches
the surface. The transition into the 1s state occurs before these electrons relax to an
intermediate bound state [194]. Due to the high K-shell fluorescence yield the np → 1s
transitions are observed efficiently.

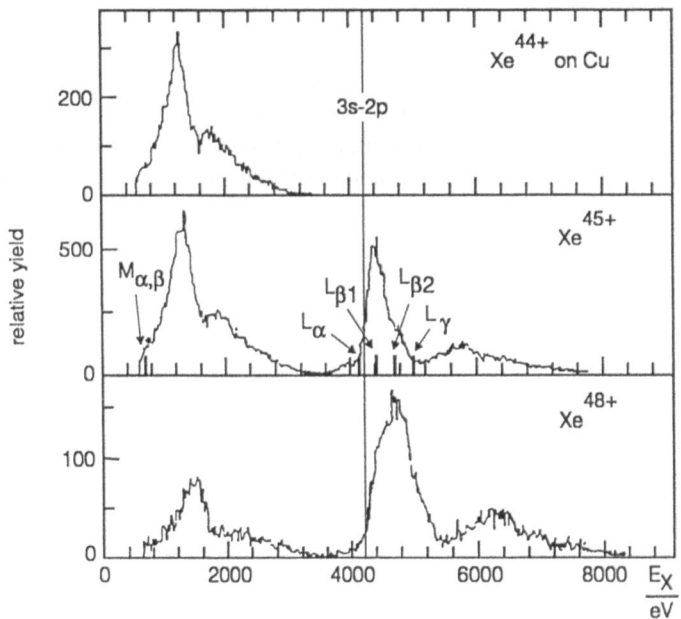

Figure 7.20   Xenon M and L x-ray emission from Xe^{44+}, Xe^{45+} and Xe^{48+} ion impact on a
Copper surface ($45°$ [419].

In the upper part of Fig.**7.19** a spectrum produced with 180 keV Ar^{18+} ions is shown.
The spectrum shows a low energy peak that matches the peak for Ar^{17+} and a second
more intense peak from the $K_{\alpha''}$ hypersatellite transition. Following the electron capture
into high-n levels (n>20) radiative and Auger decay fill an inner-shell vacancy, leading to
X-ray emission from ions with one or two initial K vacancies. The comparable intensities
reflect the similar mean K-shell fluorescence yield for single and double K-vacancy ions
with partly filled L and M shells; the mean K-shell fluorescence yield for such a case
is about 15% [422]. The third peak mainly results from K_β satellite and hypersatellite
lines.

Xenon M- and L-spectra. Fig.**7.20** shows M and L X-ray spectra from different
Xenon ion impact on Copper. The Xe^{44+} spectrum consist of the M X-ray spectrum since
there is no L vacancy in the ion. Xe^{45+} and Xe^{48+} spectra in addition show contributions
from L X-rays. For comparison recently reported line energies for the neon-like doublet
in $Xe(2p_{3/2}^5 3s_{1/2})_{J=1,2} \rightarrow 2p^6$ are plotted [368, 423].The spectra show an energy shift
towards higher energies with increasing charge states in the L X-ray spectra due to the
increasing number of L vacancies. For the Xe^{45+} the M X-ray intensity is somewhat
shifted towards higher values.

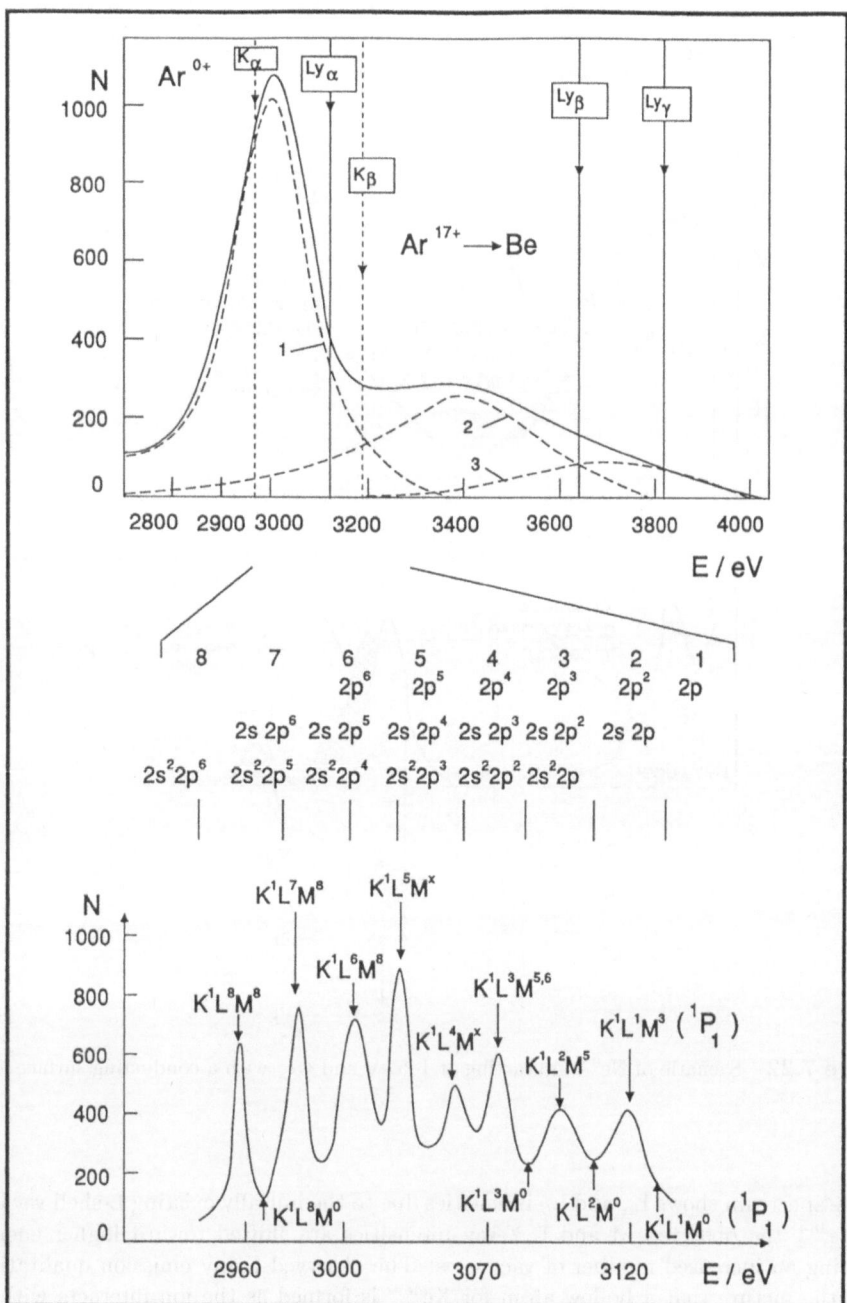

Figure 7.21 X-ray spectra of Ar^{17+} ions impinging on a solid state target. upper part: X-ray spectra measured with a Ge detector for the interaction of Ar^{17+} ions with Be [420]; middle part: Interpretation of K$_\alpha$ satellites of Ar^{17+} interacting with a solid at high energy; lower part: high resolution wavelength dispersive measured X-ray spectrum from Ar^{17+} at 340 keV interacting with a Silver target [403].

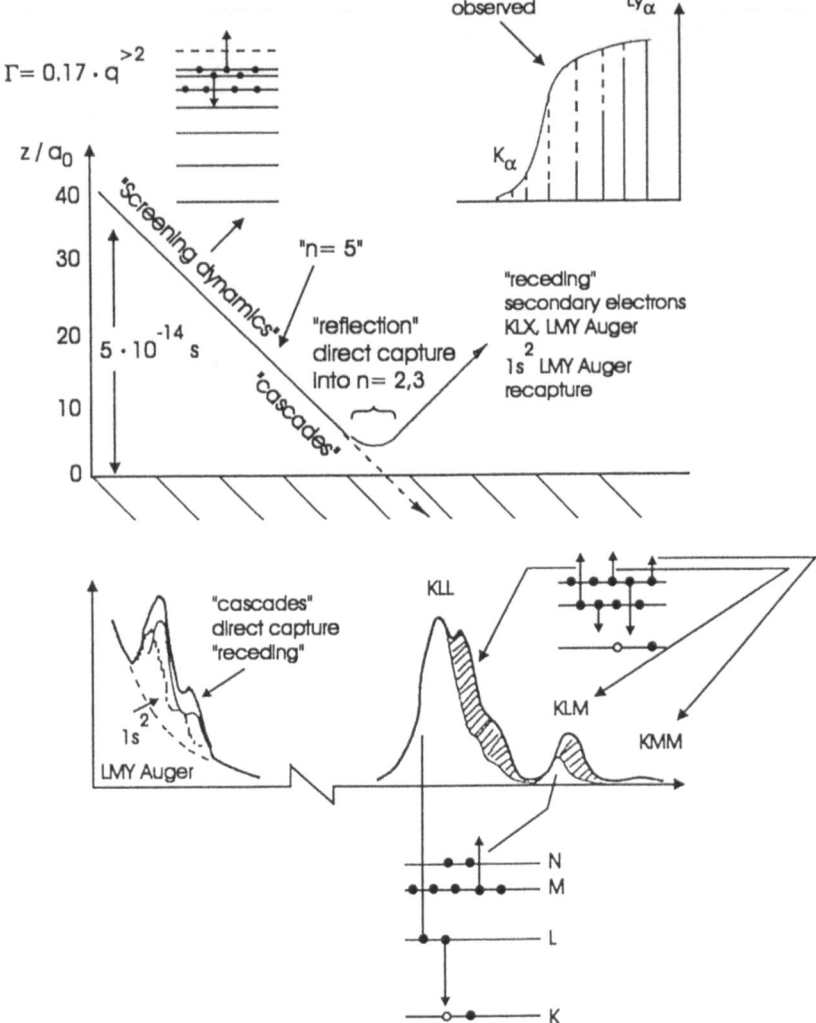

Figure 7.22   Szenario of Ne^{9+} interacting at 145 eV and 45 $^{\circ}$ with a conducting surface [424].

The spectrum shows L_{α} and L_{β} intensities due to the initially existing L-shell vacancy. For Xe^{48+} the observed M and L X-ray intensities are shifted toward higher energies reflecting an increased number of vacancies. The observed X-ray emission qualitatively infers the picture that a hollow atom for Xe^{48+} is formed as the ion interacts with the Copper surface. Further it was observed that the relative intensities between L and M X-rays are changing with increasing charge states. The increase in the L X-ray intensity relative to the M X-rays is due to an increase in the fluorescence yield with an increasing number of L vacancies. For $0Xe^{48+}$ it was observed an increased width in the L X-ray emission peak if compared to Xe^{45+} which is due to the increased number of L vacancies.

Experiments with incident multiply charged ions at high energies in earlier works. When a hydrogenic multiply charged ion (MCI) approaches a surface with high energy, its approach time is so short that the population of the n=2 manifold via cascading from $n \leq 5$ becomes negligible. The MCI thus penetrates into the solid with high probability and with a nearly empty L shell. The electrons in manifolds $n \geq 3$ are considered to be shaken off when they experience a sudden change of their effective potential when the projectile is diving into the screening electron density of the solid [424].

In these conditions, the X-ray emission can yield information on what happens to the K and L shells of the MCI inside the bulk, while the information carried by Auger electrons is limited to the history of the MCI in the first few atomic layers due to the short range escape probability of Auger electrons as summarized by Ley et al. [426]. Donets et al. [420] were the first to observe with a Germanium detector the X-ray emission of Ar^{17+} impinging at 18 keV on Be. The measured spectrum is shown in the upper part of Fig.**7.21** compared to the K_α and K_β lines of singly ionized Argon and the Ly-α, Ly-β and the Ly-γ lines of helium-like Ar^{16+}. The dominant peak is attributed to K_α satellites while the high energy tail arises from K_β and K_γ satellites. As shown in the middle part of Fig.**7.21** the region between K_α and Ly-α can be decomposed into 8 K_α satellites which differ in energy due to the screening by different numbers of L electrons. The K_α-line corresponds to a completely filled L shell while the Ly-α line is due to a single 2p electron in the L shell. Neglecting the M shell populations from the peak position in the spectrum an average of 5 electrons in the L shell is deduced when these X-rays are emitted. An analysis with correct fluorescent yields [7] yields 4-5 electrons. This interpretation has been well confirmed by high resolution measurements [403] shown in the lower part of Fig.**7.21** where the X-rays from Ar^{17+} at 340 keV on Silver were analyzed by a plane Graphite crystal. The measured spectrum shows 8 resolved K_α satellites with an intensity distribution which corresponds very well to the interpretation from Ref.[420].

For more detailled discussion of the interaction of MCI at high and low energies with solid state targets we recommend for instance the review from Andrä et al. [424].

In the mentioned above review the interaction of Ne^{9+} ions at very low energies with a conducting solid state target is also described (see also Fig.**7.22**). At rather long distances $38 a_0 < z_c < 54 a_0$ in front of the surface, a hyperexcited neutral Neon with 9 electrons in high lying Rydberg levels is created by resonant capture of electrons from the conduction band of the surface during the further approach of the Neon atom towards the surface, the screening dynamics transfers these electrons to the n=5 Rydberg manifold at a distance $Z \approx 10.8 a_0$. This screening dynamics is accompanied by autoionization cascading which produces an important fraction of the secondary electron yield Γ at very low energies.

At distances $< 10.8 a_0$ Auger cascading becomes the dominant channel of population of still lower shells which are accesible to observation via Auger electron or X-ray spectroscopy. It is during this phase that the typical ion-surface interaction spectra are emitted with a small population in the L shell and the rest of the electrons in M and N shells. Representative of this phase are the blank parts of the K Auger spectrum on the lower right of Fig.**7.22** and the full line part of the X-ray spectrum on the upper right of Fig.**7.22**.

At still smaller distances the combined action of an effective image charge and of the repulsive planar potential modify the projectile such that at 145 eV a reflection on the first or second atomic layer of the surface takes place with a probability close to one. During the closest approach on the order of $0.5\ldots2$ a_0, direct capture of electrons due to inner-atomic Auger-transitions or via quasi-resonant capture from inner shells of the surface atoms into the L and M shells takes place abundantly due to the long time spent close to the these atoms (head-on collisions). This interaction leads during the reflection and while the projectile is receding from the surface to the emission of spectra with a medium population of the L shell and the rest of the electrons in the M and N shells. Representative for this phase are the dashed part of the K Auger spectrum on the lower right of Fig.**7.22** and the dashed line part of the X-ray spectrum on the upper right of Fig.**7.22**.

All along the sections of the trajectory called "cascades", "reflection" and "receding" in Fig.**7.22**, emission of L Auger electrons takes place. In the section "cascades" the emission from configurations with one K hole dominates, but its contribution is exponentially reduced in favor of the emission from configurations with a full K shell which probably dominates in the "receding" section. The observation of L Auger emission with a full K shell implies that a resonant electron capture or near resonant Auger capture from the conduction band to continuously lowering n manifolds takes place in the "receding" section. The distance from the surface up to which this neutralization will take place is small compared to z_c. The capture of a final n=2 electron, the only channel for the completion of the neutralization, can take place up to $z \leq 1.6 < r_{2s} = 10$ a_0 only. Thus, it is very likely that the neutralization will not be completed on a receding trajectory.

X-ray emission from hydrogen-like Argon ions approaching a Germanium surface was also investigated at the CRYEBIS of the Kansas State University [68]. Thereby, the obtained data indicate an increasing fraction of X-rays emitted above the surface with decreasing vertical collision velocity. Here we gave only some examples for studies of X-ray and Auger electron emission at the interaction of highly charged ions with surfaces. Other examples for instance one can find in Refs.[403, 429, 430, 431, 432].

7.3.3.2   Electron Emission Yields following Highly Charged Ion Impact on Surfaces

Ion-surface interaction studies by means of electron emission measurements address the fundamental question of how fast ions that carry up to about 200 keV of potential energy lose their energy in the neutralization process. Corresponding beams are produced in EBIS and EBIT devices. Thus, in Ref.[476] measurements of the electron emission yield as a function of the total potential energy for extracted EBIT ions ranging in Z from 10 to 90 and charges up to 75+ on Copper and Gold targets are reported.

It has been shown that slow ions capture electrons efficiently into high-n states at relatively large distances depending on the ion charge [194]. X-ray and Auger electron spectroscopy are used to study the dynamics of the decay of these states via decay cascades. Total yield measurements [430, 434] of the emitted electrons as a function of the projectile velocity and charge as well as Auger electron spectroscopy have been performed so far using ions up to Ar^{16+}. These data help to answer the question of to what extent the ion neutralization occurs prior to penetration of the surface and how much occurs after as well as the energy loss mechanisms [435].

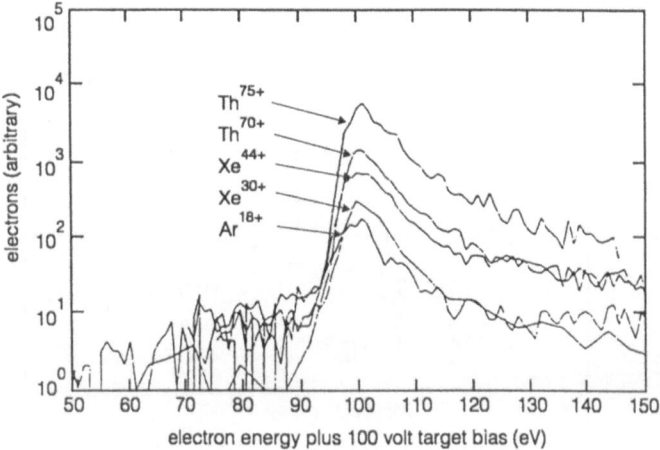

Figure 7.23   Electron emission spectra following Ar^{18+}, Xe^{30+}, Xe^{44+}, Th^{70+} and Th^{75+} impact at normal incidence on Gold [476]. Inset: Target area geometry.

In Ref.[476] an experimental setup was used to study the electron emission in the backward direction from the target. A schematic representation of the experimental geometry is shown in the inset of Fig.**7.23**. A series of low energy electron spectra is presented in Fig.**7.23** for several ions incident on the Gold target. Thereby, the spectra are shifted by the target bias of -100 V.

In [476] an analysis and comparison of model calculations and measurements of the total electron emission yield in dependence from the total ion potential energy and from the ion velocity is given. Here we follow in some details this discussion. Thus, in [436] model calculations and measurements of the velocity dependence of Auger electron emission for N^{6+} ion impact on Gold surfaces have been reported. The calculations are based on the classical over-the-barrier model where image charge, screening effects and a so-called "peeling off" of electrons in high-n-states or loss to the conduction band were taken into account. The data from N^{6+} incident on Gold targets demonstrate the appearance of an "above the surface" component in the Auger spectrum at sufficiently low ion velocities [436].

From the studies described it follows that the electron emission observed in [476] stems predominantly from neutralization processes below the surface. The electron yield increases drastically with decreasing incident ion velocity due to the wider time window available for neutralization processes to take place above the surface. Thus, the measured electron yield in Ref.[476] is representative of electrons that escape from the surface or below and they do not reflect the total yield due to the neutralization processes below the surface. A rough estimate for the fraction of electrons produced via neutralization below the surface compared to those above the surface can be deduced from a comparison of the measured yield curve to calculated values for slower ion impact using the Bardsley model [476].

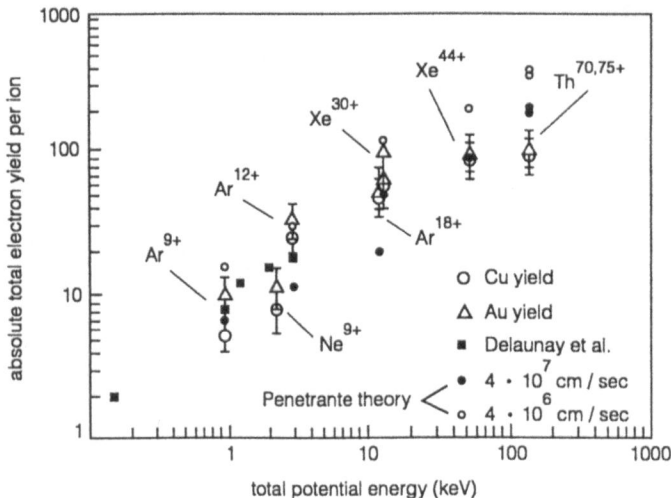

Figure 7.24   Absolute total electron emission yield as a function of total potential energy [476]. Additional data from Ref.[430] and [182] are also shown.

The total electron emission yield shown in Fig.**7.24** increases from about 10 electrons per ion for Ne^{9+} to about 100 electrons per ion for Th^{75+} incident. The increase of the measured yields with increasing total potential energy of the ions is found to be nonlinear at an ion velocity of $3.95 \cdot 10^7$ cm/sec. This corresponds with the prediction, that the proportionality of the electron yield with total potential energy is only valid up to certain charge stages, above which the electron yield increases more slowly with potential energy than for lower charge states [431]. Experimental data from Ref.[430] indicate a linear rise in electron emission with increasing ion potential for velocities up to $0.4 \cdot 10^7$ cm/sec. Extrapolation of these results yields about 1600 electrons per incident ion for Th^{75+}. The results from Ref.[476] are considerably lower. Since the high-Z highly charged ions carry inner-shell vacancies, it can be assumed that the emission of much more energetic Auger electrons or X-rays occurs with the loss of a substantial fraction of the available potential energy. For the case of Th^{75+} only about 2 keV of the available 152 keV potential energy would be released via low-energy electrons.

The collected data show that even for fast highly charged ions incident on metal surfaces the total electron emission is dominated by low-energy electron emission ($<$ 20 eV). The increased total electron yield shows a nonlinear dependence from 1 keV to about 200 keV. The number of emitted electrons per ion indicates that much of the potential energy is maintained until the ions actually reaches the surface.

In Ref.[437] total low energy yields for the normal incidence of slow highly charged ions (O^{8+}, Ne^{10+}, Ar^{18+}, $^{136}Xe^{i+}$, $21 \le i \le 51$, $^{232}Th^{i+}$, $51 \le i \le 80$) with a Copper and clean Gold surface have been determined from integrated emission yields and measured electron emission statistics. As also stated above the emission yield was found to increase proportionally with the increasing projectiles charge state in all cases studied. Results of a quantitative study of the emission yields are in good agreement with those of theoretical models based on classical over the barrier transitions [438] (for lower charge rates) and field emission theory [439] (for higher charge states).

Measurements of ion induced electron emission from metal surfaces have shown that the resulting total electron yield rises with falling ion impact velocity. This behavior has been explained by the longer times available in slow collisions for autoionization which relax the multi-excited hollow atoms that are formed in front of the surface [440]. However, the velocity dependence of the highly charged ion impact on metal surfaces is complicated due to image charge acceleration of the ions. For an ion with initial charge i the classical over-barrier model predicts a total kinetic energy gain $\Delta E_{i,im}$ until surface impact

$$\Delta E_{i,im} = 1.2 \cdot i^{3/2} \quad [\text{eV}] \ . \tag{7.2}$$

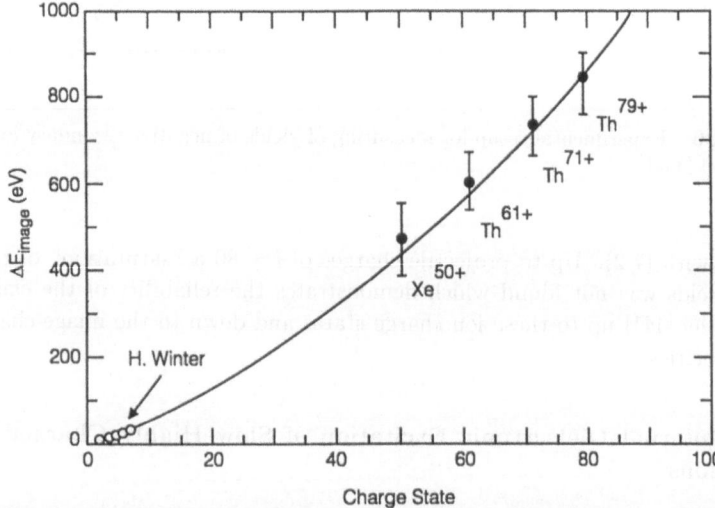

Figure 7.25   Energy gains ΔE_{image} due to image charge attraction for Xe^{i+} and Th^{i+} projectiles versus charge state i [440].

In Ref.[440] equation (**7.2**) was tested up to the highest accessible charge states i, both to check the validity of (7.2) on the basis of the classical over-barrier model and to evaluate the "image-charge limit" placed on the lowest possible kinetic energy, which determines the highest possible total electron yield. In this measurements, projectile ions Ar^{i+} ($i \leq 18$), Xe^{i+} ($i \leq 51$) and Th^{i+} ($i \leq 80$) with a nominal velocity below $2 \cdot 10^4$ m/s, corresponding to a kinetic energy as low as 2.5 eV/amu, which is much less than the potential energies of these projectiles (about 100 keV for Xe^{50+} and about 250 keV for th^{80+}) are used. From these measurements potential emission yields for slow highly charged ions are clearly derived demonstrating the effect of image charge acceleration. As shown in Fig.7.25 a quantitative evaluation of this data is in good

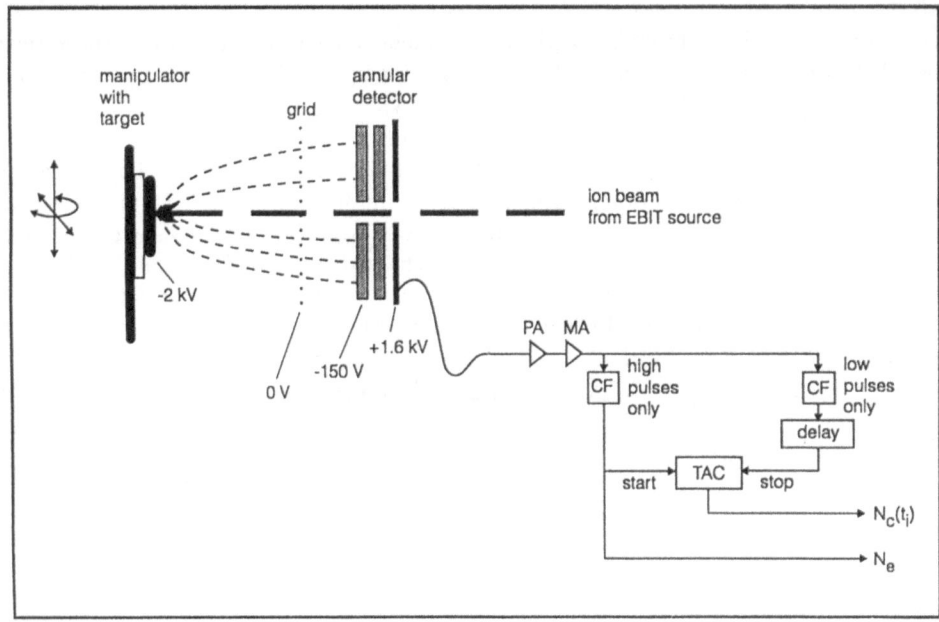

Figure 7.26 Experimental setup for measuring of yields of negative secondary ions using the TOF method [444].

agreement with (7.2). Up to projectile charges of $i = 80$ a "saturation" of the electron emission yields was not found which demonstrates the reliability of the classical over-barrier model [441] up to these ion charge states and down to the image-charge limited impact velocities.

7.3.3.3   Internal Dielectronic Excitation of Slow Highly Charged Uranium Ions

In Refs.[442, 443] X-ray emission following radiative neutralization of U^{i+} ions with charge states from 61+ up to 73+ and energies of $7i$ keV on a Be foil has been investigated. Thereby, an increasing M X-ray intensity with increasing number of M-shell vacancies in the incident ions was found. For charge states with no M-shell vacancies a significant intensity of M X-rays was also observed. This result was interpreted as evidence of an internal dielectric excitation process occuring during the neutralization of the ions in the surface.

The mechanism for the production of M-shell vacancies by internal dielectronic excitation was proposed as the following: If a $3l\,nl$ state energetically degenerate with a doubly excited $4l\,4l'$ state, then a resonant transition between the states is possible. The resulting intermediate $4l\,4l'$ state may decay either radiatively to a $3l\,4l'$ state or nonradiatively back to a $3l\,nl$ state. In the former case a M X-ray is observed. Thus, the rate for formation of the intermediate state with an M-vacancy is expected to be quite large compared to other decay rates in the hollow atom. In the in Ref.[442] studied system the formation of a $4l\,5l'$ or a higher intermediate state was not energetically allowed. So it was concluded that internal dielectronic processes must be included in any complete description of the neutralization of ions on surfaces.

7.3.3.4 Measurements of the Negative Secondary Ion Yield for Slow Highly Charged Ions Incident on Surfaces

Measurements of the yield of negative ions due to slow highly charged ions incident on insulating surface (i.e. SiO_2) have been performed at the LLNL EBIT II facility [444]. The secondary ions were identified using a single annular detector based time-of-flight (TOF) arrangement as shown in Fig.7.26. Typical TOF spectra from Ref.[444] are shown in Fig.7.27. Measurements with different ions and charge states have shown that a clear increase in secondary ion yields as well as a heavy cluster formation with increasing incident charge state is evident. Further investigations of the angular distribution and the initial energy of the positive and negative ions distributions are foreseen.

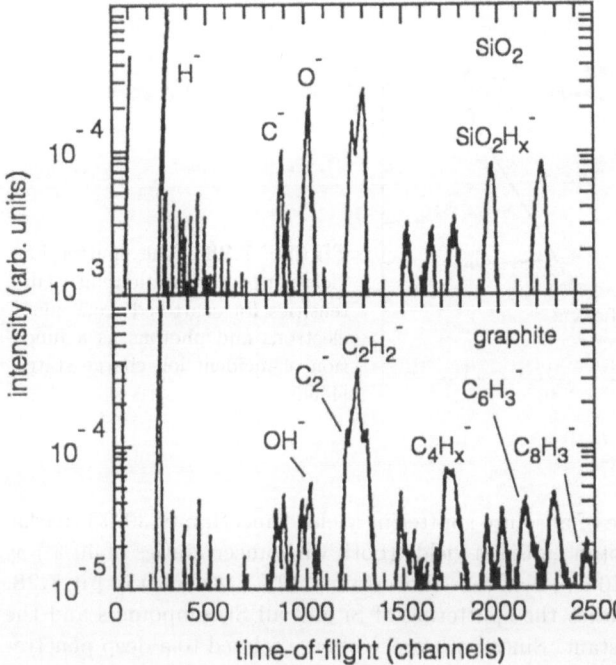

Figure 7.27 TOF spectra for 390 keV Xe^{44+} ions incident on various targets [444]. a: on SiO_2; b: on Graphite.

7.3.3.5 Particle Emission by the Interaction of Xenon Ions with a SiO_2 Surface

Data for sputtering, ion backscattering, electron and photon emission from a SiO_2 surface induced by highly charged Xenon ions (i=30-50) are reported [445, 443]. Thereby, a 300 keV ^{136}Xe beam of about 10^3 ions/sec was focussed onto the target at 25° with respect to the surface and ejected recoil ions were accelerated perpendicular to the surface into a channel-plate detector by voltages between 15-2000 V. Electrons were deflected into a channel-plate detector behind the target to produce fast start pulses for a TOF measurement.

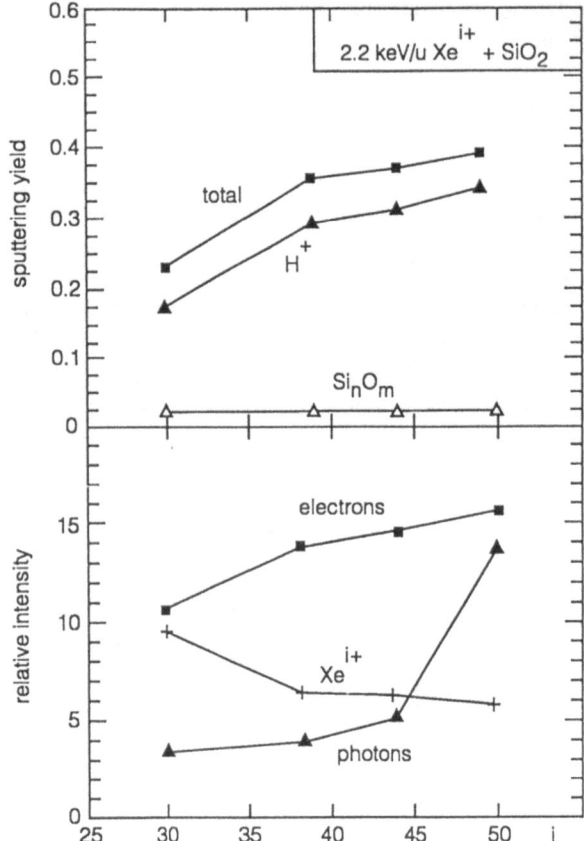

Figure 7.28 The sputter ion yield per projectile and relative intensities for scattered Xe^{i+} ions, electrons and photons as a function of incident ion charge states [445].

The charge state dependence of absolute sputtering yields (uncertainty 30%), backscattering yields (Xe^{i+}) and photon emission yields (both with uncertainties of 40%) as well as the yield of ejected ($\geq$ 10 eV) electrons (uncertainty 10%) is shown in Fig.7.28. Within the estimated uncertainties the sputtering of Si and all Si-compounds and the projectile backscattering is constant. Since both processes are related to a deep penetration of the solid, charge-state equilibrium might influence the results. All other observed yields are increasing for higher charge states. The yield of emitted electrons is linearly dependent to the ion charge state and therefore less than linear to the total ionic potential energy.

Experiments have shown that a TOF analysis of ejected particles is an efficient method to gain information on different processes in one experiment with incident ion currents of only about 1000 particles per second. Because the described experiments were conducted under non-UHV conditions, experiments under ultra-high vacuum environment with well defined surface conditions will give information on sputter ion and neutral atom and molecule yields.

Table 7.6   Measured asymtotic final charge state distribution for 49 keV O^{7+} incident on different species [447].

material	O^-/O^0	O^+/O^0	O^{2+}/O^0
C(0001)	0.40	0.09	0.01
Mica	0.59	0.09	< 0.01
Au(111)	0.35	0.19	0.03

7.3.3.6   Charge Exchange of Slow Few Electron High-Z Ions During Grazing Incidence Scattering on an Atomically Flat Low Z Surface

Measurements of the total effective charge exchange occurring during grazing incidence scattering ($\Theta < 1^o$) of highly charged ions on the surface of atomically flat highly oriented pyrolytic Graphite (mosaicity of 3.5^o) were performed at the LLNL EBIT [447] in an UHV of better than 10^{-8} Torr. Details on the experimental arrangement are given in Ref.[447]. Asymptotic final charge state distributions for a pyrolytic Graphite, a freshly cleaved Mica and an epitaxially grown Gold film on Mica (ca. 24 nm thick) are analyzed for Kr^{34+} ions at an incident angle of 0.5^o and a nominal detection angle of 1.0^o. Thereby, it is interesting to note the result that the major final states are neutral atoms or singly charged ions for all cases studied. This suggests that the neutralization of the incident highly charged ions is complete during the ion-surface interaction time of less than some 10^{-13} s. No significant effect on the final charge state distributions has been found due to the presence of a K shell vacancy. Furthermore, the probability of a residual inner-shell vacancy, present in the outgoing ion, as a function of the incident charge state was found to be linear with charge. Evidence for the formation of negatively charged scattered ions has been found for incident highly charged O^{i+} projectiles [447]. Examples for the asymptotic final charge state distribution for O^{7+} incident on an epitaxial Gold film on Mica, on pyrolitic Graphite and on Mica are tabulated in Table 7.3.3.6. The high probability for the negative ion formation indicates that the ions are in the ground state prior to leaving the surface interaction region.

7.3.3.7   Atomic Displacement Due to the Electrostatic Ion Potential at Surfaces

The occurence of single ion defects as a result of the interaction of slow highly charged ions (e.g. Xe^{44+} and U^{70+}) with large Coulombic potentials and solid insulating surfaces without any influence of secondary effects was demonstrated in [448]. Defect production in Mica has been investigated as a function of ion potential energy using an Atomic Force Microscope. Thereby, it has been found an increase in the area of blister-like defects parallel to the surface to be proportional to the total potential of the ions. On the other hand, singly and lower charged ions do not produce "blisters".

The growth of the defect volume with increasing incident ion charge and potential energy is shown in Fig.**7.29**. It is inferred that a potential energy deposition in the

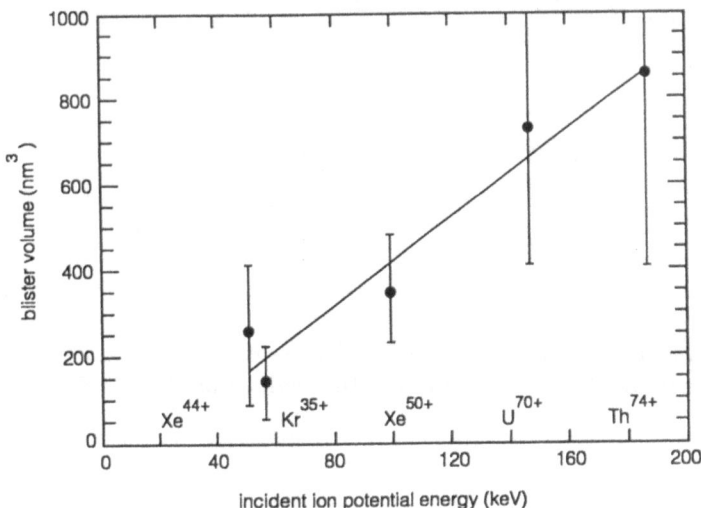

Figure 7.29   Blister volume as a function of the total potential energy of the incident ions [448].

first few atomic layers of an insulating surface due to the impact of a single slow highly charged ion is observed via the formation of nanosize defects. The defects are caused by electronic processes, primarily via emission of electrons due to the high charge of the incident ions and via acting repulsive electrostatic forces [449]. Processes causing defects are effective within the first monolayers of the surface region. The size of the defects are strongly dependent upon the total potential energy of the incident ions [450].

7.3.3.8   Projected Range of Xenon Ions in Silicon Dioxide

In Ref.[451] measurements of the effect of the incident charge state of slow highly charged ions on their projected range in insulating materials have been reported. Early estimates based on an unrealistic scaling of the energy loss with the square of the incident charge suggested a very short range of the ions on the order of nanometers. More realistic estimates do suggest that long range Coulomb interactions between highly charged incident ions and the ionic partners of dielectric solids should significantly increase the average energy loss of the ions within the first nanometers of the surface region [414]. Integration of this effect over the highly non-equilibrium charge state path length of the ion would produce a constant reduction in the projected range or range defect. The magnitude of this range defect depends strongly on the time scale of the charge state equilibration of the ion and is therefore an indirect measurement of the process involved. The existence of hollow atoms inside solids and their dynamic return to the ground state is of fundamental interest.

As reported in Ref.[451] Xenon ions at different energies and charge states are implanted into samples of thermal Silicon Dioxide on Silicon. Thereby, range profiles were determined through backscattering measurements of 2.4 MeV Ne^+. Investigations have

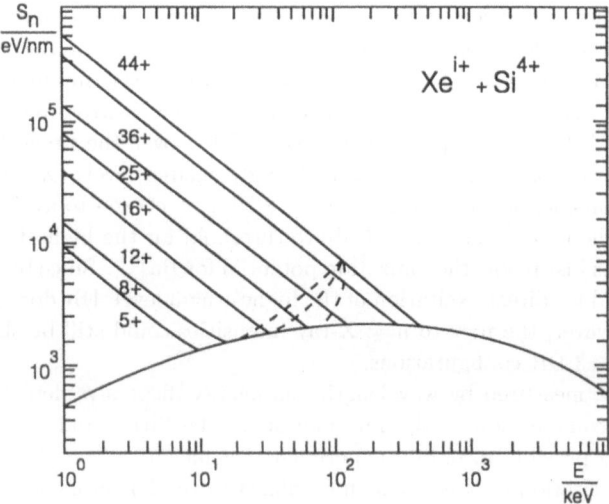

Figure 7.30   Calculation results of the effect of the incident charge state of Xe^{i+} on the component of the nuclear energy loss associated with the interaction of Si^{4+} cores with the projectile in SiO_2, assuming that the solid is completely ionic in nature [451]. The increased energy loss is due to long range Coulombic interactions with the unscreened charge of the projectile [414]. The dotted line represents the effect of the decay of the high charge state inside the solid.

shown that there is no significant effect of the incident charge on the projected range. This is in agreement with calculation results shown in Fig.7.30 for the nuclear stopping component of the interaction of Xe^{i+} and Si^{4+} cores in Silicon Dioxide, assuming that the solid is completely ionic in nature. These results indicate that either the decay time of the hollow atom inside the solid is less than 10's of femtoseconds or the interaction with the solid is significantly screened by polarization effects. Recent calculations indicate that the latter is probable, especially in covalent insulators such as SiO_2.

7.3.4   Electron-Ion Interaction Studies

7.3.4.1   Electron Impact Excitation Cross-Sections

The interaction of highly charged ions with electrons as well as their spectroscopic properties are important for the understanding of atomic structure in the high-field (high-Z) limit where relativistic and QED effects are important. In Ref.[168] a new technique was described for direct measurements of electron impact excitation (IE), dielectronic recombination (DR) and radiative recombination (RR). The technique consists of trapping ions inside an electron beam and determining cross-sections from X-ray spectroscopy of the trapped ions excited by the electron beam. Because the target ions are prepared in a single charge state and the electron beam is monoenergetic, it is possible to extract all of the separate cross-sections which contribute to X-ray emission. The method of successive ionization of ions trapped in an electron beam was also used in EBIS [158].

Excitation of neon-like Barium. In the in Ref.[168] described experiment the Barium charge state was selected by keeping the electron beam energy between the 3.66 keV and the 8.33 keV ionization potentials of Ba^{45+} and of Ba^{46+}. The neon-like Ba^{46+} configuration was selected for study because of the simplicity of a closed-shell configuration and the experimental convenience of the electron and X-ray energies involved. A crystal diffraction spectrometer was set to cover the range from 4.56 keV to 5.64 keV, where in Ba^{46+} there are 36 n=3 singly excited levels. X-ray spectra were obtained at electron energies of 5.69 keV and at 8.20 keV, chosen to avoid the strongest DR resonances. The lower energy is just above threshold for the highest n=3 level, and the higher energy is just below the ionization potential for Ba^{46+}. Since both energies are above the threshold for direct excitation of the highest n=3 level, DR does not involve an n=3 electron. However, the n=2 to n=3 X-ray intensities could still be slightly affected by cascade from n>3 DR configurations.

Typical spectra, measured by wavelength and energy dispersive detectors are shown in Fig.**7.29**. By combination of the information in the Si(Li) and crystal diffraction X-ray spectra it is possible to obtain experimental values for some of the n=2 to n=3 electron impact excitation cross-sections normalized to the RR cross-section, which can be calculated more reliably. In Ref.[168] this was done for the three strongest L X-rays since they are resolved in the Si(Li) spectrum and could extracted together with the unresolved RR n=3 lines by least-squares fitting. The contributions of the weaker Ba^{46+} L X-rays were determined from the wavelength dispersive measured spectrum and subtracted from the Si(Li) X-ray yields. The RR cross-sections were calculated using a relativistic distorted-wave code [453]. The decay scheme of the Ba^{46+} levels involved in the cross-section measurements is given in [168].

Table 7.7   Comparison of measured σ_{exp} and theoretical σ_{th} electron IE cross-sections for neon-like Ba^{46+}. Units are in 10^{-21} cm^2.

level	E/eV	$E_e = 5.69$ keV			$E_e = 8.20$ keV		
		σ_{th} [454]	σ_{th} [455]	σ_{exp}	σ_{th} [454]	σ_{th} [455]	σ_{exp}
sum J=0		2.28	2.60	2.50 ± 0.35	1.89	1.94	2.27 ± 0.32
$(2p_{3/2}^{-1} 3d_{5/2})_1$	4937	3.44	3.56	3.98 ± 0.56	2.99	3.23	3.30 ± 0.46
$(2p_{1/2}^{-1} 3d_{3/2})_1$	5295	2.42	2.00	2.12 ± 0.30	2.10	1.82	1.82 ± 0.25

Table **7.7** summarizes the results from Ref.[168] and compares them with two different theoretical calculations: a Coulomb-Born exchange method with relativistic effects treated as a pertubation [454] and a fully relativistic distorted-wave calculation [455]. The measured values in Table **7.7** are 4π sr times the differential cross-sections at an X-ray angle of 90° with respect to the electron beam. For the J=0 decays this is the total IE cross-section, but the remaining two E1 transitions in Table **7.7** may have a nonoisotropic angular distribution. The largest experimental uncertainty in the cross-sections was estimated to be about 14% due to the Tungsten background subtraction and uncertainties due to the Ba^{45+} fraction.

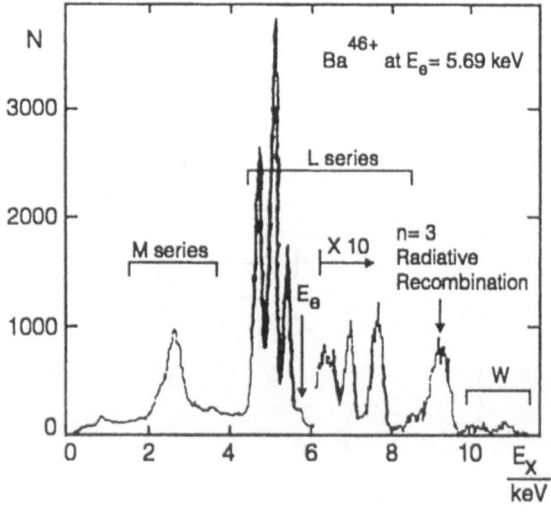

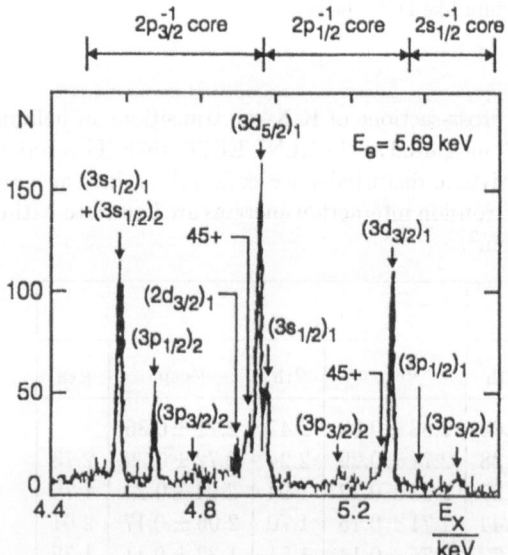

Figure 7.31 Si(Li) spectrum of Ba^{46+} at $E_e = 5.69$ keV [168] (upper part). The spectrum is cut off below 2.5 keV by absorption in the Beryllium windows. The feature w is attributed to RR onto Tungsten ions, which were contaminant in the trap. Lower part: Spectrum measured with a crystal diffraction spectrometer. Lines are isentified by (nl_j) of the excited electron and by total angular momentum J. The spectral regions indicated by bars corresponds to the three different core configurations for the identified lines.

Excitation of helium-like 3d elements. In many plasma environments electron-impact excitation is the most important line formation process. In helium-like ions it is the dominant process for exciting the $1s2p\,^1P_1$ level, whose decay results in the resonance line commonly referred as w. Direct electron collisions also excite the $1s2p\,^3P_{1,2}$ and $1s2s\,^3S_1$ levels, resulting in the intercombination lines y and x and the forbidden line z as shown schematically in Fig.7.32 for helium-like Iron. Calculations show that the electron-impact excitation cross-section of the $^1P_1$ level increases with electron energy, while those of the triplet levels decrease [274]. Consequently, the intensity ratio of the triplet lines to the single line $(x+y+z)/w$ depends on electron temperature and suggests itself as a temperature diagnostic.

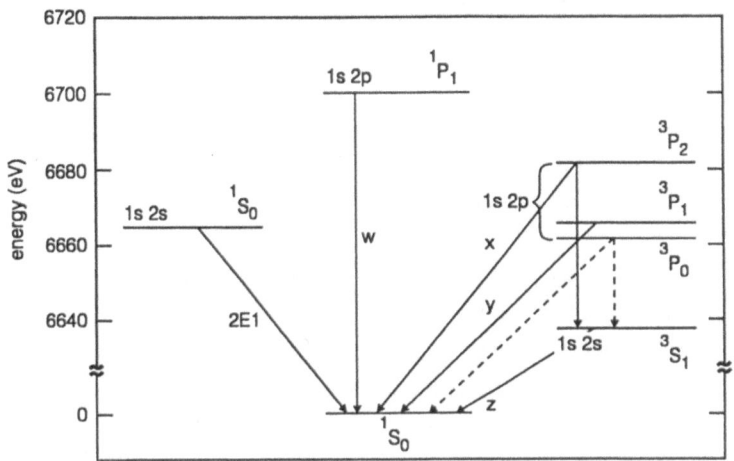

Figure 7.32   Energy level diagram of helium-like Fe^{24+} [456].

Table 7.8   Electron impact excitation cross-sections of K X-ray transitions in helium-like Ti^{20+}, V^{21+}, Cr^{22+}, Mn^{23+} and Fe^{24+} measured at the LLNL EBIT [457]. Theoretical values have been calculated with a relativistic distorted-wave code taking into account radiative cascade contributions. The electron-ion interaction energies are known to within 100 eV. All values are in units of 10^{-22} cm^2.

		w		x		y	
	E_e/keV	$\sigma_{\exp}$	σ_{th}	$\sigma_{\exp}$	σ_{th}	$\sigma_{\exp}$	σ_{th}
Ti^{20+}	4.9	7.61 ± 0.96	7.63	2.88 ± 0.38	2.47	2.71 ± 0.36	
V^{21+}	5.3	4.72 ± 0.50	6.28	2.71 ± 0.29	2.24	2.72 ± 0.29	2.73
Cr^{22+}	5.8	5.30 ± 0.46	5.27	2.03 ± 0.22	1.94	2.04 ± 0.22	1.79
Mn^{23+}	6.3	3.80 ± 0.30	4.44	1.71 ± 0.15	1.70	$2.06 \pm,0.17$	2.01
Fe^{24+}	6.8	3.17 ± 0.24	3.72	1.76 ± 0.14	1.51	$1.32 \pm,0.11$	1.38

In Ref.[456, 457, 458] electron impact ionization excitation cross-sections of K X-ray transitions in helium-like Ti^{20+}, V^{21+}, Cr^{22+}, Mn^{23+} and Fe^{24+} were determined. For comparison, Table **7.8** lists the electron-impact excitation cross-sections calculated with a relativistic distorted-wave code and experimental results. A comparison of theoretical with the measured values shows excellent agreement for the triplet lines. By contrast, the measured electron-impact excitation cross-section of w is about 20% lower than calculated what is an unexpected result. A discussion of this behaviour is given in [456].

7.3.4.2   Electron Impact Ionization Cross-Sections for Ions

In Refs.[153, 460, 461] electron impact ionization cross-sections for C, N, O, Ne, Ar, Kr and Xe are reported. Thereby, all measurements are accomplished at the Dubna EBIS. The experimental method used was a time-of-flight technique to determine abundances of differently charged ions after an concrete life-time of the ions in the EBIS. The method to determine the corresponding cross-sections is described in [153, 460, 461]. For instance, in Fig.4.7 ionization cross-sections for Neon ions at different electron energies and ion charge states are shown.

7.3.4.3   Electron Impact Ionization Cross-Section Measurements of Lithium-Like Barium

In Ref.[183] the electron impact ionization cross-section of lithium-like Ba^{53+} ($1s^2\, 2s + e^- \rightarrow 1s^2 + 2e^-$) has been measured at electron energies of 22 keV at the LLNL EBIT. The motivation of such investigations is connected with the fact, that an accurate knowledge of ionization, excitation and recombination cross-sections of highly charged ions along an isoelectronic sequence is important for checking the scaling laws of theoretical calculations because relativistic effects become important.

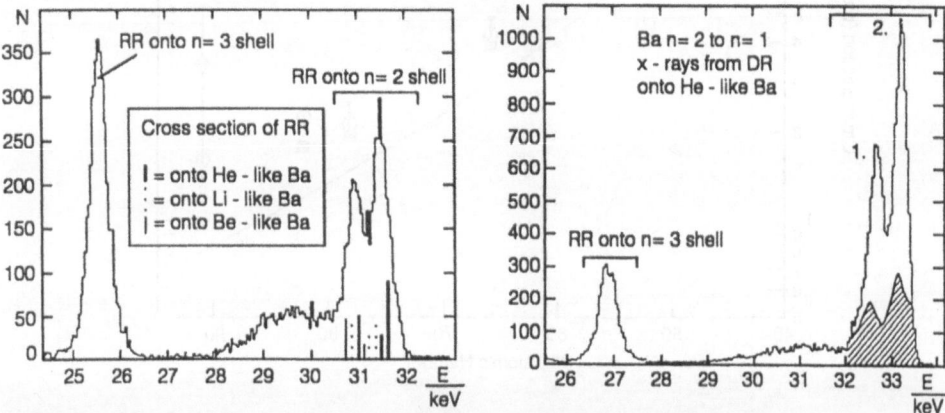

Figure 7.33   Left side: Radiative recombination (RR) X-ray spectrum at 21.2 keV electron energy [183]. The RR X-rays onto the n=2 shell are fitted to determine a charge balance. The solid and dashed lines represent the relative theoretical cross-sections of RR onto the different charge states. Right side: Dielectronic recombination (DR) X-ray spectrum at 22.4 keV electron energy. The n=2 to n=1 structure contains a contribution from X-rays due to RR onto the n=2 shell of Barium (hatched area). Peak 1 corresponds to the transition $[(1s\,2p)_1\, 2p_{3/2}]_{5/2} \rightarrow 1s^2\, 2p_{3/2}$; peak 2 to $[(1s\,2s)_0\, 2p_{3/2}]_{3/2} \rightarrow 1s^2\, 2s$.

An detailed description of the used technique is given in Ref.[183]. The experimental procedure based on measurements of RR and DR X-rays with a solid-state Germanium detector. Characteristic results are shown in Fig.7.33 for RR and DR X-ray data for Barium.

Results of lithium-like Barium are given in Table **7.9** for three energies corresponding to the energies of the dielectronic resonances in helium-like Barium. In order to determine

Table 7.9 Electron impact ionization cross-sections σ of lithium-like Barium in units of 10^{-21} cm^2 [183]. * denotes sum of adjacent resonances. CE – charge exchange recombination.

E_{beam}/keV	σ_{exp}(no CE)	σ_{exp}(with CE)	σ_{th}[463].
21.83	0.24 ± 0.16	0.36 ± 0.28	0.142
21.97	0.25 ± 0.17	0.38 ± 0.29	0.142
22.44*	0.25 ± 0.13	0.36 ± 0.24	0.142

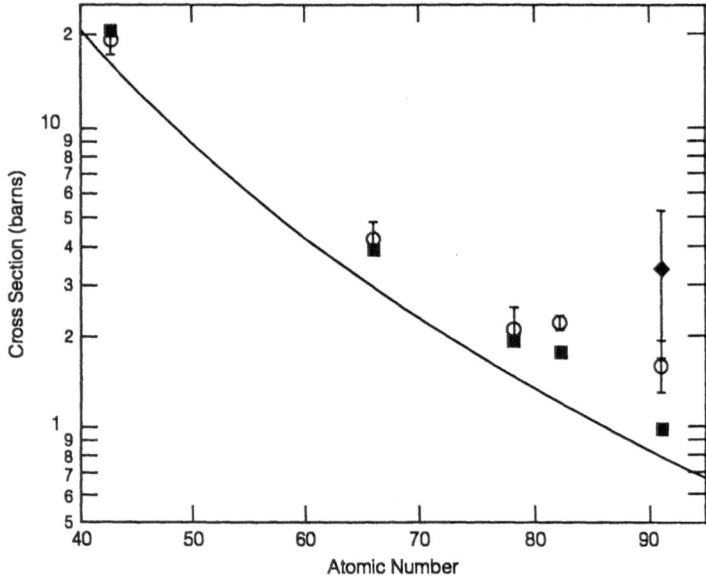

Figure 7.34 Ionization cross-sections of hydrogen-like ions [465] for electron energies approximately 1.5 times treshold. Inked line – results for the semiempirical Lotz formula; ■ – distorted wave calculations; ◊ – accelerator stripping measurements; ∘ – results from Ref.[465].

the effect of CE, the X-ray signal at energies away from the DR resonance was measured. In this regime ionization is approximately balanced by CE and RR. Knowing RR it is possible to infer CE. Since the CE contributions does not change as a function of the electron beam energy this fact can be used to correct the DR data.

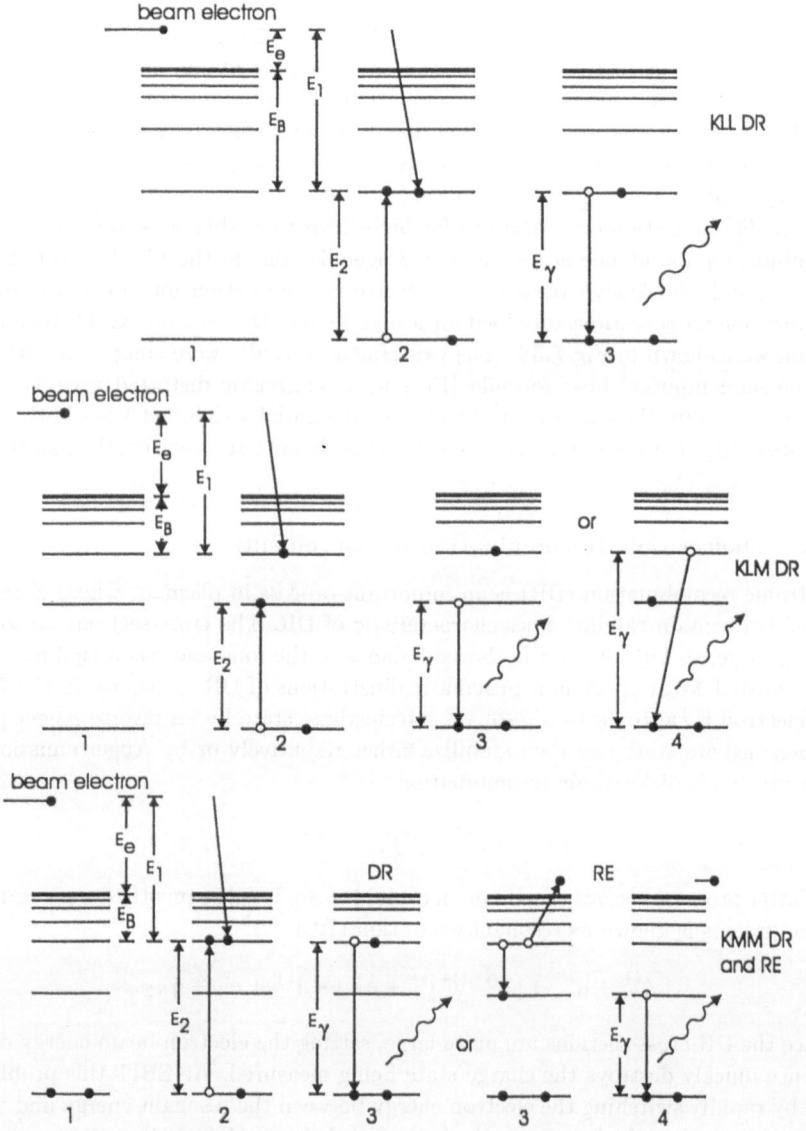

Figure 7.35 Dielectronic recombination and resonant excitation of helium-like ions [469]. a – KLL DR: A beam electron is captured into an ion exciting a bound electron. The intermediate doubly excited state can then stabilize radiatively. b – KLM DR: If the intermediate state has electrons in two states, two possible X-ray energies can result. One K X-ray always results from the DR of a helium-like ion. c – KMM DR and RE: for some resonances the intermediate state can undergo Auger decay resulting in resonant excitation.

7.3.4.4   Ionization Cross-Sections for Hydrogen-Like High-Z Atoms

Electron impact ionization cross-sections for hydrogen-like ions of high-Z elements are an important test of relativistic interactions in a simple atomic system. The interest even on this quantity is based on the fact, that the approximative size of the high-Z ionization cross-sections for 1s electrons has been uncertain in view of an accelerator stripping measurement that obtained values three to five times larger than any theory for hydrogen-like and helium-like Uranium [464]

In Ref[465] ionization cross-sections for high-Z hydrogen-like ions were obtained from the equilibrium abundance of bare and hydrogenlike ions in the LLNL EBIT using the competing and well known process of radiative recombination for normalization [166]. The ionization cross-sections obtained for Molybdenum, Dysprosium, Gold, Bismuth and Uranium were shown in Fig.**7.34**. The experimental results were compared with results from the semiempirical Lotz formula [13] and a relativistic distorted wave calculation [467]. For Uranium the experimental value is substantially different from both previous accelerator stripping measurements (Bevalac experiment) and relativistic distorted wave theory.

7.3.4.5   Dielectronic Recombination Measurements

Dielectronic recombination (DR) is an important process in plasmas, whose X-ray emission spectra contain satellite lines characteristic of DR. The cross-sections are resonant and very large, so DR affects the charge balance of the ions and has a significant effect on the emitted X-ray spectrum. Schematic illustrations of DR are shown in Fig.**7.35**. A beam electron is captured to a resonant intermediate state by an inverse Auger process. The intermediate state can then stabilize either radiatively or by Auger emission. The former process is dielectronic recombination

$$A^{i+} + e^- \rightarrow [A^{(i-1)+}]^{**} \rightarrow A^{(i-1)+} + \gamma_{DR}.$$

If the latter processes leaves the ion in an excited state, it subsequently decays radiatively and the process is known as resonant excitation (RE)

$$A^{i+} + e^- \rightarrow [A^{(i-1)+}]^{**} + e^- \rightarrow A^{i+} + e^- + \gamma_{RE}.$$

Since the DR cross-sections are quite large, setting the electron beam energy on a DR resonance quickly destroys the charge state being measured. At EBIT this problem was solved by rapidly switching the electron energy between the resonant energy and a higher nonresonant energy which restores the ionization balance [469, 184].

Typically only 10% of the time is spent in resonance. An excitation function is obtained by taking successive data runs in which the lower energy (i.e. the DR energy) is changed by small amount and normalizing each run to a common nonresonant upper energy.

Up to now, on electron impact ion sources a lot of different experiments to study DR were provided, e.g. for hydrogen-like Oxygen [470], helium-like Argon [471], hydrogen-like Argon [472], helium-like Vanadium [380], helium-like Nickel [473], lithium-like Titanium [474], for Ni^{26+}, Mo^{40+} and Au^{69+} [63], fluorine-like Xenon [475], neon-like Xenon [476] and for neon-like Gold [477].

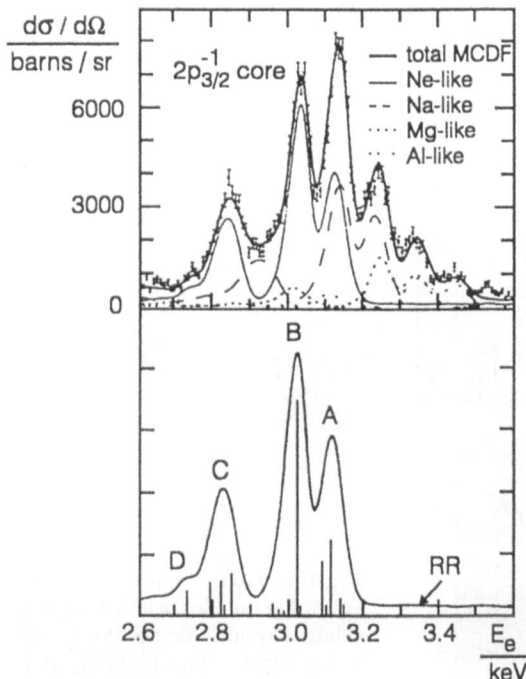

Figure 7.36   Spectrum of the LMM DR resonances of neon-like Gold [478]. Upper part: Direct excitation spectrum at $E_e = 18$ keV. The strongest excitation lines are from two E1 transitions: $(2p_{3/2}^{-1} 3d_{5/2})_{j=1} \rightarrow$ ground state near $E_e = 10.5$ keV and $(2p_{3/2}^{-1} 3s_{1/2})_{j=1} \rightarrow$ ground state near $E_e = 9.5$ keV. The other strong $n = 3 \rightarrow 2$ transitions involve the $2p_{1/2}^{-1}$ core. Lower part: X-ray density distributions versus X-ray energy and electron energy. The DR resonance are the bright spots at the intersections of the excitation X-rays in the vertical bands and the RR X-rays in the diagonal bands.

DR can be measured by detecting either the intensity of the DR X-rays or the ion abundance as a function of the interaction energy between the ion and the electron. Thereby, cross-sections were normalized to those of radiative recombination, the nonresonant recombination of electrons with ions.

For the second technique the DR cross-sections were normalized to those of electron impact ionization. DR has also been indirectly measured in resonant transfer and excitation (RTE) experiments where ions move through a gas target and interact with the target electrons. Generally, for helium-like ions all three kinds of experiment have been in good agreement with theory.

Solid state Germanium detector measurements of DR are reported in Ref.[478]. When an X-ray was detected in the Germanium detector, its photon energy, the electron beam energy and the time of the event are recorded. The data were self-normalized because the same amount of time was spent at each voltage. Data for a typical run displayed as a scatter plot are shown in Fig.7.36. Thereby, the LMM resonances occur at electron energies between 1 keV and 6 keV. An excitation function was constructed by summing all the n=3 $\rightarrow$ 2 photons and the RR to n=3 photons at each electron beam energy and correcting for dead time. The measured excitation function for the strongest region of the LMM resonances is shown in Fig.7.37. Much of the structure result from the presence of several ionization stages.

The data in Fig.7.37 were compared to multiconfiguration Dirac-Fock (MCDF) calculations of the resonance strengths for recombination onto four ionization states, neon-like to aluminium-like Gold. The ions were assumed to be initially in their ground state

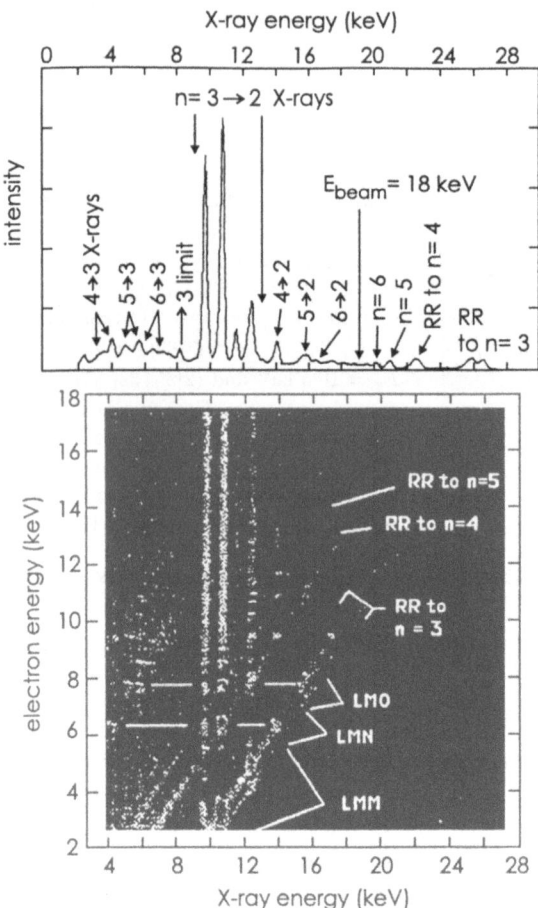

Figure 7.37 Upper part: DR excitation function and MCDF calculation results for the $2p_{3/2}^{-1}$ core region [478]. The background is from RR to n=3. Lower part: Theoretical resonance strengths for neon-like Gold target ions (vertical bars) and calculated cross-sections given the resolution of this experiment (solid line). The peaks A, B and C correspond to the second DR electron in the $3d_{5/2}$, $3d_{3/2}$ and $3p_{3/2}$ subshells, respectively.

because the electron density was in such an order that the time between electron-ion collisions was longer than the life-time of any metastable state of the ion. About 20 LMM DR resonances onto the neon-like ion make significant contributions to the DR cross-section for the data in Fig.7.37. The dominant contribution is from high angular momentum intermediate states with a $2p_{3/2}^{-1}$ core and at least one DR electron in the $3d_{5/2}$ subshell. The theoretical predictions of the resonance strengths for neon-like ions and calculated cross-sections for the energy resolution of the experiment are shown in Fig.7.37, too. All the ionization stages have a similar three-peak structure.

The excitation function in Fig.7.37 was fitted for an ionization stage distribution and an electron beam energy resolution. The fit of the data is shown in the upper part of Fig.7.37 along with the separate contributions from different ionization stages (44% neon-like, 12% magnesium-like, 35% sodium-like, 12% magnesium-like and 9% aluminium-like target ions). DR resonance strengths were derived by normalizing the data to the measured intensity of RR photons to n = 3,4,5 using calculated RR cross-sections.

The in Ref[478] measured total resonance strength for DR onto neon-like Gold between 2.6 keV and 3.6 keV was $(1.0 \pm 0.15) \cdot 10^{-17}\,\mathrm{cm^2\,eV}$; the theoretical value is $0.95 \cdot 10^{-17}\,\mathrm{cm^2\,eV}$.

Bibliography

[1] C.L.Olson; Physical Review A11 (1975) p.258

[2] Electron Impact Ionization; ed.by T.D.Mark and G.H.Dum, Springer Verlag Wien and New York 1985

[3] L.P.Presnyakov and R.K.Janev; in: Atomic Collision Processes with Multiply Charged Ions, Proc. of the II Workshop on Vinca Accelerator Installation, Belgrad 1983, p.155

[4] E.Salzborn; in: Physics of Ion-Ion and Electron-Ion Collisions, ed.by F.Brouillard and S.W.Mc Goiwal, Plenum Press, New York and London 1982

[5] A.Mueller; in: Proceedings of the Vth International Conference on the Physics of Highly Charged Ions, Giessen(Germany), Springer Verlag, Berlin, New York a.o. 1991, p.39

[6] E.D.Donets; Soviet Journal of Particles & Nuclei, 13 (1982) p.387

[7] G.Zschornack; Soviet Journal of Particles & Nuclei, 14 (1983) p.349

[8] E.A.Perelstein and G.D.Shirkov; Soviet Journal of Particles & Nuclei, 18 (1987) p.64

[9] J.J.Tompson; Phil.Mag., 20 (1912) p.752

[10] G.Zschornack, G.Musiol and W.Wagner; Report ZfK-574, Dresden-Rossendorf 1986

[11] T.A.Carlson et al.; Atomic Data Tables, 2 (1970) p.63

[12] W.Lotz et al.; Zeitschrift f. Physik, 206 (1967) p.205

[13] W.Lotz; Zeitschrift f. Physik, 216 (1968) p.466

[14] A.A.Drozdovskii; Preprint ITEF-100, Moscow 1973

[15] E.D.Donets; Physica Scripta, T3 (1983) p.11

[16] E.D.Donets and V.P.Ovsyannikov; Preprint JINR, P7-80-404, Dubna 1980 and Sov. Phys. JETP, 53 (1981) p.466

[17] M.H.Rudge and S.B.Schwartz; Proc.Phys.Soc., 88 (1966) p.563

[18] M.Gryzinski; Physical Review, 138A (1965) p.336

[19] A.Mueller and R.Frodl; Physical Review, 44 (1980) p.29

[20] A.Salop; Physical Review, A8 (1973) 3032

[21] A.Salop; Physical Review, A9 (1974) 2496

[22] H.-U.Siebert et al.; JINR Communication P9-10197, Dubna 1976

[23] F.F.Rieke and W.Prepejchal; Physical Review, A6 (1972) p.1507

[24] R.K.Janev, L.P.Presnyakov, V.P.Shevelko; Physics of Highly Charged Ions, Springer-Verlag,, Berlin, Heidelberg a.o., 1985, p.330

[25] L.P.Presnyakov, V.P.Shevelko, R.K.Janev; Elementary Processes with Multiply Charged Ions, Energoatomizdat, Moscow, 1986, p.200

[26] H.Knudsen, K.H.Haugen and P.Hvelplund; Physiacl Review A23 (1981) 597

[27] C.C.Havener, M.S.Hug, F.W.Meyer and R.A.Phaneuf; Electron Capture by Multiply Charged Ions for eV Energies, in: Proc. of the Int. Conf. on the Physics of Multiply Charged Ions, Grenoble(France), 1988 and also in: J.de Physique C1, 50 (1989) C1-7

[28] R.K.Janev and L.P.Presnyakov; Phys.Reports 70 (1981) 1

[29] R.K.Janev; in: Proc.of the 2nd Workshop on Vinca Accel. Installation, Belgrad 1983, 37

[30] L.P.Presnyakov and D.B.Uskov; in: X-Ray Plasma Spectroscopy and the Properties of Multiply Charged Ions, Vol.179, Nauka, Moscow 1987, 103

[31] A.Niehaus; in: Physics of Highly Ionized Atoms, ed.by R.Marrus, Plenum Press, New York and London 1989, 267

[32] V.I.Lendyel, V.Yu.Lasur, M.I.Karbovenec and R.K.Janev; Introduction in the Theory of Atomic Collisions, Lvov, 1989

[33] M.I.Chibisov; Soviet Phys. Tech. Phys. Lett.(USSR) 24 (1976) 56

[34] A.Mueller and V.P.Shevelko; Soviet Phys. Tech. Phys. 50 (1980) 985

[35] A.Mueller and E.Salzborn; Physics Letters 62A (1977) 391

[36] S.Bliman, M.Bonnefoy, J.J.Bonnet, S.Donsson, A.Fleury, D.Hitz and B.Jacquont; Physica Scripta (1983) 63

[37] R.K.Janev and D.S.Belic; Physica Scripta T3 (1983) 246

[38] L.Spitzer; Physics of Fully Ionized Gases, J.Wiley and Sons, New York and London 1962

[39] C.L.Langmuire; Elementary Plasma Physics, J.Wiley and Sons, New York and London 1963

[40] E.M.Lifshitz and L.P.Pitaevskii; Theoretical Physics, Vol.10: Physical Kinetics, Nauka, Moscow 1979

[41] E.D.Donets, G.D.Shirkov; Method of Highly Charged Ion Production, Soviet Invention No. 1225420, 02.07.1984, Bul. 01 No.44, p.69(1989)

[42] G.D.Shirkov; Preprint GSI-tr-89-09, GSI Darmstadt, 1989; Atomic Physics of Highly Charged Ions, Proc. of the 5th Int. Conference on the Physics of Highly Charged Ions, Springer-Verlag, Berlin, 1991, p.319

[43] B.A.Trubnikov; Ion Collisions in Fully Ionized Plasma, in: Voprosy Teorii Plasmy, Vol.1, Gosatomizdat, Moscow 1963, pp.98-182

[44] D.V.Sivuchin; Coulomb Collisions in Fully Ionized Plasma, in: Voprosy Teorii Plasmy, Vol.4, Atomizdat, Moscow 1964, pp.81-187

[45] Handbook of Mathematical Functions, ed.by M.Abramiwitz and I.A.Stegum, Nat.Bureau of Standards, Washington 1964

[46] E.Rutherford and F.Soddy; Phil. Mag., 4 (1902) 370, 569

[47] E.Rutherford and F.Soddy; Phil. Mag., 5 (1903) pp. 77, 441, 561, 576

[48] M.L.Iovnovitch and M.M.Fix; Preprint JINR P9-4849, Dubna 1969 and Atomnaya Energiya, 29 (1970) 429

[49] V.George, M.L.Iovnovitch, V.G.Novikov and V.A.Preisendorf; JINR Communication P9-6555, Dubna 1972

[50] M.L.Iovnovitch, A.B.Kusnetsov and V.A.Preisendorf; JINR Communication P9-8119, Dubna 1974

[51] D.Lehmann et al.; JINR Communication 9-10744, Dubna 1977

[52] E.A.Perelstein and G.D.Shirkov; Sov. Phys. Tech. Phys., 24 (1979) 10

[53] G.D.Shirkov; Sov. Phys. Tech. Phys., 24 (1979) 818

[54] B.S.Bochev, T.Ch.Kutsarova and V.P.Ovsyannikov; JINR Communication P9-11566, Dubna 1978

[55] B.S.Bochev, V.P.Ovsyannikov and T.Ch.Kutsarova; JINR Communication P5-11567, Dubna 1978

[56] V.P.Kutner, V.I.Kochkin, A.S.Pasuik and Yu.P.Treyakov; Preprint JINR P9-12578, Dubna 1979

[57] V.P.Kutner and V.I.Kochkin; Preprint JINR P7-12707, Dubna 1979

[58] Y.Jongen; Preprints LC-8001 and LC-8003, Univ.Catholique de Louvain, Belgium, 1980

[59] S.Bliman and N.Chang-Tung; Journ. de Physique, 42 (1981) 1241

[60] H.I.West Jr.; Preprint UCRL-53391, Lawrence Livermore National Laboratory, California, 1982

[61] G.Shirkov; JINR Communication P9-90-581, Dubna 1990

[62] G.Shirkov, C.Mühle, G.Musiol and G.Zschornack; 11. Arbeitsbericht Energiereiche atomare Stösse, Kassel, 1990, 20

[63] V.P.Kutner and G.Shirkov; Atomic Physics of Highly Charged Ions (Proc.of the Vth Int.Conf. on the Physics of Highly Charged Ions, Giessen 1990), Springer-Verlag, Berlin 1991, 323

[64] G.Shirkov, C.Mühle, G.Musiol and G.Zschornack; Nucl.Instr.Methods, A302 (1991) 1

[65] I.Steinert, G.Shirkov and G.Zschornack; 12.Arbeitsbericht Energiereiche atomare Stösse , Riezlern 1991, 125

[66] I.Steinert, G.Shirkov and G.Zschornack; Nucl.Instr.Methods, 1992,

[67] M.A.Levine, R.E.Marrs, J.R.Henderson, D.A.Knapp, M.B.Schneider; Physica Scripta, T22 (1981) 157

[68] M.A.Levine, R.E.Marrs, C.L.Bennett, J.R.Henderson, D.A.Knapp, M.B.Schneider; in: Proc. Int. Symp. on EBIS and their Applications, ed. by A.Hersheovitch, AIP Conf. Proc. No.188, (AIP, New York, 1989), p.82

[69] E.Segre; in: Experimental Nuclear Physics, Vol.III, ed.by E.Segre, New York and London 1959

[70] E.D.Donets; in: Physics and Technology of Ion Sources, ed.by L.G.Brown, J.Wiley Publ., New York 1989, 245

[71] E.D.Donets; in: Proc. of the 3rd Int.Conf. on Ion Sources, Review of Sci.Instruments, 61 (1990) 225

[72] E.A.Perelstein and G.D.Shirkov; in: Proc. Conf. on Problems of Collective Methods of Acceleration, JINR D9-82-664, Dubna 1982, 31

[73] E.A.Perelstein and G.D.Shirkov; Preprint JINR E9-85-4, Dubna 1985

[74] A.A.Vlasov; Static Distribution Functions, Nauka, Moscow, 1966

[75] Computation Methods in Plasma Physics; ed.by B.Alder, S.Fernbach and M.Rotenberg, Mir, Moscow, 1974

[76] I.M.Kapchinskii; Theory of Linear Resonance Accelerators, Energoizdat, Moscow, 1982

[77] J.D.Lawson; The Physics of Charged Particle Beams, Clarendon Press, Oxford, 1977

[78] I.M.Kapchinski; Particle Dynamics in Linear Resonance Accelerators, Atomizdat, Moscow, 1966

[79] A.S.Roshal; Modelling of Charged Particle Beams, Atomizdat, Moscow, 1979

[80] L.J.Laslett; Preprint ERAN-218, LBL, Berkeley, 1972

[81] A.A.Drozdovskii; Preprint ITEF-10, Moscow, 1973

[82] I.Hofmann; Preprint IPP 0/21, IPP, Garching, 1974

[83] V.P.Sarantsev and E.A.Perelstein; Collective Acceleration of Ions with Electron Rings, Atomizdat, Moscow, 1979

[84] N.Yu.Kazarinov, E.A.Perelstein and G.D.Shirkov; Sov. Phys. Tech. Phys., 25 (1980) 1004 and Preprint JINR, P9-12719, Dubna, 1979

[85] E.A.Perelstein and G.D.Shirkov; Preprint JINR, P9-82-526, Dubna, 1982

[86] E.A.Perelstein, V.F.Shevtsov, G.D.Shirkov and B.G.Shchinov; Sov. Phys. Tech. Phys., 29 (1984) 158

[87] A.B.Kuznetsov; Preprint JINR, P9-83-349, Dubna, 1983

[88] E.A.Perelstein; in: Int.School for Young Scientists on Problems of Charged Particles Accelerators, JINR, D9-84-817, Dubna, 1984, p.90

[89] A.D.Dymnikov and E.A.Perelstein; Nucl.Instr.Methods, 148 (1978) 567

[90] F.J.Sacherer; Preprint LRL, UCRL-18454, Berkeley, 1968

[91] O.I.Yarkovoi; Sov. J. Phys. Tech. Phys., 7 (1963) 951

[92] P.M.Lapostolle; Proc. of Part. Accelerator Conf., Chicago, IEEE Trans.Nucl.Sci., NS-18 (1971) 1101

[93] F.J.Sacherer; Proc. of Part. Accelerator Conf., Chicago, IEEE Trans.Nucl.Sci., NS-18 (1971) 1105 and Preprint CERN/SI/Int. 70-12, 1970

[94] B.Bru, M.Weiss; Preprint CERN/MPS/LIN 72-4, 1972

[95] I.S.Mirer; in: Proc.of the IV National Conf. on Charged Particles Accelerators, Vol.2, Nauka, Moscow, 1975, p.87

[96] N.Yu.Kazarinov, E.A.Perelstein, V.F.Schevtsov; Particle Accelerators, 10 (1980) 1 and Preprint JINR P9-10985, Dubna, 1977

[97] L.V.Bobuleva, N.Yu.Kazarinov, E.A.Perelstein; JINR Communication P11-81-796, Dubna, 1981

[98] E.A.Perelstein, G.D.Shirkov; Preprint JINR P9-10468, Dubna, 1978 and Sov. J. Phys. Tech. Phys., 23 (1978) 149

[99] E.P.Lee, R.K.Cooper; Particle Accelerators, 7 (1976) 83

[100] S.B.Rubin, O.I.Yarkovoi; Preprint JINR 2882-2, Dubna, 1966

[101] N.Yu.Kazarinov, E.A.Perelstein; JINR Communication P9-11916, Dubna, 1978

[102] N.Yu.Kazarinov, E.A.Perelstein, G.D.Shirkov; Sov. J. Phys. Tech. Phys., 25 (1980) 330

[103] A.A.Kolomenskij; Physical Principles of Charged Particle Acceleration Methods, Moscow State University Press, Moscow, 1980 and Preprint 155, Lebedev Physics Institute, Moscow, 1970

[104] E.A.Perelstein, G.D.Shirkov; JINR Communication 9-80-124, Dubna, 1980

[105] V.S.Alexandrov, G.V.Dolbilov, E.A.Perelstein et al.; JINR Communication 9-82-709, Dubna, 1982

[106] N.Yu.Kazarinov, E.A.Perelstein; Preprint JINR P9-12441, Dubna, 1979 and Sov. J. Phys. Tech. Phys., 25 (1980) 58

[107] N.Yu.Kazarinov, E.A.Perelstein, G.D.Shirkov; in: Proc. on Problems of Collective Acceleration Methods, JINR D9-82-664, Dubna 1982, p.62

[108] G.N.Vyalov, N.Yu.Kazarinov, E.A.Perelstein et al.; JINR Communication P9-11672, Dubna, 1978

[109] E.A.Perelstein, B.G.Shinov, G.D.Shirkov; Preprint JINR 11-84-505, Dubna, 1984 and Proc. of the Conf. on Numerical Methods and its Applications, Sofia 1985, p.471

[110] S.P.Christiansen, R.W.Hockny; Comp.Phys.Commun., 2 (1971) 139

[111] E.A.Perelstein, V.F.Shevtsov, G.D.Shirkov, B.G.Shinov; Preprint JINR P9-82-532, Dubna, 1982

[112] E.A.Perelstein, V.F.Shevtsov, G.D.Shirkov, B.G.Shinov; in: Proc.of the VIIIth National Conf. on Charged Particle Accelerations, Vol.II, Dubna, 1983, p.375

[113] V.I.Veksler; in: Proc. of Symposium CERN 1956, Vol.I, p.80

[114] W.H.Bennett; Physical Review, 45 (1934) 890

[115] U.Schuhmacher; Collective Ion Acceleration with Electron Rings, in: Springer Tracts of Modern Physics – Collective Ion Acceleration, Springer Verlag Berlin, Heidelberg, New York, 1979, p.145-231

[116] B.G.Gorinov et al.; JINR Communication 9-12148, Dubna, 1979

[117] N.C.Cristofilos; Phys.Rev.Lett., 22 (1969) 830 and IEEE Trans.Nucl.Sci., NS-16 (1969) 1039

[118] L.J.Laslett, A.M.Sessler; IEEE Trans.Nucl.Sci., NS-16 (1969) 1034

[119] V.G.Novikov, V.P.Sarantsev, Z.A. Ter-Martirosyan, B.A.Shestakov; in: Proc. of Conference on Problems of the Collective Acceleration Method, Dubna, JINR D9-82-664, 1982, p.27

[120] Yu.A.Bykovskii, V.P.Sarantsev, S.M.Silnov et al.; in: Proc. of Conference on Problems of the Collective Acceleration Method, Dubna, JINR D9-82-664, 1982, p.27

[121] G.V.Dolbilov et al.; JINR Communication P9-12414, Dubna, 1979

[122] V.P.Sarantsev, V.S.Alexandrov, L.S.Barabash et al.; JINR Communication P9-10917, Dubna, 1977

[123] G.V.Dolbilov, V.I.Mironov, V.G.Novikov et al.; JINR Communication P9-11191, Dubna, 1978

[124] V.S.Alexandrov, I.I.Golubev, G.V.Dolbilov et al.; in: Proc. of the XIIIth Int. Conf. on High Energy Physics, Vol.I, Novosibirsk, 1987, p.241

[125] H.Krauth; in: Proc. of the IVth Meeting on ERA, Garching 1971, Report IPP O/3, 1971, p.86

[126] R.H.Levy; in: Proc. of the Symposium on Electron Ring Accelerators, Lawrence Radiation Laboratory, Berkeley/Cal., Report UCRL-18103, 1968, p.318

[127] G.S.Janes; Report ERAN-17, Lawrence Radiation Laboratory, Berkeley/Cal., 1968

[128] L.J.Laslett, B.S.Levine; Report ERAN-202, Lawrence Radiation Laboratory, Berkeley/Cal., 1972

[129] G.S.Janes, R.H.Levy, H.A.Bethe, B.T.Feld; Physical Review, 145 (1966) 925

[130] E.Salzborn; IEEE Trans. Nucl. Sci., NS-23 (1976) 947

[131] N.Yu.Kazarinov, E.A.Perelstein, A.P.Sumbaev et al.; JINR Communication P9-81-428, Dubna, 1981

[132] N.Yu.Kazarinov, E.A.Perelstein, A.P.Sumbaev et al.; in: Proc. of the Conference on Problems of Collective Acceleration Methods, JINR, D9-82-664, Dubna 1982, p.16

[133] H.-U.Siebert, G.Zschornack et al.; Preprint JINR, P9-9366, Dubna 1975

[134] G.N.Kulipanov; Sov. Usp. Phys., 20 (1977) 559

[135] V.G.Novikov, E.A.Perelstein; Method of Highly Charged Ion Production, USSR Inventor No. 766384 (06.03.1979), Bull. OI 4, Moscow, 1983

[136] G.D.Shirkov; in: Proc. of the IInd Seminar on Experimental Physics of JINR Young Scientists, JINR P15-85-862, Dubna, 1985, p.22

[137] C.Pellegrini; Preprint UCRL-19815, Lawrence Radiation Laboratory, Berkeley/Cal., 1970 and Proc. of the 2nd National Conf. on Charged Particles Accelerators, Vol.I, Nauka Publ., Moscow, 1972

[138] V.S.Alexandrov, V.P.Sarantsev, G.D.Shirkov; Method of Production of Electron Rings with High Density and Device for its Realization, USSR Inventor No. 797537 (25.07.1979), Bull. OI 2, Moscow, 1981, p.255

[139] V.S.Alexandrov, E.A.Perelstein, V.P.Sarantsev, G.D.Shirkov; JINR Communication P9-81-20, Dubna, 1981

[140] E.A.Perelstein, G.D.Shirkov; in: Proc. of the Intern. Seminar on High Energy Ion-Atom Collision Processes, Debrecen/Hungary, 1981, p. 207

[141] E.A.Perelstein, G.D.Shirkov; Preprint JINR E9-88-238, Dubna, 1988 and in: Proc. of the 14th Summer School and Int. Symposium on the Physics of Ionized Gases, Sarajevo/Yugoslavia, 1988, p. 148

[142] S.M.Bijski et al.; JINR Communication P9-92-431, JINR Dubna, 1992

[143] I.V.Kuznetsov, E.A.Perelstein, V.N.Rasuvakin et al.; JINR Rapid Communication No. 16-86, Dubna, 1986, p. 33

[144] N.I.Azorski et al.; JINR Communication 9-88-224, Dubna, 1988

[145] V.S.Alexandrov et al.; JINR Communication P9-88-336, Dubna, 1988

[146] W.Wagner, V.B.Dunin, G.Karrasch et al.; JINR Rapid Communication No. 6-85, Dubna, 1985

[147] G.Zschornack, G.Musiol, M.Schiekel, W.Wagner; Soviet Journal Part.Nucl. (ECAYA), 21 (1990) 1000

[148] G.Zschornack, G.Musiol, G.Müller; JINR Communication P13-12542, Dubna, 1976

[149] E.D.Donets; USSR Inventor N 248860 (16.03.1967), Bull. OIPOTZ 23, 1969, p.65

[150] EBIT – Electron Beam Ion Trap: Selected Publications from the Electron Beam Ion Trap at Lawrence Livermore National Laboratory, ed. by E.Marrs, UCRL-ID-110491, LLNL, 1992

[151] M.P.Stockli; in: Proc. of the 5th Int. Conf. on the Physics of Highly Charged Ions, Giessen (Germany) 1990, in: Atomic Physics of Highly Charged Ions, ed. by E.Salzborn, P.H.Mokler and A.Müller, Springer Verlag, Berlin 1991, p.S111

[152] M.P.Stockli R.M.Ali, C.L.Cocke et al.; in: Proc. of the 4th Int.Conf. on Ion Sources, Bensheim(Germany), ed. by B.H.Wolf, in: Rev. of Sci. Instr., 63 (1992) 2822

[153] E.D.Donets, V.P.Ovsyannikov; Preprint JINR P7-80-515; Dubna 1980

[154] D.Schneider, M.W.Clark, B.M.Penetrante et al.; Physical Review, A44 (1991) 3119

[155] A.G.Bonch-Osmolovski; Preprint JINR P9-8379, Dubna 1974

[156] M.A.Levine, R.E.Marrs, R.W.Schmieder; Nuclear Instruments & Methods, A237 (1985) 429

[157] V.V.Afrosimov et al.; JETPh Letters, 34 (1981) 165

[158] E.D.Donets, G.Zschornack, G.A.Tutin, V.P.Eismont; JINR Communication P7-83-627, Dubna 1983

[159] E.D.Donets, S.V.Kartashov, V.P.Ovsyannikov; JINR Rapid Communications 20-86, Dubna 1986

[160] W.Wagner, E.D.Donets, V.B.Dunin, S.V.Kartashov; JINR Rapid Communications 4(24)-87, Dubna 1987, p.42

[161] E.D.Donets, V.G.Dudnikov, V.B.Dunin, S.V.Kartashov; in: Proc. of the 5th Int. Conf. on the Physics of Highly Charged Ions, Giessen(Germany), 1990 in: Atomic Physics of Highly Charged Ions, ed. by E.Salzborn, P.H.Mokler, A.Müller, Springer-Verlag, Berlin 1991, p.S337

[162] R.Becker, W.Schmidt, H.Klein; in: Proc. of the 1st Workshop on EBIS and Related Topics, GSI, 1977

[163] G.Shirkov, E.D.Donets, R.Becker, M.Kleinod; Preprint JINR E9-91-382, Dubna 1991; and Proc. of the 4th Int.Conf. on Ion Sources, Bensheim(Germany) 1991, ed. by B.H.Wolf, Review of Scint. Instruments, 63 No.4 (1992) 2819

[164] E.V.Bulyak, V.I.Kurilko; Preprint TsNII AtomInform, Moscow 1987

[165] B.M.Penetrante, M.A.Levine, J.N.Bardsley; in: Proc. of the Int.Symp. on Electron Beam Ion Sources and their Applications, ed. by A.Hershcovitch, AIP Conf. Proc. No.188, AIP, New York 1989, p.145

[166] M.B.Schneider, M.A.Levine, C.L.Bennett et al; in: Proc. of the Int.Symp. on Electron Beam Ion Sources and their Applications, ed. by A.Hershcovitch, AIP Conf.Proc. No.188, AIP, New York 1989, p.158

[167] R.Becker, W.Schmidt, M.Klein; in: Proc. of the 6th Int. Conf. on the Physics of Highly Charged Ions,

[168] R.E.Marrs, M.A.Levine, D.A.Knapp et al.; Phys. Rev. Lett., 60 (1988) 1715

[169] D.A.Knapp, R.E.Marrs, M.A.Levine et al.; Phys. Rev. Lett., 62 (1989) 2104

[170] C.M.Brown, U.Feldmann, G.A.Doschek et al.; Physical Review, A40 (1989) 4089

[171] P.Beiersdorfer, M.H.Chen, R.E.Marrs et al.; Physical Review, A41 (1990) 3453

[172] J.R.Henderson, P.Beiersdorfer, C.L.Bennett et al.; Phys. Rev. Lett., 65 (1990) 705

[173] P.Beiersdorfer, A.L.Osterheld, M.H.Chen et al.; Phys. Rev. Lett., 65 (1990) 1995

[174] D.Schneider, D.DeWitt, M.W.Clark et al.; Physical Review, A42 (1990) 3889

[175] J.W.McDonald, D.Schneider, M.W.Clark et al; Phys.Rev.Lett., 68 (1992) 2297

[176] B.M.Penetrante, J.N.Bardsley et al.; Physical Review, A43 (1991) 4861

[177] B.M.Penetrante, J.N.Bardsley, D.DeWitt et al.; Physical Review, A43 (1991) 4873

[178] R.E.Marrs, P.Beiersdorfer, C.Bennett et al.; in: Proc. of the Int. Symp. on Electron Beam Ion Sources and their Applications, ed.by A.Hershcovitch, AIP Conf. Proc. No.188, AIP, New York, 1989, p.445

[179] Y.S.Kim, P.H.Pratt; Physical Review, A27 (1983) 2913

[180] V.P.Pastukhov; in: Voprosy Teorii Plasmy, 1984, p.160 and Review of Plasma Physics, ed. by B.B.Kadomtsev, New York, London, 13 (1987) 203

[181] V.P.Pastukhov; Nucl.Fusion, 14 (1974) 3

[182] B.M.Penetrante, D.Schneider, R.E.Marrs et al.; Rev. Sci. Instruments, 63 (1992) 2806

[183] K.L.Wong, P.Beiersdorfer, D.Vogel et al.; Z.Phys. D(Suppl.), 21 (1991) S197

[184] R.E.Marrs, C.Bennett, M.h.Chen et al.; in: Proc. of the Int. Conf. on the Physics of Multiply Charged Ions, Grenoble, France, 1988; Journal de Physique, Tome 50 (1989) C1-445

[185] D.A.Knapp; in: Proc. of the 6th Int. Conf. on the Physics of Highly Charged Ions

[186] H.Postma; Physics Letters, 31A (1970) 196

[187] S.Bliman, R.Geller, W.Hess, B.Jaquot; IEEE Trans. Nucl. Sci., NS-19 (1972) 200

[188] K.Bernhardi, K.Wiesemann; Plasma Physics, 14 (1972) 1073

[189] P.Briand, R.Geller, B.Jaquot, C.Jaquot; Nuclear Instruments & Methods, 131 (1975) 407

[190] F.Bourd, R.Geller, B.Jaquot, M.Pontonnier; Proc. of the 4th Int. Workshop on ECR Ion Sources and Related Topics, Grenoble, 1982, Centre de Etudes Nucleaires Grenoble Press, p.51.

[191] Y.Jongen, C.M.Lyneis; in: Physics and Technology of Ion Sources, ed. by L.G.Brown, John Wiley Publ., New York, 1989, p.207

[192] M.P.Bourgarel, M.Bisch, J.Bony et al.; in: Proc. of the 4th Int. Conf. on Ion Sources, ed. by B.H.Wolf, Bensheim/Germany, 1991; Review of Sci. Instruments, 63 (1992) 2851

[193] Y.Jongen; in: Int. School for Young Scientists on Problems of Charged Particle Accelerators, JINR Dubna, D9-84-817, 1984, p.130

[194] T.A.Antaya; in: Proc. of the Int. Conf. on the Physics of Multiply Charged Ions, Grenoble, France, 1988; Journal de Physique, C1, 50 (1989) C1-707

[195] R.Geller; in: Proc. of the Int. Conf. on the Physics of Multiply Charged Ions, Grenoble, France, 1988, Journal de Physique, C1, 50 (1989) C1-887

[196] M.P.Bourgarel, M.Bisch, J.Bony et al.; in: Proc. of the 4th Int. Conf. on Ion Sources, Bensheim, Germany, 1991, ed. by B.H.Wolf, Review of Scientific Instruments, 63(No.4) (1992) 2851

[197] Y.Jongen; in: Int. School for Young Scientists on Problems of Charged Particles Accelerators, JINR Dubna, D9-84-817, Dubna, 1984, p.130

[198] R.Geller, F.Bourg, P.Briand et al.; in: Proc. of the Int. Conf. on ECR Ion Sources, NSLC-Report MSUCP-47, East Lansing, USA, 1987, p.1

[199] G.Melin, F.Bourg, P.Briand et al.; in: Proc. of the Int. Conf. on the Physics of Multiply Charged Ions, Grenoble, France, 1988; Journal de Physique, C1, 50 (1989) C1-673

[200] K.S.Golovanivsky; in: Int. School for Young Scientists on Problems of Charged Particle Accelerators, JINR Dubna, D9-84-817,, Dubna, 1984, p.145

[201] F.Bourg, P.Briand, J.Debernardi et al.; in: Contributed Papers of the 7th Workshop on ECR Ion Sources, ed. by H.Beuscher, Juelich/Germany, Jul-Conf-57, 1987, p.187

[202] C.Barue, P.Briand, A.Girard et al.; in: Proc. of the 4th Int. Conf. on Ion Sources, Bensheim/Germany, ed. by B.H.Wolf, Review of Scientific Instruments, 63(No.4) (1992) 2844

[203] R.H.Conen, M.E.Rensink, T.A.Cutler, A.A.Mirin; Nuclear Fusion, 18 (1978) 1229

[204] G.D.Shirkov; in: Proc. of the 4th Int. Conf. on Ion Sources, Bensheim/Germany, ed. by B.H.Wolf, 1991; Review of Scientific Instruments, 63(No.4) (1992) 2894

[205] G.D.Shirkov; Nuclear Instruments & Methods, A322 (1992) 161

[206] P.G.Akishin, A.A.Efremov, V.B.Kutner; JINR Communication P9-91-555, Dubna 1991

[207] V.B.Kutner, A.A.Evremov et al.; Proc. of the 3rd Int. Conf. on Accelerators, Kiev, 1991, p.90

[208] G.Melin, F.Bourg, P.Briand et al.; in: Proc. of the 3rd Int. Conf. on Ion Sources, 1989; Review of Sientific Instruments, 61(N0.1) (1990) 236

[209] A.G.Drentje; Nuclear Instruments & Methods, B9 (1985) 526

[210] U.Ratzinger, C.Mühle, W.Bleuel, G.Jöst, K.Leible, S.Schennach, B.H.Wolf; GSI-93-35, Darmstadt, 1993

[211] H.Beuscher et al.; in: Proc. of the 6th Int. ECR Ion Sources Workshop, LBL 5143, Livermore, 1985, p.107

[212] M.Mack, J.Haveman, R.Hoekstra, A.G.Drentje; in: Contributed Papers of the 7th Workshop on ECR Ion Sources, ed. by H.Beuscher, Juelich, Jul-Conf-57, 1986, p.153

[213] R.Baskaran, J.M.Heurtier, C.E.Hill; CERN PS Report 92-42(HI)

[214] P.Sortais; in: Proc. of the 4th Int. Conf. on Ion Sources, Bensheim/Germany, ed. by B.H.Wolf, 1991; Review of Scientific Instruments, 63(No.4) (1992) 2801

[215] K.Langbein; in: Notes of the Workshop on ECR Ion Source Problems, ed. by K.Langbein, PS/HI/Note 92-05, CERN, 1992

[216] A.G.Drentje; in: Proc. of the 4th Int. Conf. on Ion Sources, Bensheim/Germany 1991, ed. by B.H.Wolf, Rev.Sci.Instr., Vol.4, No.4 (1992) 2875

[217] G.Shirkov; in: Proc. of the 11th Int. Workshop on ECRIS, ed. by A.G.Drentje, KVI-Report 996, Groningen 1993, p. 67

[218] A.Girard; Rev.Sci.Instr. 63 (1992) 2676

[219] U.Ratzinger, C.Mühle, W.Blenel et al., Proc. of the 11th International Workshop on ECR Ion Sources, ed. by A.G.Drentje, KVI-Report 996, Groningen, 1993, p.201

[220] R.W.Schmieder; Physics of the EBIS and its Ions, NATO Workshop on Physics of Highly Ionized Ions, Cargese 1988

[221] E.Grimsehl; Lehrbuch der Physik: Struktur der Materie, Bd.4, Teubner Verlagsgesellschaft, Leipzig 1988

[222] R.W.P.McWhirter; Spectral Densities, in: Plasma Diagnostics Techniques, ed. by R.H.Huddlestone and S.L.Leonhard, Academic Press, New York 1965

[223] C.Breton and J.L.Schwob; Vacuum Ultraviolet Emission from Hot Plasmas, in: N.Damany, J.Romand, B.Vodar; Vacuum Ultraviolet Radiation Physics, Pergamon Press, New York 1974

[224] C.Barue, P.Briand, A,Girard, G.Melin, G.Brifford; Rev. Sci. Instr., 63 (1992) 2844

[225] S.A.Hokin, R.S.Post, D.L.Smatlak; Phys. Fluids, B1 (1989) 862

[226] N.Hershkowitz; in: Plasma Diagnostics Discharge Parameters an Chemistry, ed. by O.Auciello and D.I.Flamm, Academic Press, New York, 1989

[227] S.Yu.Dukyanov; Hot Plasma and Controlled Nuclear Fusion (in Russian), Nauka, Moscow, 1975

[228] D.Bohm; The Characteristics of Electrical Discharges in Magnetic Fields, ed. by A.Guthrie and R.K.Wakerling, McGraw Hill, New York, 1949

[229] M.Tamba and H.Amemiya; Rev. Sci. Instruments, 61 (1990) 247

[230] J.Asmussen, R.Fritz, L.Mahoney, G.Fournier, G.Demaggio; Rev. Sci. Instruments, 61 (1990) 282

[231] L.Mahoney and J.Asmussen; Rev. Sci. Instruments, 61 (1990) 285

[232] K.N.Leving, K.W.Ehlers, R.V.Pyle; Rev. Sci. Instruments, 56 (1985) 364

[233] C.A.Anderson, M.B.Hopkins, W.G.Graham; Rev. Sci. Instruments, 61 (1990) 448

[234] H.Schulte and B.H.Wolf; IEEE Trans. Nucl. Sci., NS-23 (1976) 1061

[235] J.Asmussen, J.Hopwood, F.C.Sze; Rev. Sci. Instruments, 61 (1990) 250

[236] J.C.Dooling and K.G.Moses; Rev. Sci. Instruments, 61 (1990) 421

[237] M.Schneider, K.Wiesemann; SFB-Report 85-05-139, Bochum 1985

[238] M.Kuelshammer; SFB-Report 90-A7-04, Bochum 1990

[239] M.Mausbach, K.Wiesemann; Plasma Physics, 32 (1990) 281

[240] H.Koehler, M.Frank, B.A.Huber, K.Wiesemann; 7th Workshop on ECR Ion Sources, Juelich Report No. ISSN 0344-5798, 1986, p.215

[241] F.W.Meyer, J.W.Hale; Rev. Sci. Instruments, 61(1990)324

[242] H.Koehler, M.Frank, B.A.Huber, K.Wiesemann; Nucl. Instr. Methods, B23 (1987) 186

[243] K.Wiesemann; J.Phys., 14 (1981) 1404

[244] K.Volk, M.Sarstedt, H.Klein, A.Schempp; Rev. Sci. Instruments, 61 (1990) 493

[245] G.Gautherin, F.Lejeune, F.Prange et al.; Plasma Physics, 11 (1969) 397

[246] P.Strehl; GSI-Preprint GSI-94-27, Darmstadt, April 1994

[247] C.Sanborn; Basic Data of Plasma Physics, Massachussetts, MIT Press, 1966

[248] D.Jamba; Rev. Sci. Instruments, 49 (1978) 634

[249] C.M.McKenna; High Current Dosimetry Techniques, Radiation Effects, 44 (1979) 93, Gordon and Breach Science Publishers, Inc., Holland

[250] G.Carter, J.Colligon; Ion Bombardement of Solids, Heinemann Educational Books Ltd., London, 1969, p.310

[251] K.B.Unser; Atomenergie-Kerntechnik, 47 (1985) 48

[252] H.W.Fulbright; Focal Plane Detectors for Magnetic Spectrometers, in: Treatise on Heavy-Ion Science, Vol.7:Instrumentation and Techniques, ed. by D.A.Bromley, Plenum Press, New York and London, 1985, p.179

[253] F.Kohlrausch; Praktische Physik, B.G.Teubner Verlag, Stuttgart, 1968, p.609

[254] J.H.Moore, C.C.Davis, M.A.Coplan; Building Scientific Apparatus, Addison-Wesley Publishing Company, London a.o., 1983, p.315

[255] G.Hertz; Lehrbuch der Kernphysik, Band I, B.G.Teubner Verlagsgesellschaft, Leipzig, 1966, p.80

[256] H.A.Enge; Deflecting Magnets, in: Focussing of Charged Particles, Vol.II, ed. by A.Septier, Academic Press, New York and London, 1967

[257] H.Tyrroff, H.Wirth, G.Zschornack; Proc. of 15th Workshop "Energiereiche atomare Stösse", Riezlern/Austria, January 1994, p.176

[258] M.A.Heald, C.B.Wharton; Plasma Diagnostics with Microwaves, Wiley, New York, 1965

[259] R.C.Garner, M.E.Mauel, S.A.Hokin, R.S.Post, D.L.Smatlak; Physical Review Letters, 59 (1987) 1821

[260] R.A.Blanken, N.H.Lazar; Phys.Fluids, 13 (1970) 2752

[261] H.Ikegami, H.Ikezi, M.Hosokawa, K.Takayama, S.Tanaky; Phys. Fluids, 11 (1968) 1061

[262] R.A.James, R.F.Ellis, C.J.Lasnier; Bull. Am. Phys. Soc., 29 (1984) 1187

[263] J.H.Booske, W.D.Getty, R.M.Gilgenbach; Phys. Fluids, 28 (1985) 3116

[264] J.C.Brown; I.A.U. Symp., 68 (1975) 261

[265] A.N.Tikhonov, V.V.Alikhaev, V.Ya.Arsenin, A.S.Dumova; Sov. Phys. JETP, 28 (1969) 1006

[266] K.Bernhardi; Comp. Phys. Communications, 19 (1980) 17

[267] P.Hoyng, G.A.Stevens; Astrophys. A. Space S., 27 (1974) 307

[268] R.Friedlein, G.Zschornack; Nuclear Instruments and Methods, A349 (1994) 554

[269] R.H.Pratt et al.; Physical Review, 16 (1977) 2169

[270] H.W.Koch, J.W.Motz; Review of Modern Physics, 31 (1959) 920

[271] L.Kissel et al.; Atomic Data and Nuclear Data Tables, 28 (1983) 384

[272] N.B.Avdonina, R.H.Pratt; J. Quant. Spectrosc. Radiation Transfer, 50 (1993) 349

[273] M.Lamoureux, L.Jacquet, R.H.Pratt; Physical Review, 39 (1989) 6323

[274] K.Bernhardi, K.Wiesemann; Plasma Physics, 24 (1982) 867

[275] P.Kosmol; Methoden zur Behandlung nichtlinearer Gleichungen und Optimierungsaufgaben, Teubner Studienbcher, Stuttgart, 1993

[276] A.J.Lichtenberg, M.T.Schwarz, M.A.Lieberman; Plasma Physics, 13 (1971) 89

[277] J.L.Shohet, S.J.Gitomer; Phys. Fluids, 10 (1967) 1359

[278] P.A.Rainbault, J.L.Shohet; Phys. Fluids, 15 (1972) 1477

[279] N.Abe, T.Yamamoto, M.Kawanishi; Japanese Journal of Applied Physics, 19 (1980) 1147

[280] R.D.Evans; The Atomic Nucleus, McGraw Hill, New York, 1955, Chap.23

[281] U.Fano Jr.; J. Opt. Soc. Am., 39 (1949) 859

[282] E.G.Harris; Plasma Physics, 2 (1961) 138

[283] R.L.Gluckstern, M.H.Hull Jr., G.Breit; Physical Review, 90 (1953) 1026

[284] R.L.Gluckstern, M.H.Hull Jr.; Physical Review, 90 (1953) 1930

[285] A.C.England, G.R.Haste; Physical Review, A7 (1973) 383

[286] J.P.Briand, G.Ban, P.Briand, R.Geller, G.Melin, C.Hill, H.Haseroth; Proc. of the 11th Int. Workshop on ECR Ion Sources, Groningen, edited by A.G.Drentje, 1993, KVI Report 996,

[287] S.Herpich, H.Hiller, H.Tyrroff, G.Zschornack; 15. Arbeitsbericht "Energiereiche atomare Stösse", EAS-15, Darmstadt, 1994, p.19

[288] I.P.Vinogradov, J.Gehring, B.Jettkant, D.Meyer, K.Wiesemann; Proc. of the 11th Int. Workshop on ECR Ion Sources, Groningen, edited by A.G.Drentje, 1993, KVI Report 996, p.57

[289] D.Hitz, M.Druetta, S.Khardi; Rev. Sci. Instrum., 63 (1992) 2889

[290] W.Pöffel, K.H.Schartner, G.Mank, E.Salzborn; Rev. Sci. Instrum., 61 (1990) 613

[291] W.H.Tait; Radiation Detection, Butterworths, London a.o., 1980

[292] N.Tsoulfanidis; Measurement and Detection of Radiation, Hemisphere Publishing Corporation, New York a.o., 1981

[293] K.Kleinknecht; Detektoren für Teilchenstrahlung, B.G.Teubner Verlag, Stuttgart, 1984

[294] B.K.Agarwal; X-Ray Spectroscopy, Springer Verlag, Heidelberg a.o., 1979

[295] W.R.:Leo; Techniques for Nuclear and Particle Physics Experiments, Springer Verlag, Berlin a.o., 1987

[296] L.Kissel, C.A.Quarles, R.H.Pratt; Atomic Data and Nuclear Data Tables, 28 (1983) 381

[297] G.D.Shirkov; CERN/PS 94-33(HI), Geneva, 1994

[298] C.Zippe; Doctor Thesis, TU Dresden, Fakultät für Mathematik und Naturwissenschaften, Dresden, 1996

[299] D.Küchler; Diploma Thesis, Institut f. Kern- und Teilchenphysik der TU Dresden, Dresden, 1994

[300] J.H.Scofield; Physical Review, A9 (1974) 1041

[301] M.Lamoureux; Resourcefulness of the Bremsstrahlung Spectra to Investigate the Energy Distributions of the Electrons in ECRIS Plasmas, 12th Int. Workshop on ECR Ion Sources, Riken, Japan (1995)

[302] W.Barnitzke et al.; 11th Workshop on ECR Ion Sources, KVI Report 996, ed. by A.G.Drentje, Groningen, 1993, p.249

[303] R.Friedlein, S.Herpich, U.Lehnert. H.Tyrroff, H.Wirth, C.Zippe, G.Zschornack; Nuclear Instruments & Methods, B98 (1995) 585

[304] L.Friedrich, E.Huttel, R.Hentschel, H.Tyrroff; 11th Workshop on ECR Ion Sources, KVI Report 996, ed. by A.G.Drentje, Groningen (1993), p.19

[305] R.Friedlein, S.Herpich, H.Tyrroff, C.Zippe, G.Zschornack; Phys. Plasmas, 2 (1995) 2183

[306] M.A.Levine, R.E.Marrs, J.N.Bardsley, P.Beiersdorfer, C.L.Bennett, M.H.Chen, T.Cowan, D.Dietrich, J.R.Henderson, D.A.Knapp, A.Osterheld, B.M.Penetrante, M.B.Schneider, J.H.Scofield; Nuclear Instruments & Methods, B43 (1989) 431

[307] X.Chen, B.G.Lane, D.L.Smatlak, R.S.Post, S.A.Hopkin; Phys. Fluids, B1 (1989) 615

[308] W.Laux; GSI Report 93-35, Darmstadt (1993)

[309] U.Lehnert, G.Zschornack; X-Ray Imaging Spectroscopy of an ECR Discharge, 6th Int. Conference on Ion Sources, September 10-16, Whistler (1995)

[310] P.Acton, G.Agnew, R.Cotton, S.Hedges, A.K.McKemey, T.Roy, S.J.Watts, C.J.S.Damerell, R.L.English, A.R.Gillman, A.L.Lintern, D.Su, F.J.Wickens; Nuclear Instruments & Methods, A305 (1991) 504

[311] D.H.Lumb, A.D.Holland; SPIE Vol. 982, X-Ray Instrumentation in Astronomy II, 1988, p.116

[312] R.H.Huddlestone, S.L.Leonhardt; Plasma Diagnostics Techniques, Academic Press, New York, 1965

[313] G.V.Marr; Plasma Spectroscopy; Elsevier Publ. Comp., New York, 1968

[314] W.Lochte-Holtgreven; Plasma Diagnostics, North-Holland Publ. Comp., Amsterdam, 1968

[315] W.Pöffel; Doctor Thesis, University Giessen, Giessen, 1990

[316] E.H.Marlinghaus; SFB-Report 85-05-137, University Bochum, Bochum, 1985

[317] E.Marlinghaus; Doctor Thesis, University Bochum, Bochum, 1984

[318] M.Jogwich, B.Huber, K.Wiesemann; Z.Phys., D17 (1990) 171

[319] H.Van Regemorter; Astrophys. J., 136 (1962) 906

[320] Y.Itikawa et al, J. Phys. Chem. Ref. Data, 15 (1986) 985

[321] J.G.Doyle, P.L.Dufton, F.P.Keenan, A.E.Kingston; Solar Physics, 89 (1983) 243

[322] F.P.Keenan, K.A.Berrington, P.G.Burke, A.E.Kingston, P.L.Dufton; Mont. Not. R. Astr. Soc., 207 (1984) 459

[323] P.L.Dufton, K.A.Berrington, P.G.Burke, A.E.Kingston; Astron. Astrophys., 62 (1978) 111

[324] F.P.Keenan, K.A.Berrington; Solar Physics, 99 (1985) 25

[325] K.A.Berrington; J. Phys.; B18 (1985) L395

[326] W.L.Wiese, M.W.Smith, B.M.Glennon; Atomic Transition Probabilities, Vol.I, Hydrogen through Neon, NSRDS-NBS 4, Washington, 1966

[327] H.Nussbaumer, P.J.Storey; Astron. Astrophys., 64 (1987) 139

[328] U.Birke; Diploma Thesis, Fachbereich Physik, Technische Universität Dresden, Dresden, 1991

[329] W.A.Abramov, E.I.Kusnetsov, W.I.Kogan; Atomnaya Energija, 26 (1969) 516

[330] W.A.Abramov, E.I.Kusnetsov, W.I.Kogan; Atomnaya Energija, 33 (1972) 845

[331] T.Bouchama, M.Druetta; Nucl. Instrum. Methods, B40-41 (1989) 1252

[332] G.C.Tisone, K.W.Bieg, P.L.Dreike; Rev. Sci. Instrum., 61 (1990) 562

[333] R.Pal, D.Hammer; Phys. Rev. Letters, 50 (1983) 732

[334] E.Salpiter; Physical Review, 120 (1960) 1528

[335] E.Salpiter; Physical Review, 122 (1961) 1663

[336] R.E.Marrs, C.L.Bennett, M.H.Chen, T.Cowan, D.Dietrich, J.R.Henderson, D.A.Knapp, M.A.Levine, M.B.Schneider, J.H.Scofield; Preprint UCRL-99699, Livermore 1988

[337] M.A.Levine, R.E.Marrs, J.R.Henderson, D.A.Knapp, M.B.Schneider; Physica Scripta, T22 (1988) 157

[338] L.A.Levine, R.E.Marrs, J.N.Bardsley, P.Beiersdorfer, C.L.Bennett, M.H.Chen, T.Cowan, D.Dietrich, J.R.Henderson, D.A.Knapp, A.Osterheld, B.M.Penetrante, M.B.Schneider, J.H.Scofield; Nucl. Instrum. Methods, B43 (1989) 431

[339] E.B.Saloman, J.H.Hubbel, J.H.Scofield; Atomic Data Nucl. Data Tables, 38 (1988) 1

[340] G.J.Herrmann; J. Appl. Physics, 29 (1958) 127

[341] D.A.Knapp, R.E.Marrs, M.A.Levine, C.L.Bennett, M.H.Chen, J.R.Henderson, M.B.Schneider, J.H.Scofield; Phys. Rev. Letters, 62 (1989) 2104

[342] H.S.W.Massey, D.R.Bates; Rep. Progr. Physics, 9 (1942) 62

[343] J.B.A.Mitchell, C.T.Ngo, J.L.Forand, D.P.Levac, R.E.Mitchell, A.Sen, D.B.Miko, J.W.Mc-Gowan; Phys. Rev. Letters, 50 (1983) 335

[344] D.S.Belic, G.H.Dunn, T.J.Morgan, D.W.Mueller, C.Timmer; Phys. Rev. Letters, 50 (1983) 339

[345] P.F.Dittner, S.Datz, S.D.Miller, C.D.Moak, P.H.Stelson, C.Bottcher, W.P.Dress, G.D.Aston, N.Nesdkovic, C.M.Fou; Phys. Rev. Letters, 51 (1983) 31

[346] D.R.DeWitt, D.Schneider, M.Chen, M.Schneider, D.Church, G.Weinberg, M.Sakurai; PRA 47, 3 (1993) R1597

[347] J.P.Briand, P.Charles, J.Arianer, H.Laurent, C.Goldstein, J.Dubau, M.Loulergue, F.Bely-Dubau; Phys. Rev. Letters, 52 (1984) 617

[348] R.Ali, C.P.Bhalla, C.L.Cocke, M.Schulz, M.Stockli; Physical Review, A44 (1991) 223

[349] J.Arianer, A.Cabrespine, C.Goldstein, T.Junquera, A.Courtois, G.Deschamps, M.Oliver; Nucl. Instrum. Methods, 198 (1982) 175

[350] B.Feinberg, I.Brown; Proc. of the 2nd EBIS Workshop, Orsay, 1981, p.1

[351] C.Goldstein, et al.; Orsay Report, IPNO 76-07, 1976

[352] M.Malard, M.Oliver; Proc. of the 2nd EBIS Workshop, Orsay, 1981, p.137

[353] P.Strehl; Rev.Sci. Instrum., 63 (1992) 2652

[354] Y.Yamyshita,, Y.Ysoya, M.Sekiguchi; Proc. of the 11th Int. Workshop on ECR Ion Sources, ed. by A.G.Drentje, Groningen, May 6-7, 1993, p.240

[355] W.M.Law, C.D.P.Levy, P.W.Schmor, J.Uegaki; Contributed Papers of the 7th Workshop on ECR Ion Sources, ed. by H.Beuscher, Jül-Conf-57, Jülich, 22-23 May, 1986, p.258

[356] J.Arianer et al.; Part. Accel. Conf., San Franscisco (1979), publ. in IEEE Trans. on Nucl. Sci., NS-26 (1979) 3713

[357] M.P.Stockli, C.L.Cocke, P.Richard; Rev.Sci.Instrum., 61 (1990) 242

[358] H.Klein; Heavy Ion Accelerators – Past, Present and Future, Report GSI-86-19, Darmstadt, 1986, p.1

[359] R.E.Marrs, S.R.Elliott, Th.Stöhlker; EBIT Annual Report 1993, UCRL-ID-118274, ed. by D.Schneider, Lawrence Livermore National Laboratory, August 1994, p.4

[360] P.Beiersdorfer, D.Knapp, R.E.Marrs, S.R.Elliott, M.H.Chen; Phys. Rev. Letters, 71(1993)3939

[361] P.Beiersdorfer, D.Knapp, R.E.Marrs, S.R.Elliott, M.H.Chen; EBIT Annual Report 1993, UCRL-ID-118274, ed. by D.Schneider, LLNL, Livermore, August 1994, p.5

[362] S.A.Blundell; Physical Review, A46 (1992) 3762

[363] P.J.Mohr; Physical Review, A26 (1982) 2338

[364] W.R.Johnson, G.Soff; Atomic Data and Nuclear Data Tables, 33 (1985) 405

[365] B.J.McKenzie, I.P.Grant, P.H.Norrington; Comp. Phys. Commun., 21 (1980) 233

[366] M.H.Chen, B.Crasemann, M.Aoyagi, K.N.Huang, H.Mark; Atomic Data and Nuclear Data Tables, 26 (1981) 561

[367] P.Beiersdorfer, M.Bitter, S. von Goeler, S.Cohen, K.W.Hill, J.Timberlake, R.S.Walling, M.H.Chen, P.L.Hagelstein, J.H.Scofield; Physical Review, A34 (1986) 1297

[368] P.Beiersdorfer, S. von Goeler, M.Bitter, E.Hinnov, R.Bell, S.Bernabei, J.Felt, K.W.Hill, R.Hulse, J.Stevens, S.Suckewer, J.Timberlake, A.Wouters, M.H.Chen, J.H.Scofield, D.D.Dietrich, M.Gerassimenko, E.Silver, R.S.Walling, P.L.Hagelstein; Physical Review, A37 (1988) 4153

[369] D.D.Dietrich, G.A.Chandler, P.O.Egan, K.P.Ziock, P.H. Mokler, S.Reusch, D.H.H.Hoffmann; Nucl. Instr. Methods, B 24/25 (1987) 301

[370] G.A.Chandler, M.H.Chen, D.D.Dietrich, P.O.Egan, K.P.Ziock, P.H.Mokler, S.Reusch, D.H.H.Hoffmann; Physical Review, A39 (1989) 565

[371] M.Loulergue, H.Nussbaumer; Astron.Astrophys., 24 (1973) 209

[372] B.K.F.Young, A.L.Osterheld, R.S.Walling, W.H.Goldstein, T.W.Phillips, R.E.Stewart, G.Charatis, G.E.Busch; Phys. Rev. Letters, 62 (1989) 1266

[373] I.P.Grant, B.J.McKenzie, P.H.Norrington, D.F.Mayers, N.C.Pyper; Comp. Phys. Commun., 21 (1980) 207

[374] P.Beiersdorfer, M.Bitter, S.von Goeler, S.Cohen, K.W.Hill, J.Timberlake, R.S.Walling, M.H.Chen, P.L.Hagelstein, J.H.Scofield; Physical Review, A37 (1986) 1297

[375] E.Träbert; Nucl. Instr. Methods, B31 (1988) 233

[376] E.Träbert; Z.Phys., 9 (1988) 143

[377] R.Hutton, L.Engström, E.Träbert; Phys. Rev. Letters, 60 (1988) 2469

[378] R.W.Dunford, M.Haas, E.Bakke, H.G.Berry, C.J.Lui, M.L.A. Raphaelian, L.J.Curtis; Phys. Rev. Letters, 62 (1989) 2809

[379] R.Marrus, P.Charles, P.Indelicato, L. de Billy, C.Tazi, J.-P.Briand, A.Simionovici, D.D.Dietrich, F.Bosch, D.Liesen; Physical Review, A39 (1989) 3725

[380] P.Beiersdorfer, M.H.Chen, S.MacLaren, R.E.Marrs, D.A.Vogel, K.Wong, R.Zasadzinski; Physical Review, A44 (1991) 4730

[381] F.Bely-Dubau, A.H.Gabriel, S.Volonté; Mon. Not. R. Astron. Soc., 186 (1979) 405

[382] M.Bitter, K.W.Hill, N.R.Sauthoff, P.C.Efthimion, E.Merservey, W.Roney, S. von Goeler, r.Horton, M.Goldman, W.Stodiek; Phys. Rev. Letters, 43 (1979) 129

[383] TFR Group, J.Dubau, M.Loulergue; J.Phys., B15 (1982) 1007

[384] F.Bombarda, R.Gianella, E.Källne, G.J.Tallents, F.Bely-Dubau, P.Faucher, M.Cornille, J.Dubau, A.H.Gabriel; Physical Review, A37 (1988) 504

[385] A.H.Gabriel; Mon. Not. R. Astronom. Soc., 160 (1972) 99

[386] J.F.Seely, U.Feldman, U.I.Safronova; Astrophys. J., 304 (1986) 838

[387] P.Beiersdorfer, R.E.Marrs, J.R.Henderson, D.A.Knapp, M.A.Levine, D.B.Platt, M.B. Schneider, D.A.Vogel, K.L.Wong; Rev. Sci. Instrum., 61 (1990) 2338

[388] P.Beiersdorfer, M.H.Chen, R.E.Marrs, M.B.Schneider, R.S.Walling; Physical Review, A44 (1991) 396

[389] J.Nilsen; Atomic Data and Nuclear Data Tables, 38 (1988) 339

[390] L.A.Vainshtein, U.I.Safronova; Atomic Data and Nuclear Data Tables, 25 (1978) 49

[391] M.Klapisch et al.; Phys. Rev. Letters; 41 (1978) 403

[392] E.Källne, J.Källne, J.E.Rice; Phys. Rev. Letters, 49 (1982) 330

[393] A.H.Gabriel, C.Jordan; Nature(London), 221 (1969) 947

[394] H.R.Griem; Astrophys. J., 156 (1969) L103

[395] A.H.Gabriel, C.Jordan; Phys. Lett., 32A (1970) 166

[396] J.-C.Gauthier et al.; J. Phys., B19 (1986) L385

[397] J.-F.Wyart et al.; Physical Review, A34 (1986) 701

[398] C.L.Cocke et al.; Physical Review, A12 (1975) 2413

[399] D.D.Dietrich et al.; Physical Review, 54 (1985) 1008

[400] P.H.Mokler et al.; Phys. Rev. Letters, 65 (1990) 3108

[401] P.Beiersdorfer, A.L.Osterheld, J.Scofield, B.Wargelin, R.E.Marrs; Phys. Rev. Letters, 67 (1991) 2272

[402] A-Bar-Shalom, M.Klapisch, J.Oreg; Physical Review, A38 (1988) 1773

[403] P.Beiersdorfer, M.H.Chen, R.E.Marrs; Physical Review, A41 (1990) 3453

[404] D.A.Knapp, R.E.Marrs, M.A.Levine, C.L.Bennett, M.H.Chen, J.R.Henderson, M.B.Schneider, J.H.Scofield; Phys. Rev. Letters, 62 (1989) 2104

[405] N.K.Del Grande, P.Beiersdorfer, J.R.Henderson, A.L.Osterheld, J.H.Scofield; Nucl. Instr. Methods, B56/57 (1991) 227

[406] S.VonGoeler, P.Beiersdorfer, M.Bitter, R.Bell, K.Hill, P.LaSalle, L.Ratzan, J.Stevens, J.Timberlake, S.Maxon, J.H.Scofield; J.Phys. (Paris), 49 (1988) C1-181

[407] P.Burkhalter, D.Nagel, R.Whitlock; Physical Review, A9 (1974) 2331

[408] M.Klapisch, A.Bar-Shalom, P.Mandelbaum, J.L.Schwob, A.Zigler, H.Zmora, S.Jackel; Physics Letters, 79A (1980) 67

[409] C.Bauch-Arnoult et al.; Nucl. Instr. Methods, B31 (1988) 153

[410] B.J.Wargelin, P.Beiersdorfer, S.M.Kahn; Phys. Rev. Letters, 71 (1993) 2196

[411] B.J.Wargelin, P.Beiersdorfer, S.M.Kahn; EBIT Annual Report 1993, UCRL-ID-118274, ed. by D.Schneider, LLNL, Livermore, August 1994, p.9

[412] J.D.Silver et al; Rev. Sci. Instrum., 65 (1994) 1072

[413] K.Widmann, P.Beiersdorfer, V.Decaux, S.Elliott, M.Bitter, A.Smith; EBIT Annual Report 1993, UCRL-ID-118274, ed. by D.Schneider, LLNL, Livermore, August 1994, p.16

[414] M.Bitter, H.Hsuan, K.W.Hill, R.Hulse, M.Zarnstorff, P.Beiersdorfer; in: Atomic and Plasma
       Material Interaction Processes in Controlled Thermonuclear Fusion, ed. by R.K.Janev and
       H.W.Darwin, Elsevier Science Publishers B.V., 1993

[415] M.Bitter, H.Hsuan, C.Bush, S.Coen, C.J.Cummings, B.Grek, K.W.Hill, J.Schivell, M.
       Zarnstorff, P.Beiersdorfer, A.Osterheld, A.Smith, B.Fraenkel; Phys. Rev. Letters, 71 (1993)
       1007

[416] R.Mewe, J.Schrijver; Astron. Astrophys., 87 (1980) 261

[417] V.Decaux, P.Beiersdorfer; EBIT Annual Report 1993, UCRL-ID-118274, ed. by D.Schneider,
       LLNL, Livermore, August 1994, p.14

[418] V.Decaux, P.Beiersdorfer; Physica Scripta, T47 (1993) 80

[419] D.Schneider, D.DeWitt, M.W.Clark, R.Schuch, C.L.Cocke, R.Schmieder, K.J.Reed, M.H.Chen,
       R.E.Marrs, M.Levine, R.Fortner; Physical Review, A42 (1990) 3889

[420] E.D.Donets, Physica Scripta, T3 (1983) 11

[421] J.Andrä; Nucl. Instr. Methods, B43 (1989) 3 and B43 (1989) 306

[422] K.Karim, C.P.Bhalla; Phys. Scripta, 38 (1988) 795

[423] P.Beiersdorfer, S. von Goeler, M.Bitter, E. Hinnov, R.Bell, S.Bernabei, J.Felt, K.W.Hill, R.
       Hulse, J.Stevens, S.Suckewer, J.Timberlake, A.Wouters, M.H.Chen, J.H.Scofield, D.D.Dietrich,
       M.Gerassimenko, E.Silver, R.S.Walling, P.L.Hagelstein; Physical Review, A37 (1988) 37

[424] H.J.Andrä, A.Simionovici, T.Lamy, A.Brenac, G.Lamboley, A.Pesnelle, S.Andriamonje,
       A.Fleury, M.Bonnefoy, M.Chassevent, J.J.Bonnett; Book of Invited Papers, ICPEAC XVII,
       Brisbane, July 1991

[425] J.P.Briand, L. de Billy, P.Charles, S.Essabaa, P.Briand, R.Geller, J.P.Desclaux, S.Bliman,
       C.Ristori; Phys. Rev. Letters, 65 (1990) 159

[426] L. Ley et al.; in: Photoemission in Solids II, ed. by L.Ley and M.Cardona, Springer Verlag,
       1979

[427] G.Zschornack, G.Musiol, W.Wagner; Physica Scripta, T3 (1983) 194

[428] M.Schulz, C.L.Cocke, M.Stöckli, S.Hagmann, H.Schmidt-Böcking; Supplement to Z.Phys. D,
       21 (1991) 341

[429] E.D.Donets; Nucl. Instr. Methods, B9 (1985) 522

[430] M.Delauny, M.Fehringer, R.Geller, D.Hitz, P.Varga, H.Winter; Physical Review, B35 (1987)
       4232

[431] S.T. de Zwart; Nucl. Instr. Methods, B23 (1987) 239

[432] F.W.Meyer; J. de Phys., 50 C1 (1989) 263

[433] J.W.McDonald, D.Schneider, M.W.Clark, D.Dewitt; Physical Review Letters, 68 (1992) 2297

[434] R.Köhrbrück, D.Lecler, F.Fremont, P.Roncin, K.Sommer, T.J.M.Zourus, J.Black-Neuhaus,
       N.Stolterfoht; Nucl. Instr. Methods, B56 (1991) 219

[435] F.W.Meyer, C.C.Havener, K.J.Snowden, S.H.Overbury, D.M.Zehner, W.Heiland; Physical Re-
       view, A35 (1987) 3176

[436] F.W.Meyer; Physical Review Letters, 67 (1991) 723

[437] D.Schneider, J.W.McDonald, M.A.Briere, F.Aumayr, H.Kurz, H.P.Winter; EBIT Annual Re-
       port 1993, UCRL-ID-118274, ed. by D.Schneider, LLNL, Livermore, August 1994, p.24

[438] J.Burgdörfer, F.W.Meyer; Physical Review, A47 (1993) R20

[439] N.Bradsley, B.Penetrante; Com. At. Mol. Phys., 27 (1991) 43

[440] D.Schneider, M.A.Briere, C.Cunningham, J.McDonald, F.Aumayr, H.Kurz, H.P.Winter; EBIT
       Annual Report 1993, UCRL-ID-118274, ed. by D.Schneider, LLNL, Livermore, August 1994,
       p.25

[441] J.Burgdörfer, P.Lerner, F.W.Meyer; Physical Review, A44 (1991) 5674

[442] D.Schneider, R.Schuch, D.A.Knapp, D.DeWitt, J.McDonald, M.H.Chen, M.W.Clark, R.E.Marrs; EBIT Annual Manual 1993, UCRL-ID-118274, ed. by D.Schneider, LLNL, Livermore, August 1994, p. 27

[443] R.Schuch, D.Schneider, M.Clark, D.DeWitt, M.Chen, D.Knapp, R.Marrs; Physical Review Letters, 70 (1993) 1073

[444] M.A.Briere, G.Schiwietz, J.McDonald, D.Schneider; EBIT Annual Manual 1993, UCRL-ID-118274, ed. by D.Schneider, LLNL, Livermore, August 1994, p.28

[445] D.Schneider, G.Schiwietz, M.Clark, B.Skogvall, D.DeWitt, J.McDonald, EBIT Annual Manual 1993, UCRL-ID-118274, ed. by D.Schneider, LLNL, Livermore, August 1994, p.38

[446] G.Schiwietz, D.Schneider, B.Skogvall, M.Clark, D.DeWitt; Radiation Effects and Defects in Solids, Vol.1278, No.1 , 1993

[447] M.A.Briere, D.Schneider, M.Reaves, J.McDonald; EBIT Annual Manual 1993, UCRL-ID-118274, ed. by D.Schneider, August 1994, LLNL, Livermore, p.30

[448] D.Schneider, M.A.Briere, W.Siekhaus, J.McDonald; EBIT Annual Manual 1993, UCRL-ID-118274, ed. by D.Schneider, August 1994, LLNL, Livermore, p.40

[449] J.McDonald, D.Schneider, M.Clark, D.DeWitt; Physical Review Letters, 46 (1992) 2297

[450] D.Schneider, M.A.Briere, M.W.Clark, J.McDonald, J.Biersack, W.Siekhaus; Surface Science, 294 (1993) 403

[451] M.A.Briere, J.P.Biersack, D.Schneider; EBIT Annual Manual 1993, UCRL-ID-118274, ed. by D.Schneider, August 1994, LLNL, Livermore, p. 33

[452] J.P.Biersack; Proc. of the 8th Int. Conference on Ion Beam Modifications of Materials, 1992, in: Nucl. Instrum. & Methods, B80/81 (1993) 12

[453] E.B.Saloman, J.H.Hubbell, J.H.Scofield; Atomic Data and Nuclear Data Tables, 38 (1988) 1

[454] H.Zhang, D.H.Sampson, R.E.H.Clark, J.B.Mann; Atomic Data and Nuclear Data Tables, 37 (1987) 17

[455] K.J.Reed; Physical Review, A37 (1988) 1791

[456] P.Beiersdorfer, R.Cauble, S.Chantrenne, M.Chen, D.Knapp, R.Marrs, T.Phillips, K.Reed, M.Schneider, J.Scofield, K.Wong, D.Vogel, R.Zasadzinski; EBIT: Selected Publications from the Electron Beam Ion Trap Program at Lawrence Livermore National Laboratory, ed. by R.E.Marrs,UCRL-ID-110491, LLNL, Livermore, May 1992, EBIT-236

[457] K.L.Wong, P.Beiersdorfer, K.J.Reed, D.A.Vogel; Physical Review, A51 (1995) 1214

[458] K.L.Wong, P.Beiersdorfer, M.H.Chen, R.E,Marrs, K.J.Reed, J.H.Scofield, D.A.Vogel, R.Zasadzinski; Physical Review, A48 (1993) 2850

[459] F.Bely-Dubau, P.Faucher, L.Steenman-Clark, M.Bitter, S.von Goeler, K.W.Hill, C.Cmahy-Val, J.Dubau; Physical Review, A26 (1982) 3459

[460] E.D.Donets, V.I.Ilyushenko; Preprint P7-8310, Dubna, 1974

[461] E.D.Donets, V.P.Ovsyannikov; Preprint P7-10780, Dubna, 1977

[462] E.D.Donets, V.P.Ovsyannikov; Preprint P7-80-404, Dubna, 1980

[463] W.Lotz; Z.Phys., 216 (1968) 241

[464] N.Claytor, b.Feinberg, H.Gould, C.E.Bemis, J.G.Campo, C.A.Ludemann. C.R.Vane; Physical Review Letters, 61 (1988) 2081

[465] R.E.Marrs, S.R.Elliott, J.H.Scofield; EBIT Annual Report 1993, ed. by D.Schneider, UCRL-ID-118274, LLNL, Livermor, 1994, p.46

[466] J.H.Scofield; Physical Review, A40 (1989) 3045

[467] H.L.Zhang, D.H.Sampson; Physical Review, A42 (1990) 5378

[468] R.E.Marrs, P.Beiersdorfer, C.Bennett, M.H.Chen, T.Cowan, D.Dietrich, J.R.Henderson, D.A.Knapp, A.Osterheld, M.B.Schneider, J.H.Scofield, M.A.Levine; Int. Symp. on Electron Beam Ion Sources and Their Applications, ed. by A.Hershcovitch, AIP Conference Proceedings No.188, AIP, New York, 1989

[469] D.A.Knapp; Suppl. to Z.Phys. D, 21 (1991) 143

[470] G.Kilgus, J.Berger, P.Blatt, M.Grieser, D.Habs, B.Hochadel, E.Jaeschke, D.Krämer, R.Neumann, G.Neureither, W.Ott, D.Schwalm, M.Steck, R.Stockstad, E.Szmola, A.Wolf, R.Schuch, A.Müller, M.Wagner; Physical Review Letters, 64 (1990) 737

[471] R.Ali, C.P.Bhalla, C.L.Cocke, M.Schulz, M.Stockli; Physical Review, A44 (1991) 223

[472] D.R.DeWitt, D.Schneider, M.W.Clark, M.H.Chen, D.Church; Physical Review, A44 (1991) 7185

[473] D.A.Knapp, R.E.Marrs, M.A.Levine, C.L.Bennett, M.H.Chen, J.R.Henderson, M.B.Schneider, J.H.Scofield; Physical Review Letters, 62 (1989) 2104

[474] P.Beiersdorfer, S.Chantrenne, M.H.Chen, R.E.Marrs, D.A.Vogel, K.L.Wong, R.Zasadzinski; Z. Phys. D (Suppl.), 21 (1991) S209

[475] D.DeWitt, D.Schneider, M.H.Chen, M.B.Schneider, D.Church, G.Weinberg, M.Sakurai; EBIT Annual Report 1993, ed. by D.Schneider, LLNL, Livermore, 1994, p. 51

[476] D.DeWitt, D.Schneider, M.H.Chen, M.W.Clark, J.W.McDonald, M.B.Schneider; Physical Review Letters, 68 (1992) 1694

[477] M.B.Schneider, D.A.Knapp, M.H.Chen, J.H.Scofield, P.Beiersdorfer, C.L.Bennett, J.R.Henderson, R.E.Marrs, M.A.Levine; Physical Review, A45 (1992) R1291

[478] M.B.Schneider, D.A.Knapp, M.H.Chen, J.H.Scofield, P.Beiersdorfer, C.L.Bennett, J.R.Henderson, R.E.Marrs; Physical Review, A45 (1992) 1291

Index